Modelling Binary Data

OTHER STATISTICS TEXTS FROM
CHAPMAN & HALL

Practical Statistics for Medical Research
D. Altman
The Analysis of Time Series
C. Chatfield
Problem solving – A statisticians guide
C. Chatfield
Statistics for Technology
C. Chatfield
Introduction to Multivariate Analysis
C. Chatfield and A. J. Collins
Applied Statistics: Principles and Examples
D. R. Cox and E. J. Snell
An Introduction to Generalized Linear Models
A. J. Dobson
Introduction to Optimization Methods and their Application in Statistics
B. S. Everitt
Multivariate Statistics – A Practical Approach
B. Flury and H. Riedwyl
Readings in Decision Analysis
S. French
Multivariate Analysis of Variance and Repeated Measures
D. J. Hand and C. C. Taylor
Multivariate Statistical Methods – A Primer
Bryan F. Manly
Statistical Methods in Agriculture and Experimental Biology
R. Mead and R. N. Curnow
Elements of Simulation
B. J. T. Morgan
Probability: Methods and Measurement
A. O'Hagan
Essential Statistics
D. G. Rees
Foundations of Statistics
D. G. Rees
Decision Analysis: A Bayesian Approach
J. Q. Smith
Applied Statistics: A Handbook of BMDP Analyses
E. J. Snell
Elementary Applications of Probability Theory
H. C. Tuckwell
Intermediate Statistical Methods
G. B. Wetherill

Further information on the complete range of Chapman & Hall *statistics books is available from the publishers*

Modelling Binary Data

D. Collett

Department of Applied Statistics,
University of Reading, UK

CHAPMAN & HALL
London · New York · Tokyo · Melbourne · Madras

Published by Chapman & Hall, 2-6 Boundary Row, London SE1 8HN

Chapman & Hall, 2-6 Boundary Row, London SE1 8HN, UK

Blackie Academic & Professional, Wester Cleddens Road, Bishopbriggs, Glasgow G64 2NZ, UK

Chapman & Hall GmbH, Pappelallee 3, 69469 Weinheim, Germany

Chapman & Hall USA., One Penn Plaza, 41st Floor, New York, NY10119, USA

Chapman & Hall Japan, ITP - Japan, Kyowa Building, 3F, 2-2-1 Hirakawacho, Chiyoda-ku, Tokyo 102, Japan

Chapman & Hall Australia, Thomas Nelson Australia, 102 Dodds Street, South Melbourne, Victoria 3205, Australia

Chapman & Hall India, R. Seshadri, 32 Second Main Road, CIT East, Madras 600 035, India

First edition 1991
Reprinted 1994

© 1991 D. Collett

Typeset in 9½/11½ Times by Interprint Limited, Malta
Printed in Great Britain by T.J Press (Padstow), Cornwall

ISBN   0 412 38790 5 (Hb)   0 412 38800 6 (Pb)

A Catalogue record for this book is available from the British Library

Library of Congress Cataloging-in-Publication Data available

Collett, D., 1952-
    Modelling binary data / D. Collett.
        p. cm.
    Includes bibliographical references and index.
    ISBN 0-412-38790-5. - ISBN 0-412-38800-6 (pbk.)
    1. Analysis of variance. 2. Distribution (Probability theory)
3. Linear models (Statistics) I. Title
QA279.C64 1991
519.5'38 - dc20   91-23845
                                                        CIP

∞ Printed on acid-free paper, manufactured in accordance with
    ANSI/NISO Z39.48-1992 and ANSI/NISO Z39.48-1984
    (Permanence of Paper)

To Janet

Contents

Preface

Data are said to be binary when each observation falls into one of two categories, such as alive or dead, positive or negative, defective or non-defective, and success or failure. In this book, it is shown how data of this type can be analysed using a modelling approach. There are several books that are devoted to the analysis of binary data, and a larger number that include material on this topic. However, there does appear to be a need for a textbook of an intermediate level which dwells on the practical aspects of modelling binary data, which incorporates recent work on checking the adequacy of fitted models, and which shows how modern computational facilities can be fully exploited. This book is designed to meet that need.

The book begins with a description of a number of studies in which binary data have been collected. These data sets, and others besides, are then used to illustrate the techniques that are presented in the subsequent chapters. The majority of the examples are drawn from the agricultural, biological and medical sciences, mainly because these are the areas of application in which binary data are most frequently encountered. Naturally, the methods described in this book can be applied to binary data from other disciplines.

Underlying most analyses of binary data is the assumption that the observations are from a binomial distribution, and in Chapter 2, a number of standard statistical procedures based on this distribution are described. The modelling approach is then introduced in Chapter 3, with particular emphasis being placed on the use of the linear logistic model. The analysis of binary data from biological assays, and other applications of models for binary data, are covered in Chapter 4.

A distinguishing feature of this book is the amount of material that has been included on methods for assessing the adequacy of a fitted model, and the phenomenon of overdispersion. Chapter 5 provides a comprehensive account of model checking diagnostics, while models for overdispersed data, that is, data which are more variable than the binomial distribution can accommodate, are reviewed in Chapter 6. Both of these chapters include a summary of the methods that are most likely to be useful on a routine basis.

A major area of application of the linear logistic model is to the analysis of data from epidemiological studies. In Chapter 7, we see how this model provides the basis of a systematic approach for analysing data from cohort studies and both unmatched and matched case-control studies. Chapter 8 contains a brief discussion of some additional topics in the analysis of binary data.

In order to use the methods described in this book, appropriate computer software is needed, and so Chapter 9 shows how some of the major statistical packages can be used in modelling binary data. This chapter also includes a review of the relative merits of these packages and details on how the non-standard procedures described in the earlier chapters can be implemented using statistical software.

Throughout this book, references have been kept to a minimum. However, notes on the source of material covered, and suggestions for further reading, are given in the final section of each chapter.

In writing this book, I have assumed that the reader has a basic knowledge of statistical methods, and is familiar with linear regression analysis and the analysis of variance. An understanding of matrix algebra is not an essential requirement, although some familiarity with the matrix representation of the general linear model is necessary to follow some of the details on model checking diagnostics in Chapter 5 and the algorithms described in Chapter 9 (Section 9.3) and Appendix B (Section B.1).

This book should be of interest to statisticians in the pharmaceutical industry, and those engaged in agricultural, biological, industrial and medical research, who need to analyse binary data. I also hope that by following the examples, numerate scientists in universities and research institutes will be able to process their own binary data with minimal assistance from a statistician. The content of Chapter 7 should be of particular interest to the epidemiologist who needs to use modern statistical methods to analyse data from cohort and case-control studies. The book should also appeal to undergraduate and postgraduate students in universities and polytechnics attending courses that include material on modelling binary data, such as courses on medical statistics, quantitative epidemiology, and the generalized linear model.

For permission to reproduce previously published data, I am grateful to the Academic Press, American Association for the Advancement of Science, American Chemical Society, American Society of Tropical Medicine and Hygiene, Association of Applied Biologists, Biometric Society, Elsevier Applied Science Publishers Ltd., Entomological Society of America, Headley Bros. Ltd., Pergamon Press PLC, Royal Statistical Society, Society for Epidemiologic Research, and John Wiley and Sons Inc. I also thank Conrad Latto and Anthony Saunders-Davies for allowing me to use the data given in Tables 1.10 and 6.10, respectively, and Jean Golding for providing the data on sudden infant death syndrome, used in Section 7.8.

It is a pleasure to acknowledge the help provided by a number of friends in the preparation of this book. Robert Curnow, Mike Kenward, Alan Kimber, Wojtek Krzanowski and John Whitehead all read through initial drafts of the chapters, providing constructive comments for which I was very grateful. I also thank Joyce Snell and the late David Williams for their valuable remarks on parts of the book. Naturally, I am solely responsible for any errors that remain.

All of the figures in this book were produced using SAS/GRAPH®, and I thank Marilyn Allum for her help in their production. I would also like to record my gratitude to Toby Lewis, who not only had a great influence on my own approach to statistics, but who also introduced me to the linear logistic model and the statistical package GLIM® in the Autumn of 1974, while I was reading for a Ph.D. under his supervision. Finally, I thank my wife, Janet, for her constant support over the period in which this book was written.

D. Collett
Reading, UK.

1

Introduction

In many areas of application of statistical principles and procedures, from agronomy through to zoology, one encounters observations made on individual experimental units that take one of two possible forms. For example, a seed may germinate or fail to germinate under certain experimental conditions; an integrated circuit manufactured by an electronics company may be defective or non-defective; a patient in a clinical trial to compare alternative forms of treatment may or may not experience relief from symptoms; an insect in an insecticidal trial may survive or die when exposed to a particular dose of the insecticide. Such data are said to be **binary**, although an older term is **quantal**, and the two possible forms for each observation are often described generically by the terms success and failure.

In some circumstances, interest centres not just on the response of one particular experimental unit (seed, integrated circuit, patient or insect) but on a group of units that have all been treated in a similar manner. Thus a batch of seeds may be exposed to conditions determined by the relative humidity and temperature, for example, and the proportion of seeds germinating in each batch recorded. Similarly, the individual responses from each of a number of patients in a clinical trial receiving the same treatment, and who are similar on the basis of age, sex, or certain physiological or biochemical measures, may be combined to give the proportion of patients obtaining relief from symptoms. The resulting data are then referred to as **grouped binary data**, and represent the number of 'successes' out of the total number of units exposed to a particular set of experimental conditions.

Data in the form of proportions are often, but not exclusively, modelled using the binomial distribution while binary data may be assumed to have the Bernoulli distribution, a special case of the binomial distribution. Methods for analysing what are often referred to as binary or binomial data are the subject of this book.

1.1 SOME EXAMPLES

A number of examples are now given which are designed to illustrate the variety of circumstances in which binary data are encountered. These examples will be used to illustrate the application of the statistical techniques presented in subsequent chapters. The description of each example will concentrate mainly on the background to the data, although some remarks will be made on those aspects of the design of the studies which have a bearing on the analysis of the data. In each case, a reference to the initial source of the data will be given, although some of the data sets have since been discussed by other authors.

Example 1.1 Smoked mice
Before clear evidence was obtained about the extent to which tobacco smoke is carcinogenic for the lungs in human beings, a number of experiments were performed using animals. Essenberg (1952) described one such experiment in which thirty-six albino mice of both sexes were enclosed in a chamber that was filled with the smoke of one cigarette every hour of a twelve-hour day. A comparable number of mice were kept for the same period in a similar chamber without smoke. After one year, an autopsy was carried out on those mice in each group which had survived for at least the first two months of the experiment. Data on the numbers of mice which had developed lung tumours amongst 23 'treated' mice and 32 'control' mice are summarized in Table 1.1.

Table 1.1 Number of mice developing lung tumours when exposed or not exposed to cigarette smoke

Group	Tumour present	Tumour absent	Total
Treated	21	2	23
Control	19	13	32

In this experiment, the allocation of mice to the two treatment groups has been randomised to ensure that any differences between tumour incidence in each group of mice can only be attributed to the smoke. The mice in the two groups are also assumed to react to the treatment independently of each other.

The aim of this experiment was to investigate whether or not exposure to cigarette smoke encourages the development of a tumour, and the binary observation recorded on each mouse refers to whether or not a lung tumour is present at autopsy. The proportion of mice which develop a tumour in the treated group is $21/23 = 0.91$, which is considerably greater than $19/32 = 0.59$, the corresponding proportion for the control group. The statistical problem is to determine whether as big a difference as this could have arisen by chance alone if the cigarette smoke was not toxic.

The data of this example constitute a 2×2 table of counts known as a **contingency table**. Our enquiry about the equality of the two proportions of affected mice is formally equivalent to assessing whether there is an association between exposure to the cigarette smoke and the subsequent development of a tumour.

Example 1.2 Propagation of plum root-stocks
The data given in Table 1.2 were originally reported by Hoblyn and Palmer (1934) and were obtained from an experiment conducted at the East Malling Research

Station on the vegetative reproduction of root-stocks for plum trees from cuttings taken from the roots of older trees. Cuttings from the roots of the variety *Common Mussel* were taken between October, 1931 and February, 1932. Half of these were planted as soon as possible after they were taken while the other half were bedded in sand under cover until spring when they were planted. Two lengths of root cuttings were used at each planting time: long cuttings 12 cm in length and short cuttings 6 cm long. A total of 240 cuttings were taken for each of the four combinations of time of planting (at once or in spring) and length of cutting (short or long) and the condition of each plant in October, 1932 (alive or dead) was assessed.

Table 1.2 Survival rate of plum root-stock cuttings

Length of cutting	Time of planting	Number surviving out of 240	Proportion surviving
Short	At once	107	0.45
	In spring	31	0.13
Long	At once	156	0.65
	In spring	84	0.35

The binary response variable here is the ultimate condition of the plant, which has been assessed for all 960 cuttings. In presenting the results of the experiment in Table 1.2, the individual binary observations have been aggregated to obtain the proportion of cuttings surviving for each of the four combinations of planting time and length of cutting.

In analysing these data, we would first examine whether the difference in the survival probabilities of long and short cuttings is the same at each planting time, that is, whether there is an interaction between length of cutting and time of planting. If there is not, we would go on to investigate whether there is an overall difference in survival rate due to the main effects of the length of cutting and the time of planting. In spite of the fact that there is no replication of the proportions, it is possible to test whether there is no interaction when the data can be assumed to have a binomial distribution. This assumption will depend critically on whether or not the individual binary responses can be assumed to be independent. It is implausible that the 960 root cuttings were originally taken from 960 different parent trees, and so the individual responses of each cutting are unlikely to be independent of each other. Unfortunately, no information is available on this aspect of the design and so there will be little alternative to assuming that the individual binary responses are independent.

Example 1.3 Germination of Orobanche
Orobanche, commonly known as broomrape, is a genus of parasitic plants without chlorophyll that grow on the roots of flowering plants. In the course of research into factors affecting the germination of the seed of the species *Orobanche aegyptiaca*, a batch of seeds was brushed onto a plate containing a 1/125 dilution

of an extract prepared from the roots of either a bean or a cucumber plant. The number of seeds which germinated was then recorded. The data in Table 1.3, which were originally presented by Crowder (1978), are the results of experiments with two varieties of *Orobanche aegyptiaca*, namely *O. aegyptiaca* 75 and *O. aegyptiaca* 73.

Table 1.3 Number of *Orobanche* seeds germinating y, out of n, in extracts of bean and cucumber roots

O. aegyptiaca 75				O. aegyptiaca 73			
Bean		Cucumber		Bean		Cucumber	
y	n	y	n	y	n	y	n
10	39	5	6	8	16	3	12
23	62	53	74	10	30	22	41
23	81	55	72	8	28	15	30
26	51	32	51	23	45	32	51
17	39	46	79	0	4	3	7
		10	13				

Notice that different numbers of batches have been used for each combination of the two seed varieties and the two root extracts, and that the batch sizes, n, themselves vary considerably, in that they range from 4 to 81. Those proportions that are based on larger batches of seeds will have a greater precision and it will be important to take account of this when the data are analysed to determine the extent to which there are differences in the germination probabilities for the two varieties of seed in each type of root extract used. An additional feature of these data is that the existence of different batches of seed assigned to the same experimental conditions provides information about the extent to which the proportions are homogeneous within each combination of variety and root extract.

Example 1.4 A toxicological study
In an article on the design of toxicological experiments involving reproduction, Weil (1970) describes a study in which female rats are fed different diets during pregnancy. One group of sixteen pregnant rats were fed a diet containing a certain chemical while a similar control group received the same diet without the addition of the chemical. After the birth of the litters, the number of pups which survived the twenty-one day lactation period, expressed as a fraction of those alive four days after birth, was recorded; the data obtained are reproduced in Table 1.4.

The object of this experiment was to investigate differences between the control and treated rats in terms of mortality over the lactation period. In this experiment, the pregnant female rats were independently assigned to the two treatment groups, although a response variable is measured on each of the offsprings rather than on the mother. Because of genetic, environmental and social influences, the offsprings in a particular litter will tend to be very similar to one another. The binary

Table 1.4 Proportions of rat pups surviving in the litters of rats assigned to two treatment groups

Treated rats	13/13, 9/10,	12/12, 9/10,	9/9, 8/9,	9/9, 11/13,	8/8, 4/5,	8/8, 5/7,	12/13, 7/10,	11/12, 7/10
Control rats	12/12, 8/9,	11/11, 4/5,	10/10, 7/9,	9/9, 4/7,	10/11, 5/10,	9/10, 3/6,	9/10, 3/10,	8/9, 0/7

responses from the animals of a given litter will therefore be correlated and unless proper account is taken of this, the precision of the estimated treatment difference is likely to be overestimated.

Experimental work in a variety of areas gives rise to problems similar to those raised by Example 1.4. Particular examples include the measurement of the proportion of fertile eggs produced by hens receiving different volumes of inseminate, binary measurements made on individual members of a household in a community health project, and dental surveys that involve counting the number of individual teeth that have caries in children using different types of toothpaste.

These first four examples all concern binary data from one-factor or two-factor experiments. The introduction of additional design factors into any of these experiments to give factorial designs with three or more factors is readily envisaged. The next three examples concern studies where interest focuses on the relationship between a binary response variable and one or more explanatory variables.

Example 1.5 Aircraft fasteners
Montgomery and Peck (1982) describe a study on the compressive strength of an alloy fastener used in the construction of aircraft. Ten pressure loads, increasing in units of 200 psi from 2500 psi to 4300 psi, were used with different numbers of fasteners being tested at each of these loads. The data in Table 1.5 refer to the number of fasteners failing out of the number tested at each load.

Table 1.5 Number of fasteners failing out of a number subjected to varying pressure loads

Load	Sample size	Number failing	Proportion
2500	50	10	0.20
2700	70	17	0.24
2900	100	30	0.30
3100	60	21	0.35
3300	40	18	0.45
3500	85	43	0.51
3700	90	54	0.60
3900	50	33	0.66
4100	80	60	0.75
4300	65	51	0.78

Here, the binary outcome variable is whether or not a particular aluminium fastener fails when a given load is applied to it, and the investigator will be interested in the pattern of the response to changing loads. A **statistical model** will be needed to describe the relationship between the probability of a component failing and the load applied to it. At low loads, none of the components is likely to fail whereas at relatively high loads, all of them will fail. Of course, the chance of a fastener failing lies between zero and unity, and so the relationship between the probability of failure and load will certainly not be a straight line over the entire range of loads applied. Typically, a graph of the probability of failure against increasing load will have an elongated S-shape, commonly called a **sigmoid**. Such a curve, in the context of this example, is presented in Fig. 1.1.

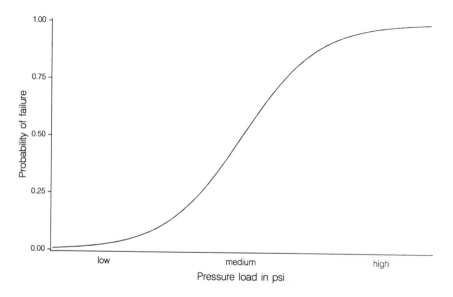

Fig. 1.1 A hypothetical load–response curve for the data from Example 1.5.

Apart from being interested in the general form of this response curve, the experimenter may want to be able to estimate the failure rate at a particular load. Alternatively, interest may centre on the load which leads to a failure rate greater than some specified percentage. In either case, a measure of the precision of the resulting estimate would be necessary; this would commonly be provided through the **standard error** of the estimate, which is defined to be the square root of the estimated variance of the estimate.

Example 1.6 Anti-pneumococcus serum
Historically, the type of study that prompted much research into methods of analysing binary data was a **bioassay**, an experiment designed to assess the potency

of a compound by means of the response produced when it is administered to a living organism. The assay which forms the basis of this example is taken from Smith (1932) who described a study of the protective effect of a particular serum, 'Serum number 32', on pneumococcus, the bacterium associated with the occurrence of pneumonia. Each of forty mice was injected with a combination of an infecting dose of a culture of pneumococci and one of five doses of the antipneumococcus serum. For all mice which died during the seven-day period following inoculation, a blood smear taken from the heart was examined to determine whether pneumococci were present or absent. Mice which lived beyond the seventh day were regarded as survivors and not further examined. The binary response variable is therefore death from pneumonia within seven days of inoculation. The numbers of mice succumbing to infection out of forty exposed to each of five doses of the serum, measured in cc, are given in Table 1.6.

Table 1.6 Number of deaths from pneumonia amongst batches of forty mice exposed to different doses of a serum

Dose of serum	Number of deaths out of 40
0.0028	35
0.0056	21
0.0112	9
0.0225	6
0.0450	1

Although a plot of the proportion of deaths against dose of serum indicates that as the dose increases, the proportion of mice which die of pneumonia decreases, an objective summary of the relationship between probability of death and dose of serum is provided by a statistical model. This model could then be used to estimate the dose of serum which would protect a given proportion of mice from the development of pneumonia following exposure to the pneumococcus bacterium.

One objective of the studies described in Examples 1.5 and 1.6 is to estimate the value of a continuous variable (pressure load or dose of serum) which would be expected to lead to a particular response probability. In Example 1.5, this might be the load above which more than 5% of the fasteners are likely to fail, while in Example 1.6 it might be the dose that would be expected to protect more than 90% of mice. If it were possible to observe the actual load at which a particular fastener fails, or the dose of serum that is just sufficient to protect a particular mouse, such values would be recorded for a number of experimental units. Information about the value of the load which causes a certain proportion of fasteners to fail, or the dose which protects a certain proportion of mice, could then be obtained directly from the distribution of the recorded observations.

Even when it is possible to observe the value of a continuous response variable, a binary outcome may be recorded when the values of the continuous variable cannot reliably be obtained. A classic example is the determination of the age at menarche for a group of adolescent girls. While the girls may not be able to recall the exact date of menarche sufficiently accurately, it is straightforward to record whether or not the girls have begun to menstruate. After classifying the girls according to age, the age–response curve can be modelled. From this, the median age at menarche can be estimated by the age at which 50% of the girls have passed the onset of menstruation.

In the following example, a continuous response variable is replaced by a binary one on the grounds that interest lies in whether or not the continuous response exceeds a certain threshold value. In general, the replacement of a continuous variable by a binary one results in a loss of information and so this practice should only be followed when the circumstances demand it.

Example 1.7 Determination of the ESR
The erythrocyte sedimentation rate (ESR) is the rate at which red blood cells (erythrocytes) settle out of suspension in blood plasma, when measured under standard conditions. The ESR increases if the levels of certain proteins in the blood plasma rise, such as in rheumatic diseases, chronic infections and malignant diseases; this makes the determination of the ESR one of the most commonly used screening tests performed on samples of blood. One aspect of a study carried out by the Institute of Medical Research, Kuala Lumpur, Malaysia, was to examine the extent to which the ESR is related to two plasma proteins, fibrinogen and γ-globulin, both measured in gm/l, for a sample of thirty-two individuals. The ESR for a 'healthy' individual should be less than 20 mm/h and since the absolute value of the ESR is relatively unimportant, the response variable used here will denote whether or not this is the case. A response of zero will signify a healthy individual (ESR < 20) while a response of unity will refer to an unhealthy individual (ESR $\geqslant$ 20). The original data were presented by Collett and Jemain (1985) and are reproduced in Table 1.7.

Table 1.7 The levels of two plasma proteins and the value of a binary response variable that denotes whether or not the ESR for each individual is greater than or equal to 20

Individual	Fibrinogen	γ-globulin	Response
1	2.52	38	0
2	2.56	31	0
3	2.19	33	0
4	2.18	31	0
5	3.41	37	0
6	2.46	36	0
7	3.22	38	0
8	2.21	37	0
9	3.15	39	0

Table 1.7—*contd*

Individual	Fibrinogen	γ-globulin	Response
10	2.60	41	0
11	2.29	36	0
12	2.35	29	0
13	5.06	37	1
14	3.34	32	1
15	2.38	37	1
16	3.15	36	0
17	3.53	46	1
18	2.68	34	0
19	2.60	38	0
20	2.23	37	0
21	2.88	30	0
22	2.65	46	0
23	2.09	44	1
24	2.28	36	0
25	2.67	39	0
26	2.29	31	0
27	2.15	31	0
28	2.54	28	0
29	3.93	32	1
30	3.34	30	0
31	2.99	36	0
32	3.32	35	0

The aim of a statistical analysis of these data is to determine the strength of any relationship between the probability of an ESR reading greater than 20 mm/h and the levels of two plasma proteins. Specifically, does this probability depend on the levels of both proteins, on the level of just one of them, or on neither of them? Since the concentrations of fibrinogen and γ-globulin are commonly elevated in inflammatory diseases, if no relationship were found between the probability of an ESR value less than 20 mm/h and the protein levels, determination of the ESR would not be considered to be worthwhile for diagnostic purposes.

The remaining examples give rise to somewhat larger data sets that include both categorical variables (factors) and continuous variables (variates). Data sets such as these are typical of those that arise in many areas of scientific enquiry.

Example 1.8 Women's role in society
This example relates to data compiled from two general social surveys carried out in 1974 and 1975 by the National Opinion Research Center, University of Chicago, Illinois. Part of each survey was concerned with the relationship of education and sex to attitudes towards the role of women in society, and each respondent was

asked if he or she agreed or disagreed with the statement 'Women should take care of running their homes and leave running the country up to men.' Since no individual took part in both surveys, and in view of the small time interval between them, data from both sets of respondents have been combined. Of the 2927 individuals who participated in the surveys, three did not give their educational level and 53 were not sure if they agreed with the statement presented. The responses for the remaining 1305 male respondents and 1566 females were given by Haberman (1978) and are reproduced in Table 1.8.

Table 1.8 Subjects in two general social surveys classified by attitude towards women staying at home, sex and years of completed education

Years of education	Responses of males		Responses of females	
	Agree	Disagree	Agree	Disagree
0	4	2	4	2
1	2	0	1	0
2	4	0	0	0
3	6	3	6	1
4	5	5	10	0
5	13	7	14	7
6	25	9	17	5
7	27	15	26	16
8	75	49	91	36
9	29	29	30	35
10	32	45	55	67
11	36	59	50	62
12	115	245	190	403
13	31	70	17	92
14	28	79	18	81
15	9	23	7	34
16	15	110	13	115
17	3	29	3	28
18	1	28	0	21
19	2	13	1	2
20	3	20	2	4

A quick look at these data suggests that for each sex, the proportion of respondents who are in agreement with the controversial statement decreases as the number of years of education increases. We might have anticipated that those who are more highly educated would generally tend to disagree with the statement, but what is the form of this relationship? To what extent do the males and females respond differently? Does any such difference depend on the amount of education received? To analyse these data, the nature of the relationship between the proportion of respondents in agreement with the statement and number of years of education would be determined and compared between the two sexes.

Example 1.9 Prostatic cancer
The treatment regime to be adopted for patients who have been diagnosed as having cancer of the prostate is crucially dependent upon whether or not the cancer has spread to the surrounding lymph nodes. Indeed, a laparotomy (a surgical incision into the abdominal cavity) may be performed to ascertain the extent of this nodal involvement. There are a number of variables that are indicative of nodal involvement which can be measured without the need for surgery, and the aim of a study reported by Brown (1980) was to determine whether a combination of five variables could be used to forecast whether or not the cancer has spread to the lymph nodes. The five variables were: age of patient at diagnosis (in years), level of serum acid phosphatase (in King–Armstrong units), the result of an X-ray examination (0 = negative, 1 = positive), the size of the tumour as determined by a rectal examination (0 = small, 1 = large) and a summary of the pathological grade of the tumour determined from a biopsy (0 = less serious, 1 = more serious). The values of each of these five variables were obtained for 53 patients presenting with prostatic cancer who had also undergone a laparotomy. The result of the laparotomy is a binary response variable where zero signifies the absence of, and unity the presence of nodal involvement. The data are reproduced in Table 1.9.

Table 1.9 Data on five predictors of nodal involvement and whether or not there is such involvement in prostatic cancer patients

Case number	Age of patient	Acid level	X-ray result	Tumour size	Tumour grade	Nodal involvement
1	66	0.48	0	0	0	0
2	68	0.56	0	0	0	0
3	66	0.50	0	0	0	0
4	56	0.52	0	0	0	0
5	58	0.50	0	0	0	0
6	60	0.49	0	0	0	0
7	65	0.46	1	0	0	0
8	60	0.62	1	0	0	0
9	50	0.56	0	0	1	1
10	49	0.55	1	0	0	0
11	61	0.62	0	0	0	0
12	58	0.71	0	0	0	0
13	51	0.65	0	0	0	0
14	67	0.67	1	0	1	1
15	67	0.47	0	0	1	0
16	51	0.49	0	0	0	0
17	56	0.50	0	0	1	0
18	60	0.78	0	0	0	0

Table 1.9—*contd*

Case Number	Age of patient	Acid level	X-ray result	Tumour size	Tumour grade	Nodal involvement
19	52	0.83	0	0	0	0
20	56	0.98	0	0	0	0
21	67	0.52	0	0	0	0
22	63	0.75	0	0	0	0
23	59	0.99	0	0	1	1
24	64	1.87	0	0	0	0
25	61	1.36	1	0	0	1
26	56	0.82	0	0	0	1
27	64	0.40	0	1	1	0
28	61	0.50	0	1	0	0
29	64	0.50	0	1	1	0
30	63	0.40	0	1	0	0
31	52	0.55	0	1	1	0
32	66	0.59	0	1	1	0
33	58	0.48	1	1	0	1
34	57	0.51	1	1	1	1
35	65	0.49	0	1	0	1
36	65	0.48	0	1	1	0
37	59	0.63	1	1	1	0
38	61	1.02	0	1	0	0
39	53	0.76	0	1	0	0
40	67	0.95	0	1	0	0
41	53	0.66	0	1	1	0
42	65	0.84	1	1	1	1
43	50	0.81	1	1	1	1
44	60	0.76	1	1	1	1
45	45	0.70	0	1	1	1
46	56	0.78	1	1	1	1
47	46	0.70	0	1	0	1
48	67	0.67	0	1	0	1
49	63	0.82	0	1	0	1
50	57	0.67	0	1	1	1
51	51	0.72	1	1	0	1
52	64	0.89	1	1	0	1
53	68	1.26	1	1	1	1

In this example, the five explanatory variables or **prognostic factors** are a mixture of factors (X-ray result, size and grade of tumour) and variates (age of patient and level of serum acid phosphatase). The problem here is to identify whether all or just a subset of these variables are required in a model that could be used to predict nodal involvement. Additional variables defined by products of the explanatory variables, or by quadratic or higher-order powers of the continuous variables, may also be included in the model. Similar problems in model selection such as are encountered in multiple linear regression analysis might be anticipated.

Example 1.10 Diverticular disease
Diverticular disease and carcinoma of the colon are amongst the most common diseases of the large intestine in the population of the United Kingdom and North America. Diverticular disease is estimated to be present in one third of the population over forty years of age, while carcinoma of the colon accounts for about 10 000 deaths in the United Kingdom per annum. It has been found that there is a geographical association between the prevalence of the two diseases in that their world-wide incidence tends to go hand in hand. In order to investigate whether the two conditions tend to be associated in individual patients, a case-control study was carried out and reported by Berstock, Villers and Latto (1978).

A total of eighty patients, admitted to the Reading group of hospitals between 1974 and 1977 with carcinoma of the colon, were studied and the incidence of diverticular disease in each patient was recorded. A control group of 131 individuals were randomly selected from a population known not to have colonic cancer, and were similarly tested for the presence of diverticular disease. Table 1.10 gives the proportions of individuals with diverticular disease, classified by age, sex and according to whether or not they had colonic cancer.

Table 1.10 Proportion of individuals with diverticular disease (DD) classified by age, sex and the presence of colonic cancer

Age interval	Midpoint of age range	Sex	Proportion with DD	
			Cancer patients	Controls
40–49	44.5	M	0/3	0/7
		F	0/6	1/15
50–54	52.0	M	1/2	1/7
		F	0/0	0/0
55–59	57.0	M	2/5	3/15
		F	1/7	4/18
60–64	62.0	M	1/5	5/18
		F	0/2	2/8
65–69	67.0	M	1/4	6/11
		F	0/5	7/17
70–74	72.0	M	0/5	1/4
		F	3/13	2/6
75–79	77.0	M	1/3	0/0
		F	5/9	0/0
80–89	84.5	M	1/2	4/5
		F	4/9	0/0

By classifying the individuals according to a number of distinct age groups, the relationship between the proportion of individuals with diverticular disease and the midpoint of the age range can easily be examined using graphical methods. One could then go on to investigate whether the relationship is the same for patients of each sex, with and without carcinoma of the colon.

An epidemiological study is usually designed to examine which of a number of possible factors affect the risk of developing a particular disease or condition. In this study, the disease of primary interest is colonic cancer, and the epidemiologist would want to look at whether the occurrence of diverticular disease in a patient increased or decreased the risk of colonic cancer, after making any necessary allowances for the age and sex of the patient. When looked at in this way, the probability that a given individual has colonic cancer cannot be estimated. This is because the experimental design that has been used does not ensure that the relative numbers of cases and controls of a given sex in a particular age group match the corresponding numbers in the population being sampled. For example, the overall proportion of colonic cancer patients in the case–control study is 80/211 which is a gross overestimate of the proportion of individuals with colonic cancer in the general population. Nevertheless, it is possible to estimate the risk of colonic cancer in a patient with diverticular disease relative to a patient without diverticular disease, and to explore the strength of the association between the occurrence of colonic cancer and age, sex and the presence of diverticular disease.

In analysing these data, one feature that might be anticipated to cause problems is the occurrence of responses of 0/0, which arise for females in the 50–54 age group and several other groups. Responses such as these provide no information about associations between colonic cancer and age, sex and the occurrence of diverticular disease. Such observations will need to be omitted from any analysis of these data.

1.2 THE SCOPE OF THIS BOOK

The ten examples that have just been presented range in level of complexity from those where the data have a relatively simple structure to those with more highly structured data. For data where the interest is in estimating a proportion, or in comparing two proportions, such as in Examples 1.1 and 1.2, relatively straightforward statistical techniques may be adopted. These techniques are outlined in Chapter 2. However, for the vast majority of applications, the data structure will have to be oversimplified in order to use these methods; important information in the data may then be lost.

In each of the examples of section 1.1, had a continuous observation been made on the experimental unit, instead of a binary one, the standard techniques of analysis of variance and linear regression analysis may have been directly applicable. This putative response variable might then have been assumed to have a constant variance for each observation. Moreover, for the purpose of significance testing, an assumption about the form of an underlying probability distribution would also be made, possibly taking it to be the ubiquitous normal distribution. When the data are binary or binomial, these standard methods cannot immediately be applied. Instead, different forms of model will be required, which will often be based on an assumed underlying binomial distribution. It turns out that many of the ideas and principles that are used in conjunction with modelling a continuous response variable carry over to modelling binary or binomial data. The models for binary data which are introduced in Chapters 3 and 4, particularly the **linear logistic regression model**, are very flexible and can be used to model the data presented in all of the examples of section 1.1. Through the use of these models, the analysis of binary data is unified, just as methods for analysing normally distributed data are unified by the general linear model.

Although methods for fitting models to binary data are well established, methods for checking whether a particular model fits the observed data are not as well known and are certainly not as widely used as they ought to be. In part, this is because techniques for detecting observations that appear to be surprising, labelled **outliers**, and those that have a severe impact on the inference to be drawn from a data set, labelled **influential values**, as well as techniques for generally assessing the adequacy of the structure of the model, have only recently been proposed. Chapter 5 provides a comprehensive account of those diagnostic methods which do seem to be useful in practice.

It sometimes happens that data in the form of proportions appear to be more variable than an underlying binomial distribution can accommodate. This phenomenon is referred to as **extra binomial variation** or **overdispersion**. A number of models have been proposed for overdispersion from which a judicious selection of that most appropriate to the context of the data is necessary. Possible models are described in Chapter 6, and a review of the utility of these models, illustrated by examples of their use, is included.

In the analysis of data from epidemiological studies designed to highlight factors that are associated with the risk of a certain disease or condition, it is appropriate to regard the presence or absence of the disease in an individual as a binary outcome. Three designs that are commonly used in such investigations are the cohort study and the unmatched or matched case-control study. It turns out that the linear logistic regression model described in Chapter 3 can be used as a basis for the analysis of data from all three types of study. Details on this are given in Chapter 7.

Models for the analysis of percentages and rates, the analysis of binary data from cross-over trials, the analysis of binary time series, and other miscellaneous topics, will be discussed briefly in Chapter 8.

1.3 USE OF STATISTICAL SOFTWARE

Most of the methods for analysing binary data that are detailed in subsequent chapters can only be carried out using a mainframe or personal computer, in conjunction with appropriate software. Some of the currently available software is described in Chapter 9 and the relative merits of GLIM, Genstat, SAS, BMDP, SPSS and EGRET for modelling binary data are reviewed.

It will often be convenient to illustrate the methods of analysis that are presented in this book using the output from a particular computer package, and for this the package GLIM will be used. This widely available package is well suited to the modelling of binary data, as indeed it is to many other types of response variable. An introduction to GLIM is given by Healy (1988) while the formal description of the language is included in Payne (1986). Full details on the use of this package in the analysis of binary data are included in Chapter 9.

In computer-based data analysis, it often happens that the values of certain statistics are required that are not part of the standard output of a particular package. Additional calculations will then need to be performed either within the package, using its arithmetic and macro facilities, or perhaps outside it, using different software or an electronic calculator. When any such statistics cannot be obtained directly using GLIM, a GLIM **macro** may be written; those used during the course of this book are given in Appendix C. Machine-readable versions of these macros, and all the data sets used in this book, are available from the author.

FURTHER READING

Texts which are devoted specifically to the analysis of binary data include those of Cox and Snell (1989), Finney (1971), Ashton (1972) and Hosmer and Lemeshow (1989). Cox and Snell (1989) give a concise, up-to-date account of the theory and although they include a number of practical examples, their book is a research monograph written primarily for post-graduate and academic statisticians. Finney (1971) deals comprehensively with the application to bioassay, although it now has something of an old-fashioned look to it. Ashton (1972) focuses on the linear logistic model for binary data, and concentrates on methods of fitting this model to data. Hosmer and Lemeshow (1989) present a practical guide to the use of this model in analysing data from the medical sciences.

There are a number of other books which include material on the analysis of binary data. Fleiss (1981) describes relatively straightforward methods for analysing contingency tables with a dichotomous variable, using illustrative examples drawn from medicine and epidemiology. Dobson (1990) includes some material on modelling binary data in her compact introduction to generalized linear modelling. The book by McCullagh and Nelder (1989) is an excellent reference book for the generalized linear model and covers a lot of ground in the thirty or so pages devoted to the analysis of binary data. Aitkin et al. (1989) emphasise the use of the statistical package GLIM in their account of the principles and practice of statistical modelling, and include a wide range of illustrative examples.

Other books that discuss models for binary data, but which are primarily devoted to the analysis of contingency tables, are those by Agresti (1990), Bishop, Fienberg and Holland (1975), Everitt (1977), Fienberg (1980) and Plackett (1981). Breslow and Day (1980, 1987) provide a comprehensive account of the application of methods for analysing binary data to case–control and cohort studies, respectively.

2
Statistical inference
for binary data

When binary data, or grouped binary data expressed as proportions, have a relatively simple structure, such as the data sets in Examples 1.1 and 1.2 of Chapter 1, questions of practical interest can usually be answered using standard statistical methods for estimating and testing hypotheses about a single proportion, and for comparing two or more proportions. Underlying these methods is the assumption that the observations have a binomial distribution, and so this distribution has a central role in the analysis of binary data. The binomial distribution was first derived by James Bernoulli, although it was not published until 1713, eight years after his death.

This chapter begins with the derivation of the binomial distribution and an outline of some of its properties. Inferential techniques associated with a single proportion, and for comparing two or more proportions are then described and illustrated in Sections 2.2–2.4. In analysing binary data, approximate rather than exact methods of inference are widely used. In part, this is because the exact methods are much more complicated, but it is often the case that inferences obtained using the approximate methods will be very similar to those obtained from the exact methods. Consequently, the use of the approximate methods is emphasised in this chapter, and subsequent chapters of this book.

2.1 THE BINOMIAL DISTRIBUTION

Consider a particular binary response that is either a success or a failure. To fix ideas, suppose that this response concerns whether or not a seed of a particular type germinates successfully under certain conditions. The germination potential of a particular seed will depend on a very large number of factors associated with the physiological and biochemical processes involved in the process of germination. These very complex circumstances cannot be known precisely, but their effect can be modelled by supposing that there is a probability, p, that the seed will germinate. This probability is termed the **success probability** or **response probability**, but it will of course be unknown. However, if it is small, the seed will be more likely to fail to germinate, while if p is close to one, the seed will be expected to germinate. Whether or not the seed germinates can be described in terms of a quantity known as a **random variable**, which will be denoted by R. This random variable may take one of two values, corresponding to successful germination and failure to germinate. It is convenient to code the two possible values that R can take as 0 and 1, where a successful germination corresponds to $R = 1$ and failure to $R = 0$. The probability that R takes the value one is the success probability, p, and so this probability can be written $P(R = 1) = p$. The corresponding probability of a failure is $P(R = 0) = 1 - p$. It is useful to be able to express these two probabilities in a single expression and so if r is the observed value of the random variable R, where r is either 0 or 1, we can write

$$P(R = r) = p^r(1 - p)^{1-r}, \quad r = 0, 1$$

This expression defines how the probabilities of the two events, $R = 0$ and $R = 1$, are distributed, and so it expresses the **probability distribution** of R. This particular probability distribution is known as the **Bernoulli distribution**.

The **mean**, or **expected value**, of the random variable R is, by definition, given by $E(R) = 0 \times P(R = 0) + 1 \times P(R = 1) = p$. The **variance** of R, a measure of dispersion of the random variable, can be found using the result that $\mathrm{Var}(R) = E(R^2) - [E(R)]^2$. Now, $E(R^2) = 0^2 \times P(R = 0) + 1^2 \times P(R = 1) = p$, and so $\mathrm{Var}(R) = p - p^2 = p(1 - p)$.

In order to throw light on the value of p for a particular batch of seeds, a number of seeds that are so similar that they can be assumed to have the same underlying germination probability are exposed to a particular set of conditions. The observations on whether or not each seed germinates will lead to a sequence such as SSFFFSFFSF..., where S and F denote success and failure, or in terms of the binary coding where 1 corresponds to a success, 1100010010.... Suppose that the germination test involves n seeds and let R_i be the random variable associated with whether or not the ith seed germinates, $i = 1, 2,..., n$. Since each of the n seeds is assumed to have the same germination probability, p, the random variables $R_1, R_2,..., R_n$ each have a Bernoulli distribution where

$$P(R_i = r_i) = p^{r_i}(1 - p)^{1 - r_i}$$

and r_i is the observed binary response for the ith seed, which takes the value zero or one, for $i = 1, 2,..., n$.

Now let y be the total number of successful germinations in the batch of n seeds, and assume that the seeds germinate independently of one another. The value of y can be regarded as the observed value of a random variable, Y, associated with the total number of successes out of n. In view of the coding used for the values taken by each of the n random variables, $R_1, R_2, ..., R_n$, it follows that $y = r_1 + r_2 + \cdots + r_n$, the sum of the n binary observations, is the observed value of the random variable $Y = R_1 + R_2 + \cdots + R_n$. The next step is to obtain the probability distribution of the random variable Y.

Suppose that a given sequence of n binary observations contains y successes, so that the sequence consists of y ones and $n - y$ zeros. Assuming that each of the n binary responses is independent of the others, the probability of any particular arrangement of y ones and $n - y$ zeros is $p^y(1 - p)^{n-y}$, for $y = 0, 1, \ldots, n$. The total number of ways in which a sequence of y ones and $n - y$ zeros can occur is given by

$$\binom{n}{y} = \frac{n!}{y!\,(n - y)!}$$

The term $\binom{n}{y}$ is the number of different combinations of n items taken y at a time and is read as '$n\,C\,y$'. In older textbooks it is often written as ^nC_y. The term $n!$ is read as 'factorial n' and is obtained from

$$n! = n \times (n - 1) \times (n - 2) \times \cdots \times 3 \times 2 \times 1$$

when n is a positive integer. For example, if there are four binary observations consisting of three successes and one failure, the number of ways in which the three ones and one zero can be arranged is

$$\binom{4}{3} = \frac{4!}{3! \times 1!} = \frac{4 \times 3 \times 2 \times 1}{(3 \times 2 \times 1) \times 1} = 4$$

The four possible arrangements are of course $1\,1\,1\,0$, $1\,1\,0\,1$, $1\,0\,1\,1$ and $0\,1\,1\,1$.

When $y = 0$, $\binom{n}{0}$ is the number of ways that n zeros can occur in n observations. This is just one, since the sequence of observations is $0\,0\,0\,0\ldots$. Now,

$$\binom{n}{0} = \frac{n!}{0! \times n!} = \frac{1}{0!}$$

and for this to be equal to one, $0!$ must be defined to be equal to one. This extends the definition of the factorial function to all non-negative integers.

The total probability of obtaining a sequence of y ones and $n - y$ zeros, independent of the ordering of the binary observations, will be $\binom{n}{y}$ multiplied by the probability of any particular arrangement of the y ones and $n - y$ zeros. The probability of y successes in n observations is therefore given by

$$P(Y = y) = \binom{n}{y} p^y(1 - p)^{n-y}$$

for $y = 0, 1 \ldots, n$. The random variable Y is said to have a **binomial distribution**. The reason for this name is that the probability $P(Y = y)$, $y = 0, 1, \ldots, n$, is the

$(y + 1)$th term in the **binomial expansion** of $(q + p)^n$, given by

$$(q + p)^n = q^n + \binom{n}{1} pq^{n-1} + \binom{n}{2} p^2 q^{n-2} + \cdots + \binom{n}{n-1} p^{n-1}q + p^n$$

where $q = 1 - p$. This expansion can be written more compactly using the Σ-notation to denote a sum, so that

$$(q + p)^n = \sum_{y=0}^{n} \binom{n}{y} p^y q^{n-y} = \sum_{y=0}^{n} P(Y = y)$$

Since $p + q = 1$, the binomial probabilities $P(Y = y)$, for $y = 0, 1, \ldots, n$, sum to one, confirming that these probabilities define a proper probability distribution.

The binomial probabilities depend on two parameters, namely the total number of observations, n, and the success probability, p. To express that the random variable Y has a binomial distribution with parameters n and p, we write $Y \sim B(n, p)$, where the symbol '$\sim$' is read as 'is distributed as'.

2.1.1 Properties of the binomial distribution

To derive the mean and variance of a random variable, Y, that has a binomial distribution with parameters n and p, recall that $Y = \sum_{i=1}^{n} R_i$, where $R_1, R_2, \ldots, R_n$ each have a Bernoulli distribution with success probability p, so that $E(R_i) = p$ and $\text{Var}(R_i) = p(1 - p)$. The expected value of Y is then $E(Y) = \sum E(R_i) = np$, and, because the R_i are independent random variables, the variance of Y is given by $\text{Var}(Y) = \sum \text{Var}(R_i) = np(1 - p)$.

The distribution of Y is symmetric when $p = 0.5$; when p is less than 0.5, it is positively skew, and if p is greater than 0.5, it is negatively skew. This feature is shown in Fig. 2.1, which gives the form of the distribution for the case where $n = 10$ and $p = 0.1, 0.25, 0.5$, and 0.9.

An important property of the binomial distribution is that as n increases, the degree of asymmetry in the distribution decreases, even when p is close to zero or one. This feature is illustrated in Fig. 2.2, which shows the binomial distribution for $p = 0.2$ and $n = 5, 15, 25$ and 50. Notice that even though the distribution is very asymmetric when $n = 5$, when $n = 50$, it is almost symmetric. It can be shown that as n increases, the binomial distribution becomes more and more closely approximated by the **normal distribution**, a distribution that is commonly used in the analysis of continuous response data. Indeed, the fourth graph in Fig. 2.2 has the appearance of a symmetric bell-shaped curve, typical of the normal distribution.

The normal distribution to which the binomial tends as n becomes large will have the same mean and variance as the binomial distribution, that is np and $np(1 - p)$, respectively. Consequently, the random variable defined by

$$Z = \frac{Y - np}{\sqrt{[np(1 - p)]}} \tag{2.1}$$

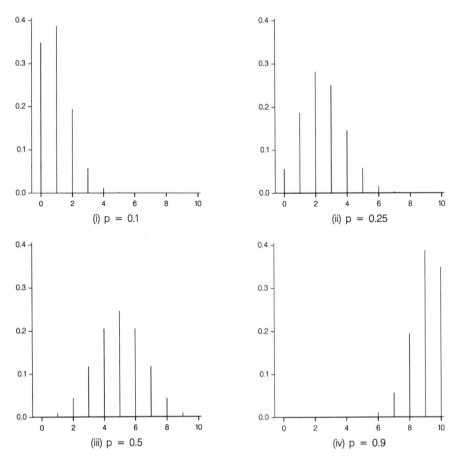

Fig. 2.1 The binomial distribution for $n = 10$ and $p = 0.1, 0.25, 0.5, 0.9$. The vertical axis exhibits the binomial probabilities for each value of y on the horizontal axis.

has an approximate normal distribution with zero mean and unit variance, known as the **standard normal distribution**. This distribution will be denoted by $N(0, 1)$, and is the distribution used in many of the approximate methods described below. The adequacy of this approximation depends on the degree of asymmetry in the binomial distribution, and hence on the value of both n and p. McCullagh and Nelder (1989) state that the approximation will be satisfactory over most of the range of the distribution of Y when $np(1 - p) \geqslant 2$. In particular, when p is close to 0.5, the approximation is very good for values of n as small as ten.

2.2 INFERENCE ABOUT THE SUCCESS PROBABILITY

Suppose that n independent binary observations have been obtained, where the underlying probability of a success is p in each case. If the n observations contain

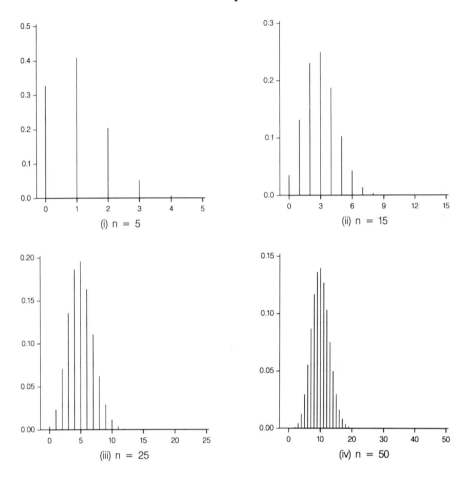

Fig. 2.2 The binomial distribution for $p = 0.2$ and $n = 5, 15, 25, 50$. The vertical axis exhibits the binomial probabilities for each value of y on the horizontal axis.

y successes, a natural estimate of the success probability, p, is the observed proportion of successes, y/n. This estimate will be denoted by $\hat{p}$.

The estimated success probability, $\hat{p}$, should become more reliable as the number of binary observations, n, increases. Consequently, a measure of the precision of the estimate is essential if the estimate is to be properly interpreted. The precision of an estimate can be assessed from its standard error, defined to be the square root of the estimated variance of the estimate. Since y is an observation on a random variable that has the binomial distribution $B(n, p)$, its variance is $np(1 - p)$. Using the result that the variance of a term such as aX is given by $\text{Var}\,(aX) = a^2\,\text{Var}\,(X)$, the variance of $\hat{p} = y/n$ is $p(1 - p)/n$. This variance can be estimated by $\hat{p}(1 - \hat{p})/n$, and so the standard error of $\hat{p}$ is given by

$$\text{s.e.}\,(\hat{p}) = \sqrt{\left(\frac{\hat{p}(1-\hat{p})}{n}\right)}$$

Estimates of the success probability that have relatively large standard errors will be less precise than estimates with smaller standard errors. The presence of the divisor n in the expression for s.e. $(\hat{p})$ shows that the precision does indeed increase with sample size.

2.2.1 Confidence intervals for p

A more informative summary of the extent to which $\hat{p}$ is reliable as an estimate of the success probability is an interval estimate for p, termed a **confidence interval**. A confidence interval for the true success probability is an interval that has a prescribed probability, usually 0.90, 0.95 or 0.99, of including p. In general, if the probability of the interval containing the true value of p is set to be $1 - \alpha$, where α is a small positive value, such as 0.1, 0.05 or 0.01, the resulting interval is called a $100(1 - \alpha)\%$ confidence interval for p.

The interpretation of a confidence interval is based on the idea of repeated sampling. Suppose that repeated samples of n binary observations are taken, and, say, a 95% confidence interval for p is obtained for each of the data sets. Then, 95% of these intervals would be expected to contain the true success probability. It is important to recognize that this does not mean that there is a probability of 0.95 that p lies in the confidence interval. This is because p is fixed in any particular case and so it either lies in the confidence interval or it does not. The limits of the interval are the random variables, and so the correct view of a confidence interval is based on the probability that the (random) interval contains p, rather than the probability that p is contained in the interval.

A $100(1 - \alpha)\%$ confidence interval for p is the interval (p_L, p_U), where p_L and p_U are the smallest and largest binomial probabilities for which the occurrence of the observed proportion y/n has a probability that is at least equal to $\alpha/2$. These values are the lower and upper limits of the confidence interval, termed the **confidence limits**. When Y has a binomial distribution with parameters n and p, $100(1 - \alpha)\%$ confidence limits for p can be found as follows. The binomial probability p_L is that for which the chance of observing y or more successes in n trials is $\alpha/2$. The probability that a binomial random variable with parameters n, p_L takes a value of y or more is

$$\sum_{j=y}^{n} \binom{n}{j} p_L^j (1 - p_L)^{n-j} \tag{2.2}$$

and so p_L is such that this probability is equal to $\alpha/2$. Similarly, the upper confidence limit is the value p_U for which the probability that a binomial random

24 Statistical inference for binary data

variable $B(n, p_U)$ takes the value y or less is also $\alpha/2$. Hence, p_U satisfies the equation

$$\sum_{j=0}^{y} \binom{n}{j} p_U^j (1 - p_U)^{n-j} = \frac{\alpha}{2} \tag{2.3}$$

The resulting interval (p_L, p_U) is often referred to as an exact confidence interval for p. The expression (2.2) and equation (2.3) are not particularly easy to manipulate but tables that give the values of p_L and p_U for given values of α, y and n are available. See, for example, Fisher and Yates (1963). Alternatively, the values of p_L and p_U which satisfy equations (2.2) and (2.3) are

$$p_L = y[y + (n - y + 1)F_{2(n-y+1),2y}(\alpha/2)]^{-1} \tag{2.4}$$

and

$$p_U = (y + 1)[y + 1 + (n - y)/F_{2(y+1),2(n-y)}(\alpha/2)]^{-1} \tag{2.5}$$

respectively, where $F_{v_1,v_2}(\alpha/2)$ is the upper $(100\alpha/2)\%$ point of the F-distribution with v_1 and v_2 degrees of freedom and y is the observed number of successes in n observations. Commonly available tables of the F-distribution only give percentage points for a relatively small range of values of v_1 and v_2, and so interpolation, more extensive tables of percentage points of the F-distribution, or computer software, will be needed to obtain an appropriate degree of accuracy. Because of this, these exact limits have not been widely used in practice, although their computation presents no real difficulties.

Confidence intervals for the true success probability are usually obtained from the normal approximation to the binomial distribution, and are constructed from percentage points of the standard normal distribution. If the random variable Z has a standard normal distribution, the upper $(100\alpha/2)\%$ point of this distribution is that value $z_{\alpha/2}$ which is such that $P(Z \geq z_{\alpha/2}) = \alpha/2$. This probability is the area under the standard normal curve to the right of $z_{\alpha/2}$, as illustrated in Fig. 2.3. By

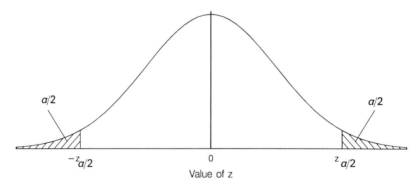

Fig. 2.3 The upper and lower $\alpha/2$ points of the standard normal distribution.

symmetry, the lower $(100\,\alpha/2)\%$ point is $-z_{\alpha/2}$, that is $P(Z \leqslant -z_{\alpha/2}) = \alpha/2$, and so $P(-z_{\alpha/2} \leqslant Z \leqslant z_{\alpha/2}) = 1 - \alpha$. The value of $z_{\alpha/2}$ for a given value of α can be obtained from widely available tables of the percentage points of the standard normal distribution, such as Lindley and Scott (1984). For example, for $\alpha = 0.1$, 0.05 and 0.01, $z_{\alpha/2} = 1.645$, 1.960 and 2.576, respectively.

From equation (2.1), it follows that

$$\frac{\hat{p} - p}{\sqrt{[p(1-p)/n]}} \tag{2.6}$$

has an approximate standard normal distribution, and so

$$P\left\{-z_{\alpha/2} \leqslant \frac{\hat{p} - p}{\sqrt{[p(1-p)/n]}} \leqslant z_{\alpha/2}\right\} = 1 - \alpha \tag{2.7}$$

The lower and upper confidence limits for p are the values of p which satisfy this probability statement, and are the two roots of the following quadratic equation in p:

$$(\hat{p} - p)^2 = z_{\alpha/2}^2 [p(1-p)/n] \tag{2.8}$$

A further approximation is to use $\hat{p}(1 - \hat{p})$ in place of $p(1 - p)$ in the denominator of expression (2.6). Then, equation (2.7) becomes

$$P\left\{-z_{\alpha/2} \leqslant \frac{\hat{p} - p}{\text{s.e.}\,(\hat{p})} \leqslant z_{\alpha/2}\right\} = 1 - \alpha \tag{2.9}$$

since s.e. $(\hat{p}) = \sqrt{[\hat{p}(1 - \hat{p})/n]}$. This leads to the most commonly used form of confidence interval for p, with confidence limits of $\hat{p} - z_{\alpha/2}\,\text{s.e.}\,(\hat{p})$ and $\hat{p} + z_{\alpha/2}\,\text{s.e.}\,(\hat{p})$.

Example 2.1 A seed germination test
In a germination test, seven seeds out of twelve were observed to germinate under a particular set of experimental conditions. The germination probability is estimated by the proportion of seeds that germinate, which is $7/12 = 0.583$. The standard error of the estimate is $\sqrt{[(0.583 \times 0.417)/12]} = 0.142$.

For comparative purposes, a 95% confidence interval for the true germination probability will be obtained using each of the three methods described in section 2.2.1. We first use the exact method, for which equations (2.4) and (2.5) give the required confidence limits. The value of $F_{12,\,14}(0.025)$, the upper 2.5% point of the F-distribution of $12, 14$ degrees of freedom, is 3.050, and $F_{16,\,10}(0.025) = 3.496$. Substituting these values in equations (2.4) and (2.5) and setting $n = 12$, $\hat{p} = 0.583$ gives $p_L = 0.277$ and $p_U = 0.848$. The interpretation of this interval is that there is a 95% chance that the interval $(0.28, 0.85)$ includes the true success probability.

To construct an approximate 95% confidence interval for the true success probability using the result in equation (2.7), the quadratic equation (2.8) has to be solved for p after setting $\hat{p} = 0.583$, $z_{\alpha/2} = 1.96$ and $n = 12$. The resulting equation is

$$1.3201\, p^2 - 1.4868\, p + 0.3403 = 0$$

whose roots are 0.320 and 0.807. An approximate 95% confidence interval for the true germination probability is therefore (0.32, 0.81), which is narrower than the exact interval.

Finally, using equation (2.9), approximate 95% confidence limits for p are $\hat{p} \pm 1.96$ s.e. $(\hat{p})$, which are 0.304 and 0.862. These limits, which are symmetric about $\hat{p}$, are a little closer to the exact limits than those found using the other approximate method. However, all three intervals are very similar.

2.2.2 Tests of hypotheses about p

On some occasions, interest centres on whether or not the true success probability has a particular value. As a concrete example, suppose that a seed merchant markets seed of a particular species that is claimed to have a germination rate of 80%. A sample of seed from a particular batch is subjected to a germination test, and the proportion of seeds that germinate is recorded. The seed merchant is interested in establishing whether or not the proportion of seeds that germinate is sufficiently high to justify his claimed germination rate. In examples such as this, a hypothesis test is used to indicate whether or not the data are in accord with the pre-specified value of the success probability.

Suppose that it is desired to test the hypothesis that the true success probability, p, takes the particular value p_0. This hypothesis, which is termed the **null hypothesis**, is the hypothesis that will be acted on unless the data indicate that it is untenable. The null hypothesis is denoted by H_0, and the hypothesis itself is written as $H_0: p = p_0$. Suppose also that the **alternative hypothesis** of interest is that p is less than p_0. This hypothesis is written as $H_1: p < p_0$. For example, this would be the appropriate alternative hypothesis in the example on seed germination, since the seed merchant is unlikely to be concerned about whether the germination rate is higher than claimed.

If the data consist of y successes in n binary observations, the null hypothesis will be rejected if the observed number of successes in n trials is too small. The extent to which the data provide evidence against H_0 will be reflected in the probability that, when H_0 is true, the number of successes is less than or equal to that observed. This is

$$\sum_{j=0}^{y} \binom{n}{j} p_0^j (1 - p_0)^{n-j}$$

If this probability is reasonably large, we would conclude that the observed number of successes, or fewer, is quite likely to be recorded when H_0 is true, and there would be no reason to reject the null hypothesis. On the other hand, a small value of this probability means that there is a relatively small chance of obtaining y successes, or fewer, in n observations when $p = p_0$, and the null hypothesis would be rejected in favour of the alternative. If the null hypothesis H_0 is to be rejected when the probability of y successes, or fewer, under H_0, is less than α, then α is called the **significance level** of the hypothesis test. If H_0 is rejected at this significance level, the difference between $\hat{p}$ and its hypothesised value, p_0, is then said to be significant at the $100\alpha\%$ level of significance.

An alternative to specifying a particular significance level for the hypothesis test is to determine the actual probability of y successes, or fewer, under H_0. This is called the **significance probability, probability value** or **P-value** for short. The P-value is a summary measure of the weight of evidence against H_0. It is not sensible to have a formal rule for deciding whether or not H_0 should be rejected on the basis of the P-value. However, by convention, if $P > 0.1$, there is said to be no evidence to reject H_0; if $0.05 < P \leqslant 0.1$, there is slight evidence against H_0; if $0.01 < P \leqslant 0.05$, there is moderate evidence against H_0; if $0.001 < P \leqslant 0.01$, there is strong evidence against H_0, and if $P \leqslant 0.001$, the evidence against H_0 is overwhelming. These guidelines are not hard and fast rules and should not be interpreted rigidly. For example, there is no practical difference between a P-value of 0.044 and 0.056, even though only the former indicates that the difference between $\hat{p}$ and p_0 is significant at the 5% level.

The procedure that has been described in the preceding paragraphs is said to be a **one-sided hypothesis test** since the alternative to the null hypothesis is that the success probability is less than p_0. One-sided tests are only appropriate when there is no interest whatsoever in values of p in the direction opposite to that specified in H_1. When H_1 is the hypothesis that p is less than p_0, for example, values of $\hat{p}$ larger than p_0 will not be regarded as evidence against $H_0: p = p_0$. Instead, it would be assumed that the large value of $\hat{p}$ is simply the result of chance variation. More usually, an appropriate alternative hypothesis to H_0 is that p is not equal to p_0, that is it is either greater than p_0 or less than p_0. A **two-sided hypothesis test** is needed in this situation to determine the probability that the observed number of successes in n trials is either too large or too small to have come from a binomial distribution with parameters n and p_0. The appropriate P-value for this test will be double the minimum of the two P-values obtained from the one-sided tests of the hypothesis $H_0: p = p_0$ against the alternatives $H_1: p < p_0$ and $H_1: p > p_0$. Because the use of one-sided tests is somewhat controversial, the hypothesis tests used in this book will be two-sided, unless otherwise stated.

The results of hypothesis testing procedures based on the binomial distribution are often very similar to those based on the normal approximation to the binomial distribution. Consider the null hypothesis $H_0: p = p_0$ and suppose that the alternative is the one-sided alternative hypothesis, $H_1: p < p_0$. The null hypothesis will then be rejected if $\hat{p}$ is much smaller than p_0. From expression (2.6), if the null hypothesis $H_0: p = p_0$ is true, $(p - p_0)/\sqrt{[p_0(1 - p_0)/n]}$ will have a standard normal distribution. The null hypothesis will therefore be rejected if

$$z = \frac{\hat{p} - p_0}{\sqrt{[p_0(1 - p_0)/n]}} \qquad (2.10)$$

has so large a negative value as to cast doubt on whether z could have come from a standard normal distribution. To judge if $\hat{p}$ is significantly lower than the hypothesized success probability p_0, we can compute the probability that a standard normal random variable, Z, has a value less than or equal to the value of z found using equation (2.10). When the alternative hypothesis is $H_1: p < p_0$, this probability is $P(Z \leqslant z)$, which can be obtained from tables of the standard normal distribution. The null hypothesis will be rejected if this P-value is sufficiently small. Similarly, if the alternative hypothesis is $H_1: p > p_0$, the null hypothesis is rejected if z in equation (2.10) is large, that is if $P(Z \geqslant z)$ is sufficiently small.

If the alternative to the null hypothesis is the two-sided alternative hypothesis $H_1: p \neq p_0$, large negative and large positive values of z will suggest that the null hypothesis is false. Consequently, we now require the probability that a standard normal random variable has a value as extreme as or more extreme than

$$|z| = \frac{|\hat{p} - p_0|}{\sqrt{[p_0(1 - p_0)/n]}}$$

where the notation $|z|$ denotes the **modulus** or **absolute value** of z, that is the unsigned value of z. This probability is $P(Z \leqslant -|z|) + P(Z \geqslant |z|)$, which is equal to $2P(Z \geqslant |z|)$.

In any hypothesis testing procedure, it is not absolutely necessary to compute the exact P-value of the test. Instead, the value of $|z|$ may be compared with percentage points of the standard normal distribution. If $|z|$ is found to be greater than $z_{\alpha/2}$, the two-sided $(100\alpha/2)\%$ point of the standard normal distribution, the null hypothesis would be rejected at the $100\alpha\%$ significance level. For example, a value of $|z|$ greater than 1.960 would be significant at the 5% level, a value greater than 2.576 would be significant at the 1% level and so on. For the one-sided test, H_0 would be rejected at the 5% and 1% levels if $|z|$ were greater than 1.645 and 1.282, respectively. However, it is preferable to quote a P-value, since this conveys more information about the extent to which the data provide evidence against the null hypothesis.

In the application of hypothesis testing procedures, such as those described in this section, one must bear in mind that a result that is declared to be statistically significant may not necessarily be of practical significance. In particular, if the sample size is large, the difference between an observed proportion $\hat{p}$ and the hypothesised value p_0 may be highly significant, although the actual difference between $\hat{p}$ and p_0 is negligible. For example, a difference between an observed proportion of $\hat{p} = 0.84$ and a hypothesized value of 0.80 would be significant at the 5% level if the number of observations exceeds 385. By the same token, a non-significant difference may be of practical importance. This might be the case in a medical research programme concerning the treatment of a life-threatening disease, where any improvement in the probability of survival beyond a certain time will be important, no matter how small. Generally speaking, the result of a statistical test of the validity of H_0 will often only be part of the evidence against H_0. This evidence will usually need to be combined with other, non-statistical, evidence before an appropriate course of action can be determined.

Example 2.2 A seed germination test
Suppose that the seed merchant who carried out the germination test described in Example 2.1 claims that the germination probability is 0.8. Does the observed proportion of seven out of twelve seeds germinating cast doubt on this hypothesis? The estimated germination probability of 0.583 is not encouraging!

In order to test $H_0: p = 0.8$ against the one-sided alternative hypothesis, $H_1: p < 0.8$, the required P-value is

$$\sum_{j=0}^{7} \binom{12}{j} 0.8^j (1 - 0.8)^{12-j}.$$

which is 0.073. This is quite a small probability and leads to the conclusion that the data provide some evidence to doubt the validity of the null hypothesis.

Using the approximate test based on the statistic in equation (2.10), the null hypothesis is rejected if $z = (0.583 - 0.8)/\sqrt{[(0.8 \times 0.2)/12]}$ has too large a negative value. The observed value of z is -1.88. From tables of the standard normal distribution, $P(Z \leqslant -1.88) = 0.030$. This means that the chance of obtaining a value of z as small as -1.88 or smaller, when the true success probability is 0.8, is 0.03. Because this probability is so small, we conclude that there is strong evidence that the germination probability is less than that claimed by the seed merchant. The approximate test suggests that the data provide more evidence against the null hypothesis than the 'exact' procedure. In this example, the difference between the exact and approximate P-values is a result of the small sample size on which the estimated germination probability has been based. However, both procedures lead to a similar conclusion.

2.3 COMPARISON OF TWO PROPORTIONS

Situations where interest centres on estimating or testing hypotheses about a success probability from a single sample of binary data are comparatively rare. On the other hand, the need to estimate the extent to which two proportions differ, or to test the equality of success probabilities underlying two independent samples of binary data, is more frequently encountered. An exact test appropriate to this situation is **Fisher's exact test**, and although details of this procedure will not be given here, they can be found in Everitt (1977), for example. Instead, an approximate test based on the normal approximation to the binomial distribution will be derived.

Consider two independent samples of binary data, of sizes n_1 and n_2, and denote the success probabilities underlying each sample by p_1 and p_2. If the number of successes in the two samples are y_1 and y_2, the estimated success probabilities are given by $\hat{p}_1 = y_1/n_1$ and $\hat{p}_2 = y_2/n_2$, respectively. Since these two estimates are independent of one another, the variance of $\hat{p}_1 - \hat{p}_2$, the estimated difference between the two success probabilities, is given by

$$\text{Var}(\hat{p}_1 - \hat{p}_2) = \text{Var}(\hat{p}_1) + \text{Var}(\hat{p}_2) = \frac{p_1(1-p_1)}{n_1} + \frac{p_2(1-p_2)}{n_2} \qquad (2.11)$$

and so the standard error of the estimated difference is

$$\text{s.e.}(\hat{p}_1 - \hat{p}_2) = \sqrt{\left\{ \frac{\hat{p}_1(1-\hat{p}_1)}{n_1} + \frac{\hat{p}_2(1-\hat{p}_2)}{n_2} \right\}} \qquad (2.12)$$

When n_1 and n_2 are large enough for the binomial distribution to be adequately approximated by the normal distribution, $\hat{p}_i$ will have a normal distribution with mean p_i and variance $p_i(1-p_i)/n_i$, for $i = 1, 2$. The difference $\hat{p}_1 - \hat{p}_2$ will then have an approximate normal distribution with mean $p_1 - p_2$ and variance given by equation (2.11), and so

$$\frac{\hat{p}_1 - \hat{p}_2 - (p_1 - p_2)}{\text{s.e.}(\hat{p}_1 - \hat{p}_2)}$$

is approximately normally distributed with zero mean and unit variance. This result is used in the construction of confidence intervals for the true difference between two success probabilities and for testing hypotheses about this difference.

2.3.1 Confidence intervals for the true difference

A $100(1 - \alpha)\%$ confidence interval for the true difference between two proportions is the set of values of $p_1 - p_2$ for which

$$P\left[-z_{\alpha/2} \leqslant \frac{\hat{p}_1 - \hat{p}_2 - (p_1 - p_2)}{\text{s.e.}(\hat{p}_1 - \hat{p}_2)} \leqslant z_{\alpha/2} \right] = 1 - \alpha$$

On rearrangement, $100(1 - \alpha)\%$ confidence limits for the true difference $p_1 - p_2$ are

$$\hat{p}_1 - \hat{p}_2 \pm z_{\alpha/2} \, \text{s.e.}(\hat{p}_1 - \hat{p}_2)$$

where $\text{s.e.}(\hat{p}_1 - \hat{p}_2)$ is obtained using equation (2.12).

2.3.2 Tests of hypotheses about the true difference

In order to test the null hypothesis that the true difference between the two success probabilities, $p_1 - p_2$, takes the value d, that is $H_0: p_1 - p_2 = d$, the statistic z is calculated, where

$$z = \frac{(\hat{p}_1 - \hat{p}_2) - d}{\text{s.e.}(\hat{p}_1 - \hat{p}_2)} \qquad (2.13)$$

When the alternative hypothesis is the two-sided alternative $H_1: p_1 - p_2 \neq d$, the null hypothesis is rejected if $|z|$ is sufficiently large. As in section 2.2, the appropriate P-value for this test of significance is $2P(Z \geqslant |z|)$, where Z is a standard normal random variable. Alternatively, the hypothesis is rejected at the $100\alpha\%$ level of significance if $|z|$ exceeds the upper $(100\alpha/2)\%$ point of the standard normal distribution, $z_{\alpha/2}$.

The null hypothesis of most practical interest is often the hypothesis that there is no real difference between the two success probabilities, that is $H_0: p_1 - p_2 = 0$. Although the test statistic in equation (2.13) could still be used here, the normal approximation is more accurate when the value of the standard error of $\hat{p}_1 - \hat{p}_2$ under the null hypothesis is used in the denominator of z. To obtain this standard error, suppose that the common value of the success probability under H_0 is p. Each of the two sets of binary data provide information about this common success probability, and so it is natural to combine the data from each data set to give a pooled estimate of p. This estimate is given by

$$\hat{p} = \frac{y_1 + y_2}{n_1 + n_2}$$

On writing $p_1 = p_2 = p$ in equation (2.11), the variance of $\hat{p}_1 - \hat{p}_2$ becomes

$$\text{Var}(\hat{p}_1 - \hat{p}_2) = p(1-p)\left[\frac{1}{n_1} + \frac{1}{n_2}\right]$$

and so the standard error of $\hat{p}_1 - \hat{p}_2$ is now

$$\sqrt{\left\{\hat{p}(1-\hat{p})\left[\frac{1}{n_1} + \frac{1}{n_2}\right]\right\}}$$

The standard error calculated using this formula will usually be very similar to that obtained using equation (2.12), and will only differ appreciably when the numbers of observations in the two samples are very different, or when p_1 and p_2 differ considerably.

To test the null hypothesis that $p_1 - p_2$ is zero, we calculate the value of the statistic

$$z = \frac{\hat{p}_1 - \hat{p}_2}{\sqrt{\left\{\hat{p}(1 - \hat{p})\left[\dfrac{1}{n_1} + \dfrac{1}{n_2}\right]\right\}}} \tag{2.14}$$

If the alternative hypothesis is that the two success probabilities are unequal, significantly large values of $|z|$ lead to the rejection of the null hypothesis.

Example 2.3 Smoked mice
To illustrate the results in sections 2.3.1 and 2.3.2, the data from an experiment described in Example 1.1 will be used. In this experiment, one group of mice was exposed to cigarette smoke while another group was used as a control. After a period of one year, 21/23 mice in the group exposed to cigarette smoke had developed a lung tumour, while 19/32 mice in the control group had developed a tumour. If p_1 and p_2 are the true probabilities of developing a tumour for mice in the treated and control groups, respectively, these probabilities are estimated by $\hat{p}_1 = 0.913$ and $\hat{p}_2 = 0.594$. The estimated difference between the two probabilities is $\hat{p}_1 - \hat{p}_2 = 0.319$, which has a standard error given by

$$\text{s.e.}\,(\hat{p}_1 - \hat{p}_2) = \sqrt{\left\{\frac{0.913 \times 0.087}{23} + \frac{0.594 \times 0.406}{32}\right\}} = 0.105$$

Approximate 95% confidence limits for the true difference in the response probabilities are $\hat{p}_1 - \hat{p}_2 \pm 1.96 \text{ s.e.}\,(\hat{p}_1 - \hat{p}_2)$, that is $0.319 \pm 1.96 \times 0.105$, from which $(0.114, 0.525)$ is the required confidence interval. This means that there is a 95% chance that the interval $(0.114, 0.525)$ includes the true difference between the probabilities of mice in the treated and control groups developing a tumour. The fact that this interval does not include zero indicates that the estimated difference between the two probabilities is significant at the 5% level. To confirm this, the validity of the hypothesis $H_0: p_1 - p_2 = 0$ will be examined by calculating the value of the statistic in equation (2.14). The common estimate of the response probability is $\hat{p} = (21 + 19)/(23 + 32) = 0.727$, and so

$$z = \frac{0.913 - 0.594}{\sqrt{\left\{0.727(1 - 0.727)\left[\dfrac{1}{23} + \dfrac{1}{32}\right]\right\}}} = 2.623$$

The *P*-value for this test is $2P(Z \geqslant |z|)$, that is $2P(Z \geqslant 2.623)$, which from tables of the standard normal distribution is 0.009. The fact that this probability is so small means that it is very unlikely that a value of z of 2.623 could have arisen by chance when there is no difference between the two response probabilities. We therefore conclude that the data provide very strong evidence that the two probabilities of tumour development are different. Since $\hat{p}_1$ is very much greater that $\hat{p}_2$, mice exposed to the cigarette smoke are much more likely to develop a lung tumour than those in the unexposed group.

2.3.3 Alternative forms of the test statistic

There are a number of alternative, but equivalent, forms of the test statistic in equation (2.14) which can be used when a two-sided alternative hypothesis is contemplated. These expressions are derived from a general result which states that if the random variable Z has a standard normal distribution, then the square of that random variable, Z^2, has a chi-squared distribution on one degree of freedom, written as χ_1^2. It then follows that

$$z^2 = \frac{(\hat{p}_1 - \hat{p}_2)^2}{\hat{p}(1 - \hat{p})\left[\dfrac{1}{n_1} + \dfrac{1}{n_2}\right]} \tag{2.15}$$

has a χ_1^2-distribution. If the upper $100\alpha\%$ point of the χ^2-distribution on one degree of freedom is denoted by $\chi_1^2(\alpha)$, then the null hypothesis $H_0: p_1 - p_2 = 0$ will be rejected at the $100\alpha\%$ level of significance if z^2 exceeds $\chi_1^2(\alpha)$. For example, if $\alpha = 0.1, 0.05$ or 0.01, the values of $\chi_1^2(\alpha)$ are 2.706, 3.841 and 6.635, respectively. In fact, these are the same as the values of $z_{\alpha/2}^2$, the square of the two-sided $(100\alpha/2)\%$ points of the standard normal distribution.

There is no advantage in using the test statistic in equation (2.15) compared to that in equation (2.14). However, alternative expressions for the statistic z^2 suggest how it can be generalised to enable more than two proportions to be compared. The statistic in equation (2.15) can be written as

$$z^2 = \frac{n_1 n_2 (\hat{p}_1^2 - 2\hat{p}_1\hat{p}_2 + \hat{p}_2^2)}{(n_1 + n_2)\hat{p}(1 - \hat{p})} \tag{2.16}$$

The numerator of z^2 in equation (2.16) can be expressed as

$$(n_1 + n_2)(n_1\hat{p}_1^2 + n_2\hat{p}_2^2) - (n_1\hat{p}_1 + n_2\hat{p}_2)^2,$$

which is

$$(n_1 + n_2)(n_1\hat{p}_1^2 + n_2\hat{p}_2^2) - (n_1 + n_2)^2 \hat{p}^2,$$

since $\hat{p} = (y_1 + y_2)/(n_1 + n_2) = (n_1\hat{p}_1 + n_2\hat{p}_2)/(n_1 + n_2)$

Now,

$$n_1\hat{p}_1^2 + n_2\hat{p}_2^2 = n_1(\hat{p}_1 - \hat{p})^2 + n_2(\hat{p}_2 - \hat{p})^2 + (n_1 + n_2)\hat{p}^2$$

so that the numerator of equation (2.16) reduces to

$$(n_1 + n_2)[n_1(\hat{p}_1 - \hat{p})^2 + n_2(\hat{p}_2 - \hat{p})^2]$$

The statistic in equation (2.16) then becomes

$$z^2 = \frac{n_1(\hat{p}_1 - \hat{p})^2 + n_2(\hat{p}_2^2 - \hat{p})^2}{\hat{p}(1 - \hat{p})}$$

that is,

$$z^2 = \sum_{i=1}^{2} \frac{n_i(\hat{p}_i - \hat{p})^2}{\hat{p}(1 - \hat{p})} \tag{2.17}$$

We will return to this version of the test statistic in section 2.4.

There is a further expression for z^2 which shows that the test of the hypothesis that there is no difference between two proportions is actually equivalent to the well-known chi-squared-test of association in 2×2 tables. The term

$$\frac{n_i(\hat{p}_i - \hat{p})^2}{\hat{p}(1 - \hat{p})}$$

in equation (2.17) can be written as

$$\frac{(y_i - n_i\hat{p})^2}{n_i\hat{p}} + \frac{\{(n_i - y_i) - n_i(1 - \hat{p})\}^2}{n_i(1 - \hat{p})}$$

Under the hypothesis that the two success probabilities are equal to p, $n_i p$ and $n_i(1 - p)$ are the expected numbers of successes and failures amongst the n_i observations in the ith set of binary data, $i = 1, 2$. Hence, $n_i\hat{p}$ and $n_i(1 - \hat{p})$ are the corresponding **estimated expected** or **fitted** numbers of successes and failures.

An equivalent way of computing the estimated expected counts follows from summarizing the observed binary data to give the following 2×2 table of counts.

	Set 1	Set 2
Number of successes	y_1	y_2
Number of failures	$n_1 - y_1$	$n_2 - y_2$
Total	n_1	n_2

Writing N for the total number of binary observations, so that $N = n_1 + n_2$, the estimated expected numbers of successes and failures in the ith data set can be expressed as $n_i(y_1 + y_2)/N$ and $n_i(N - y_1 - y_2)/N$, respectively, for $i = 1, 2$. The estimated expected counts can therefore be obtained from the row and column totals of this contingency table. In particular, the estimated expected count in the ith row and the jth column, $i = 1, 2; j = 1, 2$, is

$$(\text{Row total for row } i \times \text{Column total for column } j)/N$$

If the four observed counts in this contingency table are denoted by O_1, O_2, O_3 and O_4, and the corresponding estimated expected counts are E_1, E_2, E_3 and E_4, the statistic in equation (2.17) becomes

$$\sum_{i=1}^{4} \frac{(O_i - E_i)^2}{E_i} \tag{2.18}$$

This is the statistic used in testing for the association between two factors in a 2×2 contingency table. This statistic is generally denoted by X^2 and, because it is algebraically equivalent to z^2 in equation (2.12), X^2 has a χ^2-distribution on one degree of freedom. This statistic was first proposed by Karl Pearson and is generally referred to as **Pearson's X^2-statistic**.

2.3.4 Odds and the odds ratio

It is sometimes helpful to describe the chance that a binary response variable leads to a success in terms of the **odds** of that event. The odds of a success is defined to be the ratio of the probability of a success to the probability of a failure. Thus if p is the true success probability, the odds of a success is $p/(1 - p)$. If the observed binary data consist of y successes in n observations, the odds of a success can be estimated by $\hat{p}/(1 - \hat{p}) = y/(n - y)$.

When two sets of binary data are to be compared, a relative measure of the odds of a success in one set relative to that in the other is the **odds ratio.** Suppose that p_1 and p_2 are the success probabilities in these two sets, so that the odds of a success in the ith set is $p_i/(1 - p_i)$, $i = 1, 2$. The ratio of the odds of a success in one set of binary data relative to the other is usually denoted by ψ, so that

$$\psi = \frac{p_1/(1 - p_1)}{p_2/(1 - p_2)}$$

is the odds ratio. When the odds of a success in each of the two sets of binary data are identical, ψ is equal to one. This will happen when the two success probabilities are equal. Values of ψ less than one suggest that the odds of a success are less in the first set of data than in the second, while an odds ratio greater than one indicates that the odds of a success are greater in the first set of data. The odds ratio is a measure of the difference between two success probabilities which can take any positive value, unlike the difference between two success probabilities, $p_1 - p_2$, which is restricted to the range $(-1, 1)$.

In order to estimate the ratio of the odds of a success in one data set relative to another, suppose that the binary data are arranged as in the following 2×2 contingency table.

	Number of successes	Number of failures
Data set 1	a	b
Data set 2	c	d

The estimated success probabilities in the two data sets are $\hat{p}_1 = a/(a + b)$ and $\hat{p}_2 = c/(c + d)$, and so the estimated odds ratio, $\hat{\psi}$, is given by

$$\hat{\psi} = \frac{\hat{p}_1/(1 - \hat{p}_1)}{\hat{p}_2/(1 - \hat{p}_2)} = \frac{ad}{bc} \tag{2.19}$$

This estimate is the ratio of the products of the two pairs of diagonal elements in the above 2×2 table, and for this reason $\hat{\psi}$ is sometimes referred to as the **cross-product ratio**.

In order to construct a confidence interval for the true odds ratio, we first note that the logarithm of the estimated odds ratio is better approximated by a normal distribution than the odds ratio itself, especially when the total number of binary observations is not very large. The approximate standard error of the estimated log odds ratio, $\log \hat{\psi}$, can be shown to be given by

$$\text{s.e.} (\log \hat{\psi}) \approx \sqrt{\left(\frac{1}{a} + \frac{1}{b} + \frac{1}{c} + \frac{1}{d} \right)} \tag{2.20}$$

An outline proof of this result is given in Schlesselman (1982). An approximate $100(1 - \alpha)\%$ confidence interval for $\log \psi$ ranges from $[\log \hat{\psi} - z_{\alpha/2} \text{ s.e.} (\log \hat{\psi})]$ to $[\log \hat{\psi} + z_{\alpha/2} \text{ s.e.} (\log \hat{\psi})]$, where $z_{\alpha/2}$ is the upper $(100\alpha/2)\%$ point of the standard normal distribution. These confidence limits are then exponentiated to give a corresponding interval for ψ itself. Notice that this method of constructing a confidence interval for ψ ensures that both limits will always be non-negative, which, since an odds ratio cannot be negative, is a natural requirement.

If the approximate standard error of $\hat{\psi}$ itself is required, this can be found using the result that s.e. $(\hat{\psi}) \approx \hat{\psi}$ s.e. $(\log \hat{\psi})$, and so

$$\text{s.e.} (\hat{\psi}) \approx \hat{\psi} \sqrt{\left(\frac{1}{a} + \frac{1}{b} + \frac{1}{c} + \frac{1}{d} \right)} \tag{2.21}$$

However, for the reason given in the previous paragraph, this expression should not be used as a basis for a confidence interval for ψ.

Because an odds ratio of unity will be obtained when the success probabilities in two sets of binary data are equal, the null hypothesis that the true odds ratio is equal to unity, $H_0: \psi = 1$, can be tested using the statistics in equations (2.14) and (2.17) or the X^2-statistic of equation (2.18).

Example 2.4 Smoked mice
For the data from Example 1.1, the estimated odds of a tumour occurring in mice exposed to cigarette smoke is $21/(23 - 21) = 10.50$, while the corresponding estimated odds of a tumour for mice in the control group is $19/(32 - 19) = 1.46$. The estimated ratio of the odds of a tumour occurring in the treated group relative to the control group is given by

$$\hat{\psi} = \frac{21 \times 13}{2 \times 19} = 7.184$$

The interpretation of this odds ratio is that the odds of tumour occurrence in the treated group is more than seven times that for the control group.

A 95% confidence interval for the true log odds ratio can be found from the standard error of $\log \hat{\psi}$. Using equation (2.20), this is given by

$$\text{s.e.} (\log \hat{\psi}) = \sqrt{\left(\frac{1}{19} + \frac{1}{21} + \frac{1}{13} + \frac{1}{2}\right)} = 0.823$$

A 95% confidence interval for $\log \psi$ ranges from $[\log(7.184) - 1.96 \times 0.823]$ to $[\log(7.184) + 1.96 \times 0.823]$, that is from 0.359 to 3.585. A corresponding confidence interval for the true odds ratio is the interval from $e^{0.359}$ to $e^{3.585}$, that is, from 1.43 to 36.04. This interval does not include unity, which indicates that the evidence that the odds of a tumour is greater amongst mice exposed to the cigarette smoke is certainly significant at the 5% level.

2.4 COMPARISON OF TWO OR MORE PROPORTIONS

Suppose that n samples of binary data are now available, where the ith set contains y_i successes in n_i observations, $i = 1, 2, ..., n$. Assume that the n_i observations in the ith data set have a common success probability p_i, which is estimated by $\hat{p}_i = y_i/n_i$. The statistic in equation (2.17) can readily be extended to test the hypothesis that the n success probabilities, $p_1, p_2, ..., p_n$, are all equal. The appropriate test statistic is

$$\sum_{i=1}^{n} \frac{n_i(\hat{p}_i - \hat{p})^2}{\hat{p}(1 - \hat{p})} \qquad (2.22)$$

which can be shown to have a χ^2-distribution with $(n - 1)$ degrees of freedom. If the value of this test statistic exceeds the upper $100\alpha\%$ point of the χ^2-distribution on $(n - 1)$ degrees of freedom, the null hypothesis $H_0: p_1 = p_2 = \cdots = p_n$ would be rejected at the $100\alpha\%$ level in favour of the alternative which specifies that not all of the n success probabilities are equal.

The n sets of binary data can be summarised to give the $2 \times n$ contingency table shown below.

	Set 1	Set 2	. . .	Set n
Number of successes	y_1	y_2	. . .	y_n
Number of failures	$n_1 - y_1$	$n_2 - y_2$	. . .	$n_n - y_n$
Total	n_1	n_2	. . .	n_n

Let $\hat{p}$ be the estimated common success probability across the n sets of binary data, so that $\hat{p} = \sum y_i / \sum n_i$. The estimated expected numbers of successes, under the null hypothesis that the n success probabilities are all equal, are then $n_i \hat{p}$, and the corresponding estimated expected numbers of failures are $n_i(1 - \hat{p})$, for $i = 1, 2, \ldots, n$. The estimated expected counts may also be calculated using the result that the estimated expected count in the ith row and jth column of the $2 \times n$ contingency table is

$$(\text{Row total for row } i \times \text{Column total for column } j)/N$$

for $i = 1, 2; j = 1, 2, \ldots, n$, where $N = \sum n_i$. If the $2n$ observed counts in this table are denoted $O_1, O_2, \ldots, O_{2n}$, and the estimated expected counts under $H_0: p_1 = p_2 = \cdots = p_n$, are denoted by $E_1, E_2, \ldots, E_{2n}$, the statistic in equation (2.22) is equivalent to

$$X^2 = \sum_{i=1}^{2n} \frac{(O_i - E_i)^2}{E_i} \tag{2.23}$$

This is Pearson's X^2-statistic, which has a χ^2_{n-1}-distribution under the null hypothesis of equal success probabilities.

The statistic in equation (2.22) may be further generalized to the situation where the n success probabilities are not all equal under the null hypothesis. First note that equation (2.22) can be written in the form

$$\sum_{i=1}^{n} \frac{(y_i - n_i \hat{p})^2}{n_i \hat{p}(1 - \hat{p})}$$

If now the estimated expected success probabilities under some null hypothesis are $\hat{p}_i$, rather than $\hat{p}$, the statistic becomes

$$\sum_{i=1}^{n} \frac{(y_i - n_i \hat{p}_i)^2}{n_i \hat{p}_i(1 - \hat{p}_i)} \tag{2.24}$$

which also has a χ^2-distribution. However, the number of degrees of freedom of the statistic will not now be $(n - 1)$. In this expression, $\hat{p}_i$ is not equal to y_i/n_i unless the null hypothesis specifies that the p_i are all different, in which case the value of the test statistic in equation (2.24) is zero. We will return to this statistic in section 3.8.

Example 2.5 Propagation of plum root stocks
The experiment described in Example 1.2 is concerned with the proportion of root cuttings from a particular species of plum that are still alive after a given period of time. The cuttings used in the experiment were either long or short, and were either planted as soon as they were taken or in the following spring. The results of the experiment were given in Table 1.2, but are also given in Table 2.1 in the form of a contingency table.

Table 2.1 The data from Example 1.2 arranged as a contingency table

Ultimate condition of the cutting	Planted at once		Planted in spring	
	Short	Long	Short	Long
Alive	107	156	31	84
Dead	133	84	209	156
Total	240	240	240	240

A first step in the analysis of these data might be to compare the proportions of cuttings that survive under the four experimental treatments, for which the test statistic in expression (2.22) may be used. The combined estimate of the probability of a root-cutting surviving, over the four experimental conditions, is given by

$$\hat{p} = \frac{107 + 156 + 31 + 84}{(4 \times 240)} = 0.394$$

The individual estimates of the survival probabilities are 0.446, 0.129, 0.650 and 0.350, and the resulting value of the statistic in expression (2.22) is 142.98. The same value of the test statistic can be obtained using the equivalent statistic in equation (2.23). Under the hypothesis that the four survival probabilities are equal, the estimated expected number of live cuttings at the end of the experiment is 94.5 for each of the four combinations of length of cutting and time of planting. Similarly, the estimated expected number of cuttings that fail is 145.5 for each of the four experimental conditions. Direct application of the result in equation (2.23) then gives $X^2 = 142.98$, in agreement with that using expression (2.22), as it must be. Judged against percentage points of the χ^2-distribution on three degrees of freedom, the observed value of this test statistic is highly significant ($P < 0.001$), and so we conclude that there is very strong evidence that the four proportions are not homogeneous. This feature is really quite apparent from the observed data, and so the result of the hypothesis testing procedure is not very surprising. Indeed, the hypothesis that is being tested here is too general, and for this reason it is not particularly interesting.

The data that were obtained in this experiment have a well-defined structure, in that there are two planting times for each of the two lengths of cutting used. More interesting and relevant hypotheses concern whether or not there is a difference in the survival probabilities for the two planting dates, and whether or not the two lengths of cutting lead to different survival probabilities. To compare the survival probabilities for the two planting dates, the data for each length of cutting might be combined. Then, $(107 + 156)/480$ cuttings survive when planted at once, while $(31 + 84)/480$ survive when planted in spring. These two proportions can be compared using the methods of section 2.3. A 95% confidence interval for the true difference between the survival probabilities of cuttings planted at once and in spring is $(0.250, 0.367)$. The fact that this interval does not incude zero suggests that the time of planting is important. Likewise, the overall proportions of long and short cuttings that survive can also be compared after combining the data for the two planting times. A 95% confidence interval for the true difference between the probabilities of long and short cuttings surviving is $(0.152, 0.273)$, suggesting that length of cutting is also an important factor. These results strongly suggest that the cuttings do best if long cuttings are used and planted at once.

The method of analysis described in the previous paragraph compares the survival probabilities for the two planting dates ignoring length of cutting, and those for the two lengths of cutting ignoring time of planting. This procedure can lead to misleading results as shown in Example 2.6 below. A more satisfactory way of investigating the extent to which length of cutting, say, is important, is to compare the survival probabilities of cuttings of each length when planted at once, and those of each length when planted in spring. If the difference between the survival probabilities of long and short cuttings is not consistent between the two planting times, we would say that length of cutting and time of planting interact. It would then be appropriate to present the results for the effect of length of cutting on the ultimate condition of the cutting separately for each of the two planting times. On the other hand, if the difference between the survival probabilities for long and short cuttings is consistent between the two planting times, the information about this difference at the two planting times might be combined. This would give a more appropriate summary of the effect of length of cutting on the probability that a cutting is alive at the end of the experiment. Calculated in this way, the effect of length of cutting on survival probability is said to be **adjusted** for time of planting. A methodology for combining the information in the 2×2 table of ultimate condition of the cutting by length of cutting, for each of the two planting times, is therefore required.

In the discussion of Example 2.5, it was pointed out that misleading conclusions can follow from combining data over the levels of another factor. This phenomenon is particularly well illustrated by the following example.

Example 2.6 Survival of infants
Bishop (1969) describes a study to investigate the effect of the duration of pre-natal care, and the place where that care is received, on the survival of infants. The duration of care is classified as less than one month or at least one month, and the clinic attended by the mother is labelled as clinic A or clinic B. The proportions of infants who die in the first month of life, for each combination of clinic and duration of pre-natal care, are presented in Table 2.2.

Table 2.2 Proportion of infants who die during the first month of life, classified by clinic and duration of pre-natal care

Clinic	Duration of pre-natal care	Proportion of infants who die
A	< 1 month	3/179
A	≥ 1 month	4/297
B	< 1 month	17/214
B	≥ 1 month	2/25

Consider first the data from clinic A. The value of the z^2-statistic in equation (2.17) for testing the null hypothesis that there is no difference between the probabilities of death for infants who receive more rather than less pre-natal care, is 0.08 on 1 d.f. This is clearly not significant and so for this clinic, the survival probabilities can be taken as equal. For clinic B, the value of the z^2-statistic for comparing the two probabilities of death is zero to three decimal places, and so again there is no evidence that duration of care affects the survival of an infant. Taken together, these results show that survival is unrelated to the duration of pre-natal care.

When the data from the two clinics are combined, the proportion of infants who die after the mothers have received pre-natal care for less than one month is 20/393, while the proportion who die when the duration of pre-natal care is at least one month is 6/322. Now, the value of the z^2-statistic for comparing these two proportions is 5.26, which is significant at the 2.5% level of significance. This suggests that survival is affected by the duration of pre-natal care, and that those mothers who receive less care are at greater risk of losing their babies. This conclusion conflicts with the results from separate analyses of the data from the two clinics, and illustrates the danger of ignoring the factor associated with clinic when analysing these data.

There are a number of methods which can be used to combine information from two or more 2 × 2 contingency tables, such as that due to Cochran (1954). Alternatively, the odds ratios from each of the individual 2 × 2 tables can be combined using the Mantel–Haenszel procedure (Mantel and Haenszel, 1959) to give an estimate of the common odds ratio. These methods are well-suited to hand calculation, but they have now been superceded by methods of analysis that exploit modern computing capabilities, and so fuller details will not be given here. Instead, issues such as those raised in Example 2.5 are much more conveniently handled by using a modelling approach to analyse these data. The modelling approach will be described in the next chapter, and we will return to the analysis of the data on the propagation of plum root-stock cuttings in Example 3.5.

FURTHER READING

A comprehensive summary of the properties of the binomial distribution is given by Johnson and Kotz (1969). Reviews of standard methods for estimating and comparing two or more proportions are included in Everitt (1977), Snedecor and Cochran (1980), Steel and Torrie (1980), Armitage and Berry (1987), and many other introductory textbooks on statistical methods.

Some authors recommend that **continuity corrections** be used to improve the normal approximation to the binomial distribution. Since there is no general agreement on whether or not they should be used, they have not been referred to in this chapter. Armitage and Berry (1987) and McCullagh and Nelder (1989) discuss the effect of these corrections on the normal approximation.

The relationship between the binomial distribution and the F-distribution, used in obtaining an exact confidence interval for a success probability, is discussed by Jowett (1963). Cochran's method for combining information about associations in a series of 2×2 tables is described in Everitt (1977) and Armitage and Berry (1987). The Mantel–Haenszel procedure is also described in Armitage and Berry (1987) and in textbooks on quantitative epidemiology, such as Breslow and Day (1980) and Schlesselman (1982).

3

Models for binary and binomial data

The techniques for processing binary data presented in the previous chapter can be useful when the structure of the data is not particularly complex. In particular, when the data are from a number of similarly treated units and the aim is simply to estimate a proportion and its associated standard error, or when it is desired to compare the proportions of individuals in a number of groups who respond to some stimulus, the methods described in Chapter 2 may be suitable. But when, for example, a single binary response is recorded on individuals who differ from each other on the basis of other variables, or when the responses are recorded on individuals who are grouped according to a number of different factors, the use of these relatively unsophisticated methods will generally be uninformative.

An alternative approach to analysing binary and binomial response data is based on the construction of a statistical model to describe the relationship between the observed response and explanatory variables. This approach is equally applicable to data from experimental studies where individual experimental units have been randomized to a number of treatment groups, such as in Examples 1.1–1.6 or to observational studies where individuals have been sampled from some conceptual population by random sampling, such as in Examples 1.7–1.10. Furthermore, when a modelling approach is adopted, the methods presented in Chapter 2 are both unified and extended.

3.1 STATISTICAL MODELLING

The basic aim of modelling is to derive a mathematical representation of the relationship between an observed response variable and a number of explanatory variables, together with a measure of the inherent uncertainty of any such relationship. Having obtained such a representation, there are a number of different uses to which the resulting model may be put. The objective might be to determine if there really is a relationship between a particular response and a number of other variables, or to study the pattern of any such relationship. For example, these might be the aims in Example 1.2 concerning the relationship between the survival rate of plum root-stock cuttings and the time of planting and length of the cutting. Modelling may motivate the study of the underlying reasons for any model structure found to be appropriate, which may in turn throw new light on the data generating process under study. For example, in Example 1.7, a modelling approach may highlight whether both proteins influence the probability of an ESR value in excess of 20, which may prompt a study of the physiological reasons underlying this phenomenon. Prediction may be the primary objective, such as in Example 1.9, where one wants to predict the extent to which cancer has spread to the lymph nodes in patients presenting with prostatic cancer. Another objective might be to estimate in what way the response would change if certain explanatory variables changed in value. For example, in Example 1.6, we might enquire about the effect of increasing the dose of the protective serum from 0.01 to 0.02 cc on the survival of mice exposed to pneumococci. Of course, these objectives are not mutually exclusive and one might hope to realise two or perhaps more of them simultaneously.

Statistical models constructed for response variables are at best an approximation to the manner in which some observable variable depends on other variables. No statistical model can be claimed to represent truth and, by the same token, no one model can be termed the correct model. Some models will be more appropriate than others, but typically, for any set of data, there will be a number of models which are equally well suited to the purpose in hand, and the basis for choosing a single model from amongst them will not rest on statistical grounds alone. Statistical models are essentially descriptive and, inasmuch as they are based on experimental or observational data, may be described as **empirical models**. The structure of an empirical model will therefore be crucially dependent upon the data. In contrast, **mathematical models** are largely based on the underlying subject area, but in this book, statistical models alone will be discussed.

A widely used statistical model is one where the response variable is expressed as the sum of two components as follows:

$$\text{response variable} = \text{systematic component} + \text{residual component}$$

The **systematic component** summarises how the variability in the response of individual experimental units is accounted for by the values of certain variables or the levels of certain factors measured on those units. The remaining variation is referred to as the **residual component**. This component summarises the extent to which the observed response deviates from the average or expected response, described by the systematic part of the model. In a satisfactory model, the systematic component will account for all non-random variation in the response. The unexplained variation may of course be due to variables that have not been measured – it is impossible to measure everything – but if the model is valid, this random component will be taken to be a manifestation of the inherent variability of the process that leads to the measured response. An important part of any modelling exercise will be to determine whether the residual component of variation may be regarded as random or whether it still includes systematic variation, indicating that the model is unsatisfactory in some respect. This aspect of modelling is discussed in Chapter 5.

3.2 LINEAR MODELS

When a response variable is continuous, models for n observations $y_1, y_2, \ldots, y_n$ are often of the form

$$y_i = \beta_0 + \beta_1 x_{1i} + \beta_2 x_{2i} + \cdots + \beta_k x_{ki} + \varepsilon_i$$

which expresses the fact that the response for the ith observation, y_i, $i = 1, 2 \ldots, n$, depends linearly on the values of k **explanatory variables** labelled $x_1, x_2, \ldots, x_k$, through unknown parameters $\beta_0, \beta_1, \ldots, \beta_k$. The explanatory variables are assumed to be fixed and known without error. Since the response is linear in the unknown parameters, it is called a **linear model**. The term ε_i is an unobservable random variable which represents the residual variation and will be assumed to have zero mean, implying that all systematic variation in the response has been accounted for. The further assumption that ε_i has constant variance σ^2 will often be made, and for significance testing it might also be assumed that the ε_i have a normal distribution. Under the assumption that the mean and variance of ε_i are given by $E(\varepsilon_i) = 0$ and $\mathrm{Var}(\varepsilon_i) = \sigma^2$, respectively, for $i = 1, 2, \ldots, n$, the expected value of the ith observation, y_i, is

$$E(y_i) = \beta_0 + \beta_1 x_{1i} + \beta_2 x_{2i} + \cdots + \beta_k x_{ki}$$

and the variance of that observation is $\mathrm{Var}(y_i) = \sigma^2$. Strictly speaking, it is a random variable Y_i associated with the ith observation y_i, rather than y_i itself, which has an expected value and variance, but for simplicity of presentation, this distinction will not usually be made in this book. If the ε_i are further assumed to be normally distributed with zero mean and common variance σ^2, written $N(0, \sigma^2)$, then as a consequence, y_i will have an $N(E(y_i), \sigma^2)$ distribution, where each of the n observations typically has a different mean dependent upon the values of the k explanatory variables x_{ji}, $j = 1, 2, \ldots, k$, for that observation.

The right-hand side of the expression for $E(y_i)$, the **linear systematic component** of the model, is conveniently denoted by η_i. Then, in general, $\eta_i = \sum_{j=0}^{k} \beta_j x_{ji}$, where $x_{0i} = 1$, and the model can be written as $y_i = \eta_i + \varepsilon_i$. A particular example is the simple linear regression model $y_i = \beta_0 + \beta_1 x_i + \varepsilon_i$, where the linear systematic component of the model is $\eta_i = \beta_0 + \beta_1 x_i$. By defining the k explanatory variables to be powers of a single explanatory variable x, that is, writing $x_j = x^j$, $j = 1, 2, \ldots, k$, this class of models is seen to include polynomial regression models. A linear model may also include terms corresponding to qualitative variables known as **factors** which can take a limited set of values known as the **levels** of the factor. For example, in Example 1.3 concerning the germination of seeds, there are two factors: variety of seed with two levels and root extract also with two levels.

To see how factors are included in a model, suppose that A is a factor with m distinct levels. These will be coded as 1, 2,..., m, although in practice, they will usually have names or numerical values attached to them. A linear model may then include a term α_j, $j = 1, 2,..., m$, representing the effect of the jth level of A on the expected response. The α_j are known as the **main effects** of the factor. Models that contain factors can still be expressed as a linear combination of explanatory variables by defining **indicator** (or **dummy**) **variables**.

3.2.1 Use of indicator variables

If a model includes a factor with m levels, there are various ways in which appropriate indicator variables can be defined. One way is to introduce the m indicator variables $x_1, x_2,..., x_m$ which take the values shown in the following table.

Factor level	x_1	x_2	x_3	. . .	x_m
1	1	0	0	. . .	0
2	0	1	0	. . .	0
3	0	0	1	. . .	0
. . . .					
m	0	0	0	. . .	1

Fitting the term α_j in a model is then equivalent to fitting these m indicator variables with coefficients $\alpha_1, \alpha_2, \ldots, \alpha_m$, that is, α_j is replaced by $\alpha_1 x_1 + \alpha_2 x_2 + \cdots + \alpha_m x_m$.

When data are cross-classified by a number of factors, a series of terms may be included in the model to represent individual effects for each combination of levels of the factors. Such effects are known as **interactions**. For example, if there are two factors A and B where the response at particular levels of A depends on the level of B, A and B are said to interact. It would then be appropriate to include terms representing the main effects of the two factors, say, α_j and β_k, as well as terms $(\alpha\beta)_{jk}$ representing the two-factor interaction. By defining a set of indicator variables for each of the two factors, an interaction term is represented by a product of two indicator variables, one from each set. For example, suppose that the two factors A and B have three and two levels, respectively. Define two sets of indicator variables u_1, u_2, u_3 and v_1, v_2 corresponding to the levels of A and B as in the following two tables.

Level of A	u_1	u_2	u_3	Level of B	v_1	v_2
1	1	0	0	1	1	0
2	0	1	0	2	0	1
3	0	0	1			

A linear model that includes the main effects of both A and B, that is, terms α_j and β_k, can then be expressed as $\alpha_1 u_1 + \alpha_2 u_2 + \alpha_3 u_3 + \beta_1 v_1 + \beta_2 v_2$. To incorporate parameters $(\alpha\beta)_{jk}$ in the model, the following set of terms would also be included:

$$(\alpha\beta)_{11} u_1 v_1 + (\alpha\beta)_{12} u_1 v_2 + (\alpha\beta)_{21} u_2 v_1 + \cdots + (\alpha\beta)_{32} u_3 v_2$$

The products $u_j v_k$ are now explanatory variables with coefficients $(\alpha\beta)_{jk}$.

It is often necessary to include mixed terms in a model for it to correspond to the situation where the coefficient of an explanatory variable varies according to the levels of some factor. Models that contain terms that are combinations of factors and variates are encountered when comparing regression relationships between groups of individuals. The dependence of the coefficient of x on the levels of a factor A would be depicted by including the term $\alpha_j x$ in the linear model. To include such a term, indicator variables u_j for the factor A are defined and each of these is then multiplied by x. The resulting series of explanatory variables, $u_j x$, then have coefficients α_j.

As an illustration, suppose that there are six pairs of observations (x_i, y_i), $i = 1, 2, \ldots, 6$, of which the first three are observed at the first level of a factor A, and the next three are observed at the second level of A. If the slope of a straight line relationship between y and x is to depend on the level of A, the explanatory variable x is included in the model with a different coefficient for each level of A. This is achieved by defining two indicator variables, u_1 and u_2, as in the following table. The products $u_1 x$ and $u_2 x$ in the final two columns of this table would then be included as terms in the model. The coefficients of these two terms would be the required coefficients of x at each level of the factor A.

Observation	y	x	Level of A	u_1	u_2	$u_1 x$	$u_2 x$
1	y_1	x_1	1	1	0	x_1	0
2	y_2	x_2	1	1	0	x_2	0
3	y_3	x_3	1	1	0	x_3	0
4	y_4	x_4	2	0	1	0	x_4
5	y_5	x_5	2	0	1	0	x_5
6	y_6	x_6	2	0	1	0	x_6

The particular choice of indicator variables used in the preceding paragraphs does mean that when a constant term is included in the model along with a term such as α_j, or when two or more factors are present, it is not possible to obtain unique estimates of all the parameters in the model. For example, consider the model that includes a constant and the term α_j, $j = 1, 2, \ldots, m$, representing the effect of the jth level of a factor with m levels. This model contains more terms for the effect α_j than the available number of degrees of freedom for it, and the model is said to be **overparameterised**. This difficulty can be resolved by defining just $m - 1$ indicator variables, $x_2, x_3, \ldots, x_m$, say, as in the following table.

Factor level	x_2	x_3	x_4	$\cdots$	x_m
1	0	0	0	$\cdots$	0
2	1	0	0	$\cdots$	0
3	0	1	0	$\cdots$	0
4	0	0	1	$\cdots$	0
$\cdots$					
m	0	0	0	$\cdots$	1

Fitting the term α_j is then equivalent to including $x_2, x_3, \ldots, x_m$ in the model with coefficients $\alpha_2, \alpha_3, \ldots, \alpha_m$. The value of α_1 is set to zero and the model is no longer overparameterized. We will return to this point in the discussion of Example 3.3 in section 3.7.

Most computer packages for statistical modelling generate indicator variables automatically when a term in the model has been specified as a factor, making it easy to incorporate variates, factors and combinations of the two in a model. However, because there are a number of different ways of coding these indicator variables, it is essential to know which coding has been used if the estimated values of their coefficients are to be interpreted correctly.

3.3 METHODS OF ESTIMATION

Fitting a model to a set of data first entails estimating the unknown parameters in the model. The two most important general methods of estimation used in fitting linear models are the **method of least squares** and the **method of maximum likelihood**. These two methods of estimation are summarized in this section.

3.3.1 The method of least squares

Suppose that $y_1, y_2, \ldots, y_n$ are n independent observations such that the expected response for the ith observation is $E(y_i) = \Sigma \beta_j x_{ji}$ with $\mathrm{Var}(y_i) = \sigma^2, i = 1, 2, \ldots, n$. The least squares estimates of the unknown parameters in the model are then those values $\hat{\beta}_o, \hat{\beta}_1, \ldots, \hat{\beta}_k$ which minimize the sum of squared deviations of the observations from their expected values, given by

$$S = \sum_i \{y_i - E(y_i)\}^2 = \sum_i (y_i - \beta_0 - \beta_1 x_{1i} - \cdots - \beta_k x_{ki})^2$$

In principle, these values can be obtained by differentiating S with respect to each of the unknown parameters, equating the derivatives to zero, and solving the resulting set of linear equations to give the least squares estimates. The minimized sum of squared deviations $\sum(y_i - \hat{y}_i)^2$, where $\hat{y}_i = \sum\hat{\beta}_j x_{ji}$, is generally called the **residual sum of squares** or, less appropriately, the **error sum of squares**, while $\hat{y}_i$ is the **estimated expected value**, or **fitted value** for the ith observation. The residual sum of squares divided by its corresponding number of degrees of freedom, $(n - k - 1)$ in this case, is an unbiased estimate of the variance σ^2.

There are two reasons for the almost universal use of the method of least squares in fitting linear models. The first lies in its intuitive appeal – what could be more natural than minimizing differences between observations and their expected value? The second reason, a more theoretical one, is that the parameter estimates, and derived quantities such as fitted values, have a number of optimality properties: they are unbiased and have minimum variance when compared with all other unbiased estimators that are linear combinations of the observations. Furthermore, linearity of the estimates means that if the data are assumed to have a normal distribution, the residual sum of squares on fitting a linear model has a chi-squared distribution. This provides a basis for the use of F-tests to examine the significance of a regression or for comparing two models that have some terms in common.

3.3.2 The method of maximum likelihood

While the method of least squares is usually adopted in fitting linear regression models, the most widely used general method of estimation is the method of maximum likelihood. To operate this method, we construct the **likelihood** of the unknown parameters in the model for the sample data. This is the joint probability or probability density of the observed data, interpreted as a function of the unknown parameters in the model, rather than as a function of the data. Consequently, in order to specify the likelihood function, a particular distributional form for the data must be assumed. If an observation y has a normal distribution with mean μ and variance σ^2, the probability density function of y is

$$\frac{1}{\sigma\sqrt{(2\pi)}}\exp\left\{-\tfrac{1}{2}[(y - \mu)/\sigma]^2\right\}, \qquad -\infty < y < \infty$$

and the joint probability density of the n observations $y_1, y_2, \ldots, y_n$, each with different mean $\mu_1, \mu_2, \ldots, \mu_n$, say, is the product of n such density functions. Using the symbol Π to denote a product, so that $\Pi_{i=1}^n z_i = z_1 \times z_2 \times \cdots \times z_n$, the joint probability density function is

$$\prod_{i=1}^{n} \frac{1}{\sigma\sqrt{(2\pi)}}\exp\left\{-\tfrac{1}{2}[(y_i - \mu_i)/\sigma]^2\right\}$$

which is regarded as a function of the observations $y_1, y_2, \ldots, y_n$. The likelihood of the data is the same expression, but regarded as a function of $\boldsymbol{\mu} = (\mu_1, \mu_2, \ldots, \mu_n)$ and σ, and is denoted by $L(\boldsymbol{\mu}, \sigma)$. Since the mean response μ_i for each of the n observations depends on the model parameters $\boldsymbol{\beta} = (\beta_0, \beta_1, \ldots, \beta_k)$ the likelihood function itself is a function of the β-parameters and σ, and can be written as

$$L(\beta, \sigma) = \prod_{i=1}^{n} \frac{1}{\sigma\sqrt{(2\pi)}} \exp\{-\tfrac{1}{2}[(y_i - \beta_0 - \beta_1 x_{1i} - \cdots - \beta_k x_{ki})/\sigma]^2\} \qquad (3.1)$$

The sample likelihood is interpreted as the likelihood of the parameters in the model on the basis of the observed data. The larger its value, the greater is the weight of evidence in favour of a given set of parameter values. It may also be regarded as summarizing the information in the data about the unknown parameters in the model, and so a method of estimation which seeks to maximize this has great intuitive appeal.

The maximum likelihood estimators of the parameters in β are those values $\hat{\beta}$ which maximize the likelihood function in equation (3.1), but it is usually more convenient to maximize the logarithm of the likelihood, rather than the likelihood itself. The values $\hat{\beta}$ which maximize $\log L(\beta, \sigma)$ will be the same as those which maximize $L(\beta, \sigma)$ and can generally be obtained as the solutions of the equations

$$\left. \frac{\partial \log L(\beta, \sigma)}{\partial \beta_j} \right|_{\hat{\beta}} = 0, \quad j = 0, 1, \ldots, k$$

for which

$$\left. \frac{\partial^2 \log L(\beta, \sigma)}{\partial \beta_j^2} \right|_{\hat{\beta}} < 0$$

The variance, σ^2, is estimated by setting the derivative of $\log L(\beta, \sigma)$ with respect to σ, evaluated at $\hat{\beta}, \hat{\sigma}$, equal to zero.

The advantage of this method of estimation lies in its general applicability, but since exact properties of the resulting estimators are often difficult to obtain, asymptotic results are widely used. In particular, maximum likelihood estimates are asymptotically unbiased, though they will generally be biased in small samples. The exact variance of the resulting estimators is often intractable, but large sample approximations can usually be obtained from second partial derivatives of the log-likelihood function. For large samples, it is the maximum likelihood estimate that has the smallest possible variance, and this estimator is known to be normally distributed, irrespective of the distribution from which the original data were drawn. When the data have a normal distribution, maximizing the likelihood function in equation (3.1) is equivalent to minimizing the sum of squared deviations of the observations from their expectation and so this method of estimation yields the same estimates of the β-parameters as the method of least squares. In this particular case, the maximum likelihood estimators will be unbiased, irrespective of the sample size.

3.4 FITTING LINEAR MODELS TO BINOMIAL DATA

For binary or binomial data, the response from the ith unit, $i = 1, 2, ..., n$, is a proportion y_i/n_i, which is conveniently denoted by $\tilde{p}_i$; in the particular case of binary data, $n_i = 1$ and $y_i = 0$ (failure) or $y_i = 1$ (success). As was seen in Chapter 2, an appropriate distribution for the ith observation y_i is usually the binomial distribution with parameters n_i and p_i, $B(n_i, p_i)$, where p_i is termed the **success probability**, and the n_i are sometimes referred to as the **binomial denominators**. Under this model for the distribution of the response variable, $E(y_i) = n_i p_i$ and $\mathrm{Var}(y_i) = n_i p_i(1 - p_i)$. Rather than directly modelling the dependence of $E(y_i)$ on explanatory variables, it is customary to explore how the success probability $p_i = E(y_i/n_i)$ can be described by observed explanatory variables.

One approach to modelling binary data, sadly encouraged by the widespread availability of statistical software for linear regression analysis, is to adopt the model $p_i = \beta_0 + \beta_1 x_{1i} + \cdots + \beta_k x_{ki}$ and apply the method of least squares to obtain those values $\hat{\beta}_0, \hat{\beta}_1, ..., \hat{\beta}_k$ for which

$$\sum_i \left(\frac{y_i}{n_i} - p_i \right)^2 = \sum_i (\tilde{p}_i - \beta_0 - \beta_1 x_{1i} - \cdots - \beta_k x_{ki})^2$$

is minimised.

There are a number of drawbacks to this approach. The first problem concerns assumptions made about the variance of $\tilde{p}_i$. Since y_i actually has a binomial distribution, $B(n_i, p_i)$, $\mathrm{Var}(y_i) = n_i p_i(1 - p_i)$ and so $\mathrm{Var}(\tilde{p}_i) = p_i(1 - p_i)/n_i$. Even when the denominators of the proportions, n_i, are all equal, which they certainly are for the particular case of binary data, the variance of the observed proportion of successes is not constant but depends on the true unknown success probability p_i. If these success probabilities do not vary too much, this is not a major problem. For example, if all the p_i lie in the range 0.25–0.75, then $0.19 < p_i(1 - p_i) < 0.25$, and $p_i(1 - p_i)$ is a maximum when $p_i = 0.5$. There are a number of ways of taking account of this non-constant variance when it is important to do so. One proposal, valid when the denominators n_i are equal or at least approximately equal, is to use a **variance stabilising transformation**. If $\sin^{-1} \sqrt{(\tilde{p}_i)}$ is used in place of $\tilde{p}_i$, and the angle whose sine is $\sqrt{(p_i)}$ is measured in radians, the transformed values will have a variance of $1/4n$, where n is the common value of the n_i. This is the **arc sine** or **angular transformation**. A better procedure, appropriate whether or not the n_i are equal, is to use the method of weighted least squares to minimize the function

$$\sum w_i(\tilde{p}_i - p_i)^2$$

where $p_i = \sum_{j=0}^{k} \beta_j x_{ji}$ with $x_{0i} = 1$, and the weights w_i are the reciprocals of the variance of $\tilde{p}_i$ given by $w_i = [p_i(1 - p_i)/n_i]^{-1}$. An immediate problem here is that the p_i and hence the w_i are unknown at the outset – if they were known, there would be no need to estimate them! The observed proportions could be used to calculate the weights, after replacing observed proportions of $0/n$ and n/n

by $0.5/n$ and $(n - 0.5)/n$, respectively, but a better procedure is the following. Starting with initial estimates of the p_i, $\hat{p}_{i0} = y_i/n_i$, or appropriately adjusted values if y_i is equal to zero or n_i, $w_{i0} = [\hat{p}_{i0}(1 - \hat{p}_{i0})/n_i]^{-1}$ is obtained before finding those values $\hat{\beta}_{j0}$, $j = 0, 1, ..., k$, which minimize $\Sigma\, w_{i0}(\hat{p}_i - p_i)$. From these estimates, $\hat{p}_{i1} = \hat{\beta}_{00} + \hat{\beta}_{10}x_{1i} + \cdots + \hat{\beta}_{k0}x_{ki}$ is obtained, followed by revised weights $w_{i1} = [\hat{p}_{i1}(1 - \hat{p}_{i1})/n_i]^{-1}$. This iterative scheme is continued until the parameter estimates converge to $\hat{\beta}$ and the fitted probabilities converge to $\hat{p}_i$. This procedure is referred to as **iteratively weighted least squares** method, although it is equivalent to maximum likelihood estimation.

The second difficulty is that since the assumption of a normally distributed response variable cannot be made, the elegant distribution theory associated with fitting linear models to normal data is no longer valid. When the n_i are all reasonably large, this will not be a severe restriction, in view of the fact that the binomial distribution tends to normality for large sample sizes.

A third difficulty is of fundamental importance and concerns the fitted values, $\hat{p}_i$, under the model. The $\hat{\beta}_j$ are totally unconstrained, and can take any value, positive or negative, large or small, so that any linear combination of them can in principle lie anywhere in the range $(-\infty, \infty)$. Since fitted probabilities are obtained from $\hat{p}_i = \hat{\beta}_0 + \hat{\beta}_1x_{1i} + \cdots + \hat{\beta}_kx_{ki}$, there can be no guarantee that the fitted values will lie in the interval $(0, 1)$ – a matter which could well be the cause of some embarrassment! A linear model for $\hat{p}_i$ will generally be unacceptable in any case, a point which was made earlier in the discussion of Example 1.5 on the failure of aircraft fasteners. Furthermore, bearing in mind the fact that the $\hat{p}_i$ are not bound to lie in the range $(0, 1)$, it would usually be hazardous to adopt any such model for prediction.

Example 3.1 Anti-pneumococcus serum
To illustrate this third point, consider the data from Example 1.6 on the protective effect of a particular serum on the occurrence of pneumonia. Denoting the number of deaths out of forty for the ith batch of mice by y_i, $i = 1, 2, ..., 5$, and the corresponding dose of serum by d_i, the linear regression model

$$E(y_i/n_i) = p_i = \beta_0 + \beta_1 d_i$$

is fitted to the data using the method of least squares, assuming y_i/n_i to have constant variance. The fitted (or estimated expected) probability of death at dose d_i is found to be given by

$$\hat{p}_i = 0.64 - 16.08\, d_i$$

and the fitted probability of death for mice treated with 0.045 cc of the serum is

$$\hat{p} = 0.64 - 16.08 \times 0.045 = -0.084$$

A negative probability is clearly unacceptable. A plot of the data with the fitted line superimposed is given in Fig. 3.1. It is easily seen that the predicted probability of death will not exceed one unless the dose is negative, while it will be negative for doses greater than 0.04 cc. But the most obvious feature shown by this plot is that a simple linear regression line simply does not fit the observed data. A quadratic expression in dose may be a better fit over the range of the data, but is likely to lead to fitted probabilities greater than unity for small doses. However, the main difficulty with a quadratic expression is that while the fitted probabilities will first decrease with increasing level of serum, there will come a point when the probability of death from pneumonia starts to increase with increasing dose, contrary to biological considerations. In this example, the upturn would occur at a dose of 0.034 cc.

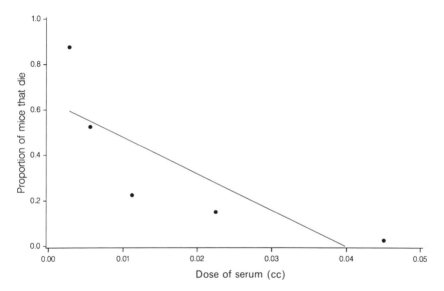

Fig. 3.1 Relationship between the proportion of mice that die from pneumonia and dose of serum.

All this suggests that fitting linear models using the method of least squares is inappropriate for modelling binary response data.

3.5 MODELS FOR BINOMIAL RESPONSE DATA

Instead of using a linear model for the dependence of the success probability on explanatory variables, the probability scale is first transformed from the range $(0, 1)$ to $(-\infty, \infty)$. A linear model is then adopted for the transformed value of the success probability, a procedure which ensures that the fitted probabilities will lie between zero and one. Some possible transformations are described below.

3.5.1 The logistic transformation

The **logistic transformation** of a success probability p is $\log\{p/(1-p)\}$, which is written as logit (p). Notice that $p/(1-p)$ is the odds of a success and so the logistic transformation of p is the log odds of a success. It is easily seen than any value of p in the range $(0, 1)$ corresponds to a value of logit (p) in $(-\infty, \infty)$. As $p \to 0$, logit $(p) \to -\infty$; as $p \to 1$, logit $(p) \to \infty$, and for $p = 0.5$, logit $(p) = 0$. Values of logit (p), for $p = 0.01(0.01)1.00$, are given in the table of Appendix A. The function logit (p) is a sigmoid curve that is symmetric about $p = 0.5$; a graph of this function is included in Fig. 3.2. The logistic function is essentially linear between $p = 0.2$ and $p = 0.8$, but outside this range it becomes markedly non-linear.

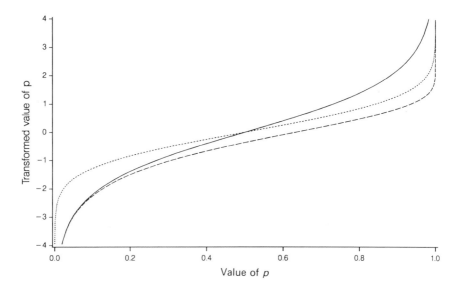

Fig. 3.2 The logistic [———], probit [······] and complementary log–log [------], transformations of p as a function of p.

3.5.2 The probit transformation

The **probit** of a probability p is defined to be that value ξ for which

$$\frac{1}{\sqrt{(2\pi)}} \int_{-\infty}^{\xi} \exp\left(-\tfrac{1}{2}u^2\right) du = p$$

This integral is the distribution function of a standard normal random variable, U, and so $p = P(U \leqslant \xi)$; an interpretation of U will be given in section 4.1. The standard normal distribution function is usually denoted by $\Phi(\xi)$, and so ξ is such that $\Phi(\xi) = p$. On rearrangement, $\xi = \Phi^{-1}(p)$, where the inverse function $\Phi^{-1}(p)$ is the probit transformation of p, written as probit(p). The original definition of a probit took probit$(p) = 5 + \Phi^{-1}(p)$, primarily to avoid having to work with negative probits; $\Phi^{-1}(p)$ will only be smaller than -5 when p is less than 2.9×10^{-7}. This definition is still used in some quarters, but in the major statistical software packages for what is referred to as **probit analysis**, probits are defined without the addition of 5.

The probit function is symmetric in p, and for any value of p in the range $(0, 1)$, the corresponding value of the probit of p will lie between $-\infty$ and ∞. When $p = 0.5$, probit$(p) = 0$, and, as shown by Fig. 3.2, the probit transformation of p has the same general form as the logistic transformation. To obtain the probit of a probability, the table in Appendix A can be used, whereas a table of the standard normal distribution function, such as that given in Lindley and Scott (1984), can be used for the reverse calculation.

3.5.3 The complementary log–log transformation

The **complementary log–log transformation** of a probability p is $\log[-\log(1-p)]$. This function again transforms a probability in the range $(0, 1)$ to a value in $(-\infty, \infty)$, but unlike the logistic and probit transformations, this function is not symmetric about $p = 0.5$. This is shown by the graph of the complementary log–log function in Fig. 3.2. This figure also shows that the complementary log–log transformation is barely distinguishable from the logistic transformation when p is small.

3.5.4 Discussion

Of the three transformations that have been reviewed in this section, the use of the complementary log–log transformation is limited to those situations where it is appropriate to deal with success probabilities in an asymmetric manner. Some areas of application where the complementary log–log transformation arises naturally are described in section 4.6. The logistic and probit transformations are quite similar to each other, but from the computational viewpoint, the logistic transformation is the more convenient. There are two other reasons why the logistic transformation is preferred to the other transformations. First, it has a direct interpretation in terms of the logarithm of the odds in favour of a success. This interpretation is particularly useful in the analysis of data from epidemiological studies, and is discussed in greater detail in Chapter 7. Second, models based on the logistic transformation are particularly appropriate for the analysis of data that have been collected retrospectively, such as in the case-control study. This point is also discussed more fully in Chapter 7.

For these reasons, the logistic transformation is the one that will be studied in greatest detail in this book. However, most of the principles and procedures used in connection with models based on the logistic transformation, to be described in subsequent sections of this chapter, apply equally to models based on the probit and complementary log–log transformations.

3.6 THE LINEAR LOGISTIC MODEL

Suppose that we have n binomial observations of the form y_1/n_i, $i = 1, 2, \ldots, n$, where $E(y_i) = n_i p_i$ and p_i is the success probability corresponding to the ith observation. The **linear logistic model** for the dependence of p_i on the values of the k explanatory variables, $x_{1i}, x_{2i}, \ldots, x_{ki}$, associated with that observation, is

$$\text{logit}(p_i) = \log(p_i/(1 - p_i)) = \beta_0 + \beta_1 x_{1i} + \beta_2 x_{2i} + \cdots + \beta_k x_{ki}$$

On some rearrangement,

$$p_i = \frac{\exp(\beta_0 + \beta_1 x_{1i} + \cdots + \beta_k x_{ki})}{1 + \exp(\beta_0 + \beta_1 x_{1i} + \cdots + \beta_k x_{ki})} \tag{3.2}$$

or, writing $\eta_i = \sum_j \beta_j x_{ji}$,

$$p_i = \frac{e^{\eta_i}}{1 + e^{\eta_i}}$$

Note that since y_i is an observation from a binomial distribution with mean $n_i p_i$, a corresponding model for the expected value of y_i is $E(y_i) = n_i e^{\eta_i}/(1 + e^{\eta_i})$.

The linear logistic model is a member of a class of models known as **generalized linear models**, introduced by Nelder and Wedderburn (1972). This class also includes the linear model for normally distributed data and the log-linear model for data in the form of counts. The function that relates p to the linear component of the model is generally known as the **link function**, so that a logistic link function is being used in the linear logistic model.

A particular application of the linear logistic model is to modelling the dependence of a proportion on a single explanatory variable. In Example 1.5, the proportion of aluminium fasteners failing is to be modelled as a function of the load applied to them and a linear logistic model for the relationship between the expected proportion of failures, p_i, at the ith load, x_i, $i = 1, 2, \ldots, 10$, is

$$p_i = \frac{\exp(\beta_0 + \beta_1 x_i)}{1 + \exp(\beta_0 + \beta_1 x_i)}$$

This relationship between p and x is sigmoidal, whereas logit (p) is linearly related to x. When a linear logistic model is fitted to explore the relationship between a binary response variable and one or more explanatory variables, as in this illustration, the model is also referred to as a **logistic regression** model.

3.7 FITTING THE LINEAR LOGISTIC MODEL TO BINOMIAL DATA

Suppose that binomial data of the form y_i successes out of n_i trials, $i = 1, 2, \ldots, n$, are available, where the logistic transform of the corresponding success probability,

p_i, is to be modelled as a linear combination of k explanatory variables, $x_{1i}, x_{2i}, ..., x_{ki}$, so that

$$\text{logit}(p_i) = \beta_0 + \beta_1 x_{1i} + \beta_2 x_{2i} + \cdots + \beta_k x_{ki}$$

The binomial observations y_i have mean $n_i p_i$ and can be expressed as $y_i = n_i p_i + \varepsilon_i$. The residual components $\varepsilon_i = y_i - n_i p_i$ have zero mean, but they no longer have a binomial distribution. In fact, the ε_i have what is termed a **shifted binomial distribution**. Although there is no correspondence between the distribution of the data and that of the residual terms here, in contrast to the situation for normally distributed data referred to in section 3.2, it is only the distribution of the data that is important in model fitting.

In order to fit a linear logistic model to a given set of data, the $k + 1$ unknown parameters $\beta_0, \beta_1, ..., \beta_k$ have first to be estimated. These parameters are readily estimated using the method of maximum likelihood. The likelihood function is given by

$$L(\boldsymbol{\beta}) = \prod_{i=1}^{n} \binom{n_i}{y_i} p_i^{y_i}(1 - p_i)^{n_i - y_i}$$

This likelihood depends on the unknown success probabilities p_i, which in turn depend on the βs through equation (3.2), and so the likelihood function can be regarded as a function of $\boldsymbol{\beta}$. The problem now is to obtain those values $\hat{\beta}_0, \hat{\beta}_1, ..., \hat{\beta}_k$ which maximize $L(\boldsymbol{\beta})$, or equivalently $\log L(\boldsymbol{\beta})$.

The logarithm of the likelihood function is

$$\log L(\boldsymbol{\beta}) = \sum_i \left\{ \log \binom{n_i}{y_i} + y_i \log p_i + (n_i - y_i)\log(1 - p_i) \right\}$$

$$= \sum_i \left\{ \log \binom{n_i}{y_i} + y_i \log \left(\frac{p_i}{1 - p_i} \right) + n_i \log(1 - p_i) \right\}$$

$$= \sum_i \left\{ \log \binom{n_i}{y_i} + y_i \eta_i - n_i \log(1 + e^{\eta_i}) \right\} \tag{3.3}$$

where $\eta_i = \sum_{j=0}^{k} \beta_j x_{ji}$ and $x_{0i} = 1$ for all values of i. The derivatives of this log-likelihood function with respect to the $k + 1$ unknown β parameters are

$$\frac{\partial \log L(\boldsymbol{\beta})}{\partial \beta_j} = \sum y_i x_{ji} - \sum n_i x_{ji} e^{\eta_i}(1 + e^{\eta_i})^{-1}, \quad j = 0, 1, ..., k$$

Evaluating these derivatives at $\hat{\boldsymbol{\beta}}$ and equating them to zero gives a set of $k + 1$ non-linear equations in the unknown parameters $\hat{\beta}_j$ that can only be solved numerically. It turns out to be more straightforward to use an algorithm known as Fisher's method of scoring to obtain the maximum likelihood estimates $\hat{\boldsymbol{\beta}}$. This procedure is equivalent to using an iteratively weighted least squares procedure in which values of an adjusted dependent variable $z_i = \eta_i + (y_i - n_i p_i)/\{n_i p_i(1 - p_i)\}$ are regressed on the k explanatory variables, $x_{1i}, x_{2i}, \ldots, x_{ki}$, using weights $w_i = n_i p_i(1 - p_i)$. Details of this algorithm are given in Appendix B.1. This algorithm is implemented in a number of widely available computer packages for fitting models to binary response data.

Once $\hat{\beta}$ has been obtained, the estimated value of the linear systematic component of the model is

$$\hat{\eta}_i = \hat{\beta}_0 + \hat{\beta}_1 x_{1i} + \hat{\beta}_2 x_{2i} + \cdots + \hat{\beta}_k x_{ki}$$

which is termed the **linear predictor**. From this, the fitted probabilities $\hat{p}_i$ can be found using $\hat{p}_i = \exp(\hat{\eta}_i)/[1 + \exp(\hat{\eta}_i)]$.

Example 3.2 Aircraft fasteners
A plausible model for the data from Example 1.5 is that there is a linear relationship between the logistic transform of the probability of a fastener failing and the load applied, that is

$$\log(p_i/(1 - p_i)) = \beta_0 + \beta_1 \text{load}_i$$

where p_i is the probability of failure at the ith load, denoted by load_i. On fitting this model to the ten observations, the parameter estimates from GLIM output are as follows.

	estimate	s.e.	parameter
1	− 5.340	0.5457	1
2	0.001548	0.0001575	LOAD

The equation of the fitted model is logit $(\hat{p}) = -5.340 + 0.00155$ load, so that the fitted probability of failure at the ith load is given by

$$\hat{p}_i = \frac{\exp(-5.340 + 0.00155 \ \text{load}_i)}{1 + \exp(-5.340 + 0.00155 \ \text{load}_i)}$$

For example, at a load of 2500 psi, logit $(\hat{p}) = -5.340 + 0.00155 \times 2500 = -1.465$; the corresponding estimated failure probability is $\hat{p} = 0.1877$ and the estimated expected or fitted number of components failing at 2500 psi is $50 \times 0.1877 = 9.39$. From the relationship between the fitted probability and the explanatory variable load, the fitted curve can be superimposed on a graph of the observed proportion of fasteners failing against load. Such a graph is given in Fig. 3.3. The data appear to be well fitted by the linear logistic model.

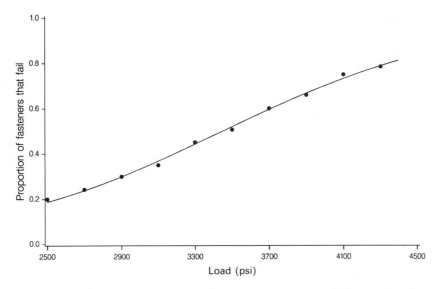

Fig. 3.3 Relationship between the proportion of aircraft fasteners that fail and load.

Some packages also give the values of the statistics $\hat{\beta}/\text{s.e.}(\hat{\beta})$, which can be used to test the hypothesis that the corresponding coefficient in the linear logistic model is zero. These statistics are often referred to as t-values, but in the analysis of binomial data they are generally taken to have a normal distribution rather than a t-distribution. The reason for this is that no scale parameter, analogous to σ^2 for normally distributed data, is being estimated here, a point which will be discussed in greater detail in Chapter 6. More specifically, under the null hypothesis that $\beta = 0$, the ratio $\hat{\beta}/\text{s.e.}(\hat{\beta})$ has a standard normal distribution.

In Example 3.2, the t-value (or more correctly the z-value) for the constant term can be used to test the hypothesis that logit $(p) = 0$, or $p = 0.5$, when the load is zero. This hypothesis is clearly untenable, and not at all relevant. In general, it will not be useful to interpret the t-value for the constant. The t-value for load tests the hypothesis that the coefficient of load is zero, that is, there is no relationship between failure probability and load. Here the observed value of the t-statistic is 9.83, which is highly significant, since it exceeds the upper 0.1% point of the standard normal distribution. We can therefore conclude that there is a relationship between failure probability and the load applied to the fasteners. Although the t-values can be used in this way to determine whether a term is needed in a model, alternative models are best compared using the approach described in section 3.9.

Example 3.3 Smoked mice
The object of the study described in Example 1.1 was to compare the proportions of treated and control mice which develop lung tumours during the course of an experiment on the carcinogenic effect of cigarette smoke. Methods for comparing

two proportions were discussed in Chapter 2, but here we see how the procedure for comparing two proportions can be formulated in terms of a modelling approach.

There are two possible models for the data. Under the first, model (1), the probability that a mouse develops a tumour is the same in each of the two groups. Under the second model, model (2), the probabilities of tumour development are different. Making comparisons between the two probabilities then amounts to comparing these two models. Let p_i be the probability of a mouse developing a lung tumour when in the ith treatment group for $i = 1$ (treated) and $i = 2$ (control), and denote the observed proportions of mice in the two groups by y_i/n_i. The two linear logistic models for the p_i are then as follows

$$\text{Model (1):} \quad E(y_i) = n_i p_i, \text{ where logit}(p_i) = \beta_0, \ i = 1, 2$$

$$\text{Model (2):} \quad E(y_i) = n_i p_i, \text{ where logit}(p_i) = \beta_0 + \gamma_i, \ i = 1, 2$$

In model (2), the term γ_i represents the effect of the ith treatment group.

On fitting model (1), an estimate of β_0, $\hat{\beta}_0$, is obtained from which $\hat{p}_i = \exp(\hat{\beta}_0)/[1 + \exp(\hat{\beta}_0)]$ is the estimated value of the common probability of a mouse developing a tumour in either treatment group.

On using GLIM to fit model (1) to the data from Example 1.1 the following parameter estimate is obtained.

	estimate	s.e.	parameter
1	0.9808	0.3028	1

From this output, the logistic transform of the estimated probability of a mouse developing a tumour in either treatment group is given by logit$(\hat{p}) = 0.981$, from which $\hat{p} = 0.727$. In fact, this is the maximum likelihood estimate of the overall probability of developing a tumour, calculated from $(21 + 19)/(23 + 32)$.

On using GLIM to fit model (2) to the data, where the identifier **GRP** denotes the treatment group, the following output is obtained.

	estimate	s.e.	parameter
1	2.351	0.7400	1
2	−1.972	0.8229	GRP(2)

In order to be able to interpret this output, it is necessary to look at the structure of the model in closer detail. Corresponding to the term γ_i in the model being fitted are two indicator variables x_1, x_2 defined by

$$x_{1i} = \begin{cases} 1 & \text{for } i = 1 \text{ (treated)} \\ 0 & \text{for } i = 2 \text{ (control)} \end{cases} \qquad x_{2i} = \begin{cases} 0 & \text{for } i = 1 \text{ (treated)} \\ 1 & \text{for } i = 2 \text{ (control)} \end{cases}$$

The model can then be expressed in the form

$$\text{logit}\,(p_i) = \beta_0 + \gamma_1 x_{1i} + \gamma_2 x_{2i}, \quad i = 1, 2$$

As written, this model contains three unknown parameters, β_0, γ_1 and γ_2, although there are only two binomial proportions to which the model is to be fitted. Because the model contains more unknown parameters than is justified by the number of observations, the model is overparameterized. The simplest solution to this problem is to omit one of the parameters, and there are a number of possibilities. For example, one of the two indicator variables, say, x_1, could be excluded from the model. The corresponding estimate of the coefficient of x_1 can then be taken to be zero. In fact, this is exactly what the package GLIM does, so that the model that is actually fitted by GLIM is

$$\text{logit}\,(p_i) = \beta_0 + \gamma_2 x_{2i}, \quad i = 1, 2$$

Then, if $\hat{\beta}_0$, $\hat{\gamma}_2$ are the estimated values of β_0, γ_2 under this model, $\text{logit}\,(\hat{p}_1) = \hat{\beta}_0$ and $\text{logit}\,(\hat{p}_2) = \hat{\beta}_0 + \hat{\gamma}_2$, from which the two fitted probabilities are $\exp(\hat{\beta}_0)/[1 + \exp(\hat{\beta}_0)]$ and $\exp(\hat{\beta}_0 + \hat{\gamma}_2)/[1 + \exp(\hat{\beta}_0 + \hat{\gamma}_2)]$, respectively.

Notice that the GLIM output gives $\hat{\beta}_0$ (the parameter labelled **1**) and $\hat{\gamma}_2$ (the parameter labelled **GRP(2)**); the label **GRP(2)** indicates that the estimate contributes only to the model for the data of the second group, the control group in this example. From these estimates, $\text{logit}\,(\hat{p}_1) = 2.351$ and $\text{logit}\,(\hat{p}_2) = 0.379$, so that $\hat{p}_1$ and $\hat{p}_2$ are 0.913 and 0.594, respectively. In model (2), two unknown probabilities are being estimated from just two binomial proportions. The model must therefore be a perfect fit to the observed data and so $\hat{p}_1$ and $\hat{p}_2$ are identical to the observed proportions of mice with tumours, 21/23 and 19/32, respectively.

From the coding that has been used for the indicator variables, $\hat{\gamma}_2$ measures the extent to which the two proportions are different; the smaller the value of $\hat{\gamma}_2$, the closer the estimated proportions. Consequently, the t-value associated with this parameter can be used to compare the two proportions. The observed value of this statistic can be compared with a standard normal distribution to examine its significance. In this example, the t-value is 2.40, which is significant at the 5% level, and so there is evidence of a difference between the responses of the two groups of mice. This test is equivalent to the test of equality of two proportions given in section 2.3.2, and although the numerical values of the two statistics usually differ, the difference is rarely of practical importance.

An alternative to leaving out one of the indicator variables is to omit the constant term β_0 from the model so that model (2) becomes

$$\text{logit}\,(p_i) = \gamma_i = \gamma_1 x_{1i} + \gamma_2 x_{2i}, \quad i = 1, 2$$

The fitted probabilities are then found from $\text{logit}\,(\hat{p}_1) = \hat{\gamma}_1$ and $\text{logit}\,(\hat{p}_2) = \hat{\gamma}_2$. In computer packages such as GLIM, a constant term, or intercept β_0, is always included in a model unless the user specifies that it is to be omitted. GLIM output obtained from this specification is given below.

	estimate	s.e.	parameter
1	2.351	0.7400	GRP(1)
2	0.3795	0.3599	GRP(2)

Notice that the values of $\text{logit}(\hat{p}_1)$ and $\text{logit}(\hat{p}_2)$ are obtained directly from here and agree with those calculated from the previous output.

In the course of Example 3.3, we have seen how two proportions can be compared on the basis of the estimated parameters in the linear part of the model. A more general approach to the comparison of different linear logistic regression models is presented in Section 3.9.

3.8 GOODNESS OF FIT OF A LINEAR LOGISTIC MODEL

After fitting a model to a set of data, it is natural to enquire about the extent to which the fitted values of the response variable under the model compare with the observed values. If the agreement between the observations and the corresponding fitted values is good, the model may be acceptable. If not, the current form of the model will certainly not be acceptable and the model will need to be revised. This aspect of the adequacy of a model is widely referred to as **goodness of fit**. This is something of a misnomer, since one is not measuring how good a model is, but how distant the model is from the data – in other words how bad it is. Rather than adopt the phrase 'badness of fit', an ill-fitting model is said to display **lack of fit**.

There are a number of summary statistics that measure the discrepancy between observed binomial proportions, y_i/n_i, and fitted proportions, $\hat{p}_i$, under an assumed model for the true success probability p_i. Of these, the most widely used is based on the likelihood function for the assumed model. Recall from section 3.3.2 that the likelihood function summarises the information that the data provide about the unknown parameters in a given model. The value of the likelihood, when the unknown parameters are set equal to their maximum likelihood estimates, can therefore be used to summarize the extent to which the sample data are fitted by this current model. This is the maximized likelihood under the **current model**, denoted $\hat{L}_c$. This statistic cannot be used on its own to assess the lack of fit of the current model, since it is not independent of the number of observations in the sample. It is therefore necessary to compare the current model with an alternative baseline model for the same data. This latter model is taken to be the model for which the fitted values coincide with the actual observations, that is, a model that fits the data perfectly. Such a model will have the same number of unknown parameters as there are observations. The model is termed the **full** or **saturated model** and the maximized likelihood under it is denoted by $\hat{L}_f$. The full model is not useful in its own right, since it does not provide a more simple summary of the data than the individual observations themselves. However, by comparing $\hat{L}_c$ with

$\hat{L}_f$, the extent to which the current model adequately represents the data can be judged.

To compare $\hat{L}_c$ and $\hat{L}_f$, it is convenient to use minus twice the logarithm of the ratio of these maximized likelihoods, which is denoted by D, so that

$$D = -2 \log (\hat{L}_c/\hat{L}_f) = -2[\log \hat{L}_c - \log \hat{L}_f] \qquad (3.4)$$

Large values of D are encountered when $\hat{L}_c$ is small relative to $\hat{L}_f$, indicating that the current model is a poor one. On the other hand, small values of D are obtained when $\hat{L}_c$ is similar to $\hat{L}_f$, indicating that the current model is a good one. The statistic D therefore measures the extent to which the current model deviates from the full model and is termed the **deviance**.

In modelling n binomial observations, where p_i is the true success probability corresponding to the ith observation y_i/n_i, $i = 1, 2, \ldots, n$, the likelihood function is

$$\prod_{i=1}^{n} \binom{n_i}{y_i} p_i^{y_i} (1 - p_i)^{n_i - y_i}$$

On fitting a linear logistic model with $k + 1$ unknown parameters $\beta_0, \beta_1, \ldots, \beta_k$, fitted values $\hat{p}_i$ are obtained, where

$$\text{logit} \, (\hat{p}_i) = \hat{\beta}_0 + \hat{\beta}_1 x_{1i} + \hat{\beta}_2 x_{2i} + \cdots + \hat{\beta}_k x_{ki}$$

From equation (3.3), the maximized log-likelihood function under this model is given by

$$\log \hat{L}_c = \sum_i \left\{ \log \binom{n_i}{y_i} + y_i \log \hat{p}_i + (n_i - y_i) \log (1 - \hat{p}_i) \right\}$$

Under the full model, the fitted probabilities will be the same as the observed proportions $\tilde{p}_i = y_i/n_i$, $i = 1, 2, \ldots, n$, and so the maximized log-likelihood function for the full model is

$$\log \hat{L}_f = \sum_i \left\{ \log \binom{n_i}{y_i} + y_i \log \tilde{p}_i + (n_i - y_i) \log (1 - \tilde{p}_i) \right\}$$

The deviance is then given by

$$D = -2[\log \hat{L}_c - \log \hat{L}_f]$$
$$= 2 \sum_i \left\{ y_i \log \left(\frac{\tilde{p}_i}{\hat{p}_i} \right) + (n_i - y_i) \log \left(\frac{1 - \tilde{p}_i}{1 - \hat{p}_i} \right) \right\}$$

If the fitted number of successes under the current model is $\hat{y}_i = n_i \hat{p}_i$, the deviance can be written as

$$D = 2 \sum_i \left\{ y_i \log \left(\frac{y_i}{\hat{y}_i} \right) + (n_i - y_i) \log \left(\frac{n_i - y_i}{n_i - \hat{y}_i} \right) \right\} \tag{3.5}$$

and it is easily seen that this is a statistic that compares the observations y_i with their corresponding fitted values $\hat{y}_i$ under the current model.

3.8.1 The deviance for binary data

In the important special case of binary data, where $n_i = 1$ for $i = 1, 2, \ldots, n$, the deviance depends only on the fitted success probabilities p_i, and so is uninformative about the goodness of fit of a model. To see this, the likelihood for n binary observations, as a function of the β parameters, is

$$L(\beta) = \prod_i p_i^{y_i} (1 - p_i)^{1 - y_i} \tag{3.6}$$

and so the maximized log-likelihood under some current model is

$$\log \hat{L}_c = \sum \{ y_i \log \hat{p}_i + (1 - y_i) \log (1 - \hat{p}_i) \}$$

For the full model, $\hat{p}_i = y_i$, and since $y_i \log y_i$ and $(1 - y_i) \log (1 - y_i)$ are both zero for the only two possible values of y_i, 0 and 1, $\log \hat{L}_f = 0$. The deviance for binary data then becomes

$$D = -2 \sum \{ y_i \log \hat{p}_i + (1 - y_i) \log (1 - \hat{p}_i) \}$$
$$= -2 \sum \{ y_i \log (\hat{p}_i / (1 - \hat{p}_i)) + \log (1 - \hat{p}_i) \} \tag{3.7}$$

Next, differentiation of $\log L(\beta)$ with respect to the jth parameter, β_j, yields

$$\frac{\partial \log L(\beta)}{\partial \beta_j} = \sum_i \left\{ \frac{y_i}{p_i} - \frac{1 - y_i}{1 - p_i} \right\} p_i (1 - p_i) x_{ji}$$

from which

$$\sum_j \beta_j \frac{\partial \log L(\beta)}{\partial \beta_j} = \sum_i (y_i - p_i) \sum_j \beta_j x_{ji}$$
$$= \sum_i (y_i - p_i) \log (p_i / (1 - p_i))$$

Because $\hat{\beta}$ is the maximum likelihood estimate of β, the derivative on the left-hand side of this equation is zero at $\hat{\boldsymbol{\beta}}$. Consequently, the fitted probabilities $\hat{p}_i$ must satisfy the equation

$$\sum (y_i - \hat{p}_i) \operatorname{logit}(\hat{p}_i) = 0$$

and so

$$\sum y_i \operatorname{logit}(\hat{p}_i) = \sum \hat{p}_i \operatorname{logit}(\hat{p}_i)$$

Finally, substituting for $\sum y_i \operatorname{logit}(\hat{p}_i)$ in equation (3.7), the deviance becomes

$$D = -2 \sum \{\hat{p}_i \operatorname{logit}(\hat{p}_i) + \log(1 - \hat{p}_i)\}$$

This deviance depends on the binary observations y_i only through the fitted probabilities $\hat{p}_i$ and so it can tell us nothing about the agreement between the observations and their corresponding fitted probabilities. Consequently, the deviance on fitting a model to binary response data cannot be used as a summary measure of the goodness of fit of a model.

3.8.2 Distribution of the deviance

In order to evaluate the extent to which an adopted model fits a set of binomial data, the distribution of the deviance, under the assumption that the model is correct, is needed. Since the deviance is the likelihood-ratio statistic for comparing a current model with the full model, the null distribution of the deviance follows directly from a result associated with likelihood-ratio testing. According to this result, under certain conditions, the deviance is asymptotically distributed as χ^2 with $(n - p)$ degrees of freedom, where n is the number of binomial observations and p is the number of unknown parameters included in the current linear logistic model.

In the analysis of continuous data assumed to have a normal distribution, the deviance turns out to be the residual sum of squares, $\sum (y_i - \hat{y}_i)^2$. In this particular situation, the exact distribution of the deviance (that is, the residual sum of squares) is known to be a multiple of a χ^2-distribution, irrespective of the sample size; in fact, it has a $\sigma^2 \chi^2$-distribution, where σ^2 is the variance of the observations. For binomial data, only an asymptotic result is available, which is used as an approximation to the (unknown) distribution of the deviance in analysing data based on finite sample sizes.

The validity of the large-sample approximation to the distribution of the deviance depends on the total number of individual binary observations $\sum n_i$ rather than on n, the actual number of proportions y_i/n_i. So even when the number of binomial observations is small, the χ^2-approximation to the distribution can be used so long as $\sum n_i$ itself is reasonably large. Although little can be said in response to the inevitable query 'How large is reasonably large?', the adequacy of the approximation appears to depend more on the individual values of the binomial denominators n_i and the underlying success probabilities p_i than on $\sum n_i$.

In the particular case where ungrouped binary responses are available, so that $n_i = 1$ for all i, the deviance is not even approximately distributed as χ^2. The reason for this is somewhat technical, but it has to do with the fact that under the full model for a set of binary outcomes, the number of parameters being estimated tends to infinity as the number of observations tends to infinity, since one parameter is being fitted for each observation. When binary responses are grouped to give a binomial response based on n_i observations ($n_i > 1$) only one parameter is being fitted to each proportion y_i/n_i in the full model. Then, as $n_i \to \infty$, the number of parameters being fitted for that proportion remains constant, a condition that is required for the valid application of the chi-squared result. As the deviance for binary data cannot be used as a measure of goodness of fit, the fact that it does not have a χ^2-distribution is not of practical importance.

Even when the n_i all exceed one, the chi-squared approximation to the null distribution of the deviance may not be particularly good. This tends to be the case when some of the binomial denominators n_i are very small, that is, when the data are sparse, and the fitted probabilities under the current model are near zero or one.

In circumstances where the deviance on fitting a particular model can be used as a summary measure of the goodness of fit of the model, it is straightforward to judge whether or not the model displays lack of fit. In view of the chi-squared result, the statistic D can be compared to tables of percentage points of the χ^2-distribution with $(n - p)$ degrees of freedom. If the observed value of the statistic exceeds the upper $100\alpha\%$ point of the χ^2-distribution on the given number of degrees of freedom, where α is sufficiently small, the lack of fit is declared significant at the $100\alpha\%$ level. When the deviance on fitting a particular model is declared to be significantly large, the model is deemed to be an inappropriate summary of the data. Since the expected value of a chi-squared random variable with v degrees of freedom is v, a useful rule of thumb is that when the deviance on fitting a linear logistic model is approximately equal to its degrees of freedom, the model is satisfactory. If the mean deviance is defined as the ratio of the deviance to its degrees of freedom, a satisfactory model will then have a mean deviance of around one.

Since the chi-squared approximation to the distribution of the deviance on fitting a model is sometimes not adequate, this goodness of fit procedure should be used with caution. A better approach to examining lack of fit is to determine whether additional terms need to be included in a model, using the method described in Section 3.9.

Example 3.4 Aircraft fasteners
After fitting the model logit $(p_i) = \beta_0 + \beta_1$ load$_i$ to the data from Example 1.5, the deviance can easily be calculated from the fitted failure probabilities $\hat{p}_i$ using equation (3.5). It is found to be 0.372, and since a model with two unknown parameters, β_0 and β_1, is being fitted to ten binomial observations, this deviance has $10 - 2 = 8$ degrees of freedom. This value is nowhere near significant at any conventional significance level and so there is no evidence of lack of fit in the logistic regression model. In fact, the deviance is suspiciously small, suggesting that the binary responses from the fasteners subjected to a given load are correlated (see Chapter 6) or that the data are artificial!

GLIM output corresponding to this aspect of the analysis is reproduced below.

scaled deviance = 0.37192 at cycle 3
d.f. = 8

In this output, the scaled deviance is identical to the quantity referred to as the deviance throughout this book, and is the deviance on fitting a linear logistic model that contains the explanatory variable load. The reason for this terminology is that after fitting a model to binomial data, GLIM gives the deviance divided by a quantity called the **scale parameter**, which is actually equal to unity for binomial data. Further information on this is given in Section 6.5.

3.8.3 Pearson's X^2-statistic

So far, the deviance has been considered exclusively as a summary measure of lack of fit. However, there are a number of alternatives, of which the most popular is Pearson's X^2-statistic defined by

$$X^2 = \sum_{i=1}^{n} \frac{(y_i - n_i \hat{p}_i)^2}{n_i \hat{p}_i (1 - \hat{p}_i)}$$

This statistic was given in expression (2.24).

Both the deviance and this X^2-statistic have the same asymptotic χ^2-distribution. The numerical values of the two statistics generally differ, but the difference is seldom of practical importance. For example, the deviance of fitting a linear regression model to the data on aircraft fasteners is 0.372 on 8 degrees of freedom, whereas the value of the X^2-statistic is 0.371. Large differences between the two statistics can be taken as an indication that the chi-squared-approximation to the distribution of the deviance or the X^2-statistic is not adequate.

Since the maximum likelihood estimates of the success probabilities maximise the likelihood function for the current model, the deviance is the goodness of fit statistic that is minimised by these estimates. On this basis, it is more appropriate to use the deviance rather than the X^2-statistic to measure goodness of fit when linear logistic models are being fitted using the method of maximum likelihood. However, this does not mean that the deviance will always be smaller than X^2, as the above example shows. There is a further reason for preferring the deviance as a measure of goodness of fit. In comparing two models, where one model includes terms in addition to those in the other, the difference in deviance between the two models can be used to judge the import of those additional terms. The X^2-statistic cannot be used in this way. A general approach to the comparison of alternative linear logistic regression models is presented in the next section.

3.9 COMPARING LINEAR LOGISTIC MODELS

Alternative linear logistic models for binomial data can be compared on the basis of a goodness of fit statistic such as the deviance. When one model contains terms that are additional to those in another, the two models are said to be **nested**. The difference in the deviances of two nested models measures the extent to which the additional terms improve the fit of the model to the observed response variable.

In examining the effect of including terms in, or excluding terms from, a model, it is important to recognise that the change in deviance will depend on what terms are in the model at the outset. For example, consider the following sequence of models:

$$\text{Model (1):} \quad \text{logit}(p) = \beta_0$$

$$\text{Model (2):} \quad \text{logit}(p) = \beta_0 + \beta_1 x_1$$

$$\text{Model (3):} \quad \text{logit}(p) = \beta_0 + \beta_2 x_2$$

$$\text{Model (4):} \quad \text{logit}(p) = \beta_0 + \beta_1 x_1 + \beta_2 x_2$$

The difference in deviance between model (4) and model (2) and between model (3) and model (1) both measure the effect of including x_2 in the model. In the first comparison, between model (4) and model (2), the effect of x_2 is adjusted for x_1, since x_2 is being added to a model that already includes x_1. In the second comparison, between model (3) and model (1), the effect of x_2 is unadjusted for x_1. The extent to which the deviances, on including x_2 in the model with and without x_1, differ depends on the extent to which x_1 and x_2 are associated. If x_1 and x_2 are acting independently, the differences in deviance will be quite similar, whereas if x_1 and x_2 are highly correlated, they will be very different.

In general, suppose that two linear logistic models, model (1) and model (2), say, are to be compared, where the two models are as follows:

$$\text{Model (1):} \quad \text{logit}(p) = \beta_0 + \beta_1 x_1 + \cdots + \beta_h x_h$$

$$\text{Model (2):} \quad \text{logit}(p) = \beta_0 + \beta_1 x_1 + \cdots + \beta_h x_h + \beta_{h+1} x_{h+1} + \cdots + \beta_k x_k$$

Denote the deviance under each model by D_1 and D_2, so that D_1 and D_2 have $(n - h - 1)$ and $(n - k - 1)$ degrees of freedom, respectively, where n is the number of binomial observations. Model (1) is nested within model (2) and since model (2) contains more terms than model (1), it will inevitably fit the data better, and D_2 will be smaller than D_1. The difference in deviance $D_1 - D_2$ will reflect the combined effect of the variables $x_{h+1}, x_{h+2}, \ldots, x_k$ after $x_1, x_2, \ldots, x_h$ have already been included in the model. This difference in deviance is described as the deviance of fitting $x_{h+1}, x_{h+2}, \ldots, x_k$ adjusted for, or eliminating, $x_1, x_2, \ldots, x_h$.

One important difference between modelling binomial data by fitting a linear logistic model and modelling a continuous response variable using a linear model relates to the concept of orthogonality. In linear modelling, when the effects of two factors can be estimated independently, they are said to be **orthogonal**. Similarly, if two explanatory variates have zero correlation, they can be termed orthogonal. In the analysis of continuous data from an experimental design with equal numbers of observations at each combination of factor levels, the factors will be orthogonal, and the design itself is termed an **orthogonal design**. It will not then make any difference whether the effect of one factor is measured with or without the other factor in the model. In processing binary or binomial data, there is no concept of orthogonality, and even when the data arise from an orthogonal design, the effect of each term in a model cannot be estimated independently of the others. The order in which terms are included in a model will therefore be important in most analyses.

To compare two nested models for binomial data, no exact distribution theory is available. However, since the deviance for each model has an approximate χ^2-distribution, the difference between two deviances will also be approximately distributed as χ^2. For example, suppose that two models are being compared, where model (1) is nested within model (2); model (1) will therefore have fewer terms than model (2). Let the deviance for model (1) be D_1 on v_1 degrees of freedom (d.f.) and that for model (2) be D_2 on v_2 d.f. Denoting the maximized likelihood under model (1) and model (2) by $\hat{L}_{c1}$ and $\hat{L}_{c2}$, respectively, the two deviances are

$$D_1 = -2[\log \hat{L}_{c1} - \log \hat{L}_f], \qquad D_2 = -2[\log \hat{L}_{c2} - \log \hat{L}_f]$$

where $\hat{L}_f$ is the maximized likelihood under the full model. Since D_1 has an approximate χ^2-distribution on v_1 d.f. and D_2 has an approximate χ^2-distribution on v_2 d.f., the difference in deviance $D_1 - D_2$ will have an approximate χ^2-distribution on $v_1 - v_2$ d.f. In the discussion of the chi-squared approximation to the distribution of the deviance in section 3.8.2, it was pointed out that for binary data the deviance cannot be approximated by a χ^2-distribution. Essentially, this is because of the inclusion of the likelihood under the full model in the expression for the deviance. But when comparing two deviances, the term involving $\hat{L}_f$ disappears and $D_1 - D_2 = -2[\log \hat{L}_{c1} - \log \hat{L}_{c2}]$. The chi-squared approximation to the difference between two deviances can therefore be used to compare nested models for binary as well as for binomial data. In other situations where the distribution of the deviance cannot be reliably approximated by a χ^2-distribution, such as when the data are sparse, the χ^2-distribution is still a good approximation to the distribution of the difference between two deviances.

Example 3.5 Propagation of plum root stocks
The aim of the experiment presented in Example 1.2 was to determine the impact of two factors, length of cutting and time of planting, on the survival of plants produced from plum root cuttings. Let p_{jk} be the probability of a cutting surviving when a cutting of length j is planted at time k, for $j = 1, 2; k = 1, 2$. There are five possible models for the survival probabilities p_{jk}, catalogued below

Model (1): $\text{logit}(p_{jk}) = \beta_0$

Model (2): $\text{logit}(p_{jk}) = \beta_0 + \lambda_j$

Model (3): $\text{logit}(p_{jk}) = \beta_0 + \tau_k$

Model (4): $\text{logit}(p_{jk}) = \beta_0 + \lambda_j + \tau_k$

Model (5): $\text{logit}(p_{jk}) = \beta_0 + \lambda_j + \tau_k + (\lambda\tau)_{jk}$

In these models β_0 is a constant term, λ_j and τ_k are the effects due to the jth length of cutting and the kth planting time and the terms $(\lambda\tau)_{jk}$ are associated with an interaction between the length of cutting and the time of planting. Under model (1), the survival probability is completely independent of both the length of the cutting and the time of planting. Model (2) and model (3) suggest that survival depends only on one of the two factors. Model (4) indicates that the outcome depends on both length of cutting and time of planting, where these two factors act independently. Under model (5), there is an interaction between length and time, so that the effect of length of cutting depends on the actual time of planting.

This sequence of possible models progresses from the most simple [model (1)] to the most complex [model (5)]. Notice also that the interaction term $(\lambda\tau)_{jk}$ is included only in a model that has both of the corresponding main effects λ_j and τ_k. This is because the terms $(\lambda\tau)_{jk}$ can be interpreted as interaction effects only when the corresponding main effect terms, λ_j and τ_k, are both contained in the model. Indeed, this is a general rule in model building: higher-order terms are included in a model only when the corresponding lower-order terms are present.

These five models can be compared using an analysis of deviance. The deviances on fitting each of the models, obtained using GLIM, are given in Table 3.1.

Table 3.1 Values of deviances for the plum root-stock data

Terms fitted in model	Deviance	d.f.
β_0	151.02	3
$\beta_0 + \lambda_j$	105.18	2
$\beta_0 + \tau_k$	53.44	2
$\beta_0 + \lambda_j + \tau_k$	2.29	1
$\beta_0 + \lambda_j + \tau_k + (\lambda\tau)_{jk}$	0.00	0

When either of the main effects is added to a model that includes a constant term alone, the reduction in deviance is highly significant. Moreover, the model that has both main effects is far better than a model with just one of them. The deviance on fitting model (5) is identically zero, since model (5) is actually the full model: four unknown parameters are being fitted to four observations. The change in deviance on introducing the interaction term into the model is 2.29 of 1 d.f. which is not significant at the 10% level; in fact, the P-value associated with this test is 0.13. This also indicates that the model with both main effects provides a satisfactory description of the observed survival probabilities.

To see that the order in which terms are included in a model is important, notice that when λ_j is added to a model that includes β_0 only, the deviance decreases from 151.02 to 105.18, a reduction of 45.84. This is the change in deviance on fitting length of cutting ignoring time of planting. When λ_j is added to a model that already includes τ_k, the deviance reduces from 53.44 to 2.29. This reduction of 51.15 is the deviance due to fitting length of cutting allowing for or adjusting for time of planting. These two reductions in deviance differ even though the design might be described as

Table 3.2 Analysis of deviance table for the plum root-stock data

Source of variation	Deviance	d.f.
Length ignoring time	45.84	1
Time ignoring length	97.58	1
Length adjusted for time	51.15	1
Time adjusted for length	102.89	1
Length × time adjusted for length and time	2.29	1

orthogonal. These results, and others based on the deviances in Table 3.1, are summarised in the analysis of deviance table given as Table 3.2

The conclusion from this analysis of deviance is that both length of cutting and time of planting affect the chance of a root cutting surviving, but the difference in the response at the two planting times does not depend on the length of cutting. GLIM output on fitting the model with the two main effects [model (4)] is given below

$$\text{scaled deviance} = 2.2938 \text{ at cycle 3}$$
$$\text{d.f.} = 1$$

	estimate	s.e.	parameter
1	−0.3039	0.1172	1
2	1.018	0.1455	LENG(2)
3	−1.428	0.1465	TIME(2)

From the parameter estimates under the fitted model, the fitted survival probabilities can be obtained. The model fitted can be written as $\text{logit}(p_{jk}) = \beta_0 + \lambda_j + \tau_k$ for $j = 1, 2; k = 1, 2$, where λ_1 is the effect of a short cutting and τ_1 is the effect of planting at once. For reasons given in Example 3.3 in section 3.7, this model is overparameterized, and GLIM sets the estimates corresponding to the first level of each factor equal to zero. From the output, $\hat{\beta}_0 = -0.304$, $\hat{\lambda}_2 = 1.018$ and $\hat{\tau}_2 = -1.428$. Both $\hat{\lambda}_1$ and $\hat{\tau}_1$ are taken to be zero. The four fitted probabilities are then obtained from

$$\text{logit}(\hat{p}_{11}) = \hat{\beta}_0 + \hat{\lambda}_1 + \hat{\tau}_1 = -0.304$$
$$\text{logit}(\hat{p}_{12}) = \hat{\beta}_0 + \hat{\lambda}_1 + \hat{\tau}_2 = -0.304 - 1.428 = -1.732$$
$$\text{logit}(\hat{p}_{21}) = \hat{\beta}_0 + \hat{\lambda}_2 + \hat{\tau}_1 = -0.304 + 1.018 = 0.714$$
$$\text{logit}(\hat{p}_{22}) = \hat{\beta}_0 + \hat{\lambda}_2 + \hat{\tau}_2 = -0.304 + 1.018 - 1.428 = -0.714$$

leading to the fitted probabilities given in Table 3.3. From this table, we see that

Table 3.3 Observed and fitted survival probabilities for the data on the survival of plum root-stocks

Length of cutting	Time of planting	Survival probability	
		observed	fitted
Short	At once	0.446	0.425
	In spring	0.129	0.150
Long	At once	0.650	0.671
	In spring	0.350	0.329

the observed survival probabilities are well fitted by the model. The difference between the fitted survival probabilities for short cuttings planted at once and short cuttings planted in spring is 0.275, which is smaller than the corresponding difference of 0.342 for long cuttings. However, from the analysis of deviance we know that this difference is not large enough to warrant the inclusion of an interaction term in the model. If the future strategy is to maximise the chance of success with the root cuttings, long cuttings should be taken and planted at once. Sixty-seven per cent of such cuttings will then be expected to survive.

Example 3.6 Determination of the ESR
The data from Example 1.7 were obtained in order to study the extent to which the disease state of an individual, reflected in the ESR reading, is related to the levels of two plasma proteins, fibrinogen and γ-globulin. Let p_i be the probability that the ith individual has an ESR reading greater than 20, so that p_i is the probability that the individual is diseased, and let f_i and g_i denote the levels of fibrinogen and γ-globulin in that individual. In modelling the dependence of the disease state on these two explanatory variables, there are a number of possible linear logistic models, four of which are:

Model (1): $\operatorname{logit}(p_i) = \beta_0$

Model (2): $\operatorname{logit}(p_i) = \beta_0 + \beta_1 f_i$

Model (3): $\operatorname{logit}(p_i) = \beta_0 + \beta_2 g_i$

Model (4): $\operatorname{logit}(p_i) = \beta_0 + \beta_1 f_i + \beta_2 g_i$

Under model (1), there is no relationship between disease state and the two plasma proteins. Models (2) and (3) indicate a relationship between disease state and the level of fibrinogen and γ-globulin, respectively, while under model (4), the binary response variable is related to the levels of both plasma proteins. This set of possible models could be enlarged by including models that contain powers of f_i and g_i such as f_i^2, g_i^2, or cross-product terms such as $f_i g_i$, but at the outset attention will be restricted to the more simple models.

Table 3.4 Deviances on fitting four linear logistic regression models to the ESR data

Terms fitted in model	Deviance	d.f.
β_0	30.88	31
$\beta_0 + \beta_1 f_i$	24.84	30
$\beta_0 + \beta_2 g_i$	28.95	30
$\beta_0 + \beta_1 f_i + \beta_2 g_i$	22.97	29

The deviance on fitting each of the four models given above to the observed data was obtained using GLIM and are summarized in Table 3.4. The change in deviance on adding f_i to a model that includes a constant term alone is $30.88 - 24.84 = 6.04$ on 1 d.f. Since the upper 5% point of the χ^2-distribution on 1 d.f. is 3.84, this is significant at the 5% level. On the other hand, when g_i is added to a model that includes only a constant, the deviance is reduced by 1.93 on 1 d.f. which is not significant. Before concluding that the only term that need be included in the model is f_i, it is necessary to look at whether g_i is needed in addition to f_i. On including g_i in the model that already has f_i, the deviance is reduced from 24.84 to 22.97, a non-significant amount when judged against a chi-squared variate on 1 d.f. In summary, the probability that an individual has an ESR reading greater than 20 mm/h seems to depend only on the level of fibrinogen in the blood plasma. GLIM output obtained on fitting this logistic regression model, in which f_i is denoted **FIB**, is reproduced below.

scaled deviance = 24.840 at cycle 4
d.f. = 30

	estimate	s.e.	parameter
1	−6.845	2.764	1
2	1.827	0.8991	FIB

The fitted model is

$$\text{logit}(\hat{p}_i) = -6.85 + 1.83 f_i$$

so that the fitted probability of an ESR reading greater than 20 mm/h for an individual with fibrinogen level f is

$$\hat{p} = \frac{\exp(-6.85 + 1.83 f)}{1 + \exp(-6.85 + 1.83 f)}$$

A graph of this function, together with the observed binary responses, is given in Fig. 3.4. Although the deviance on fitting this model is less than its corresponding

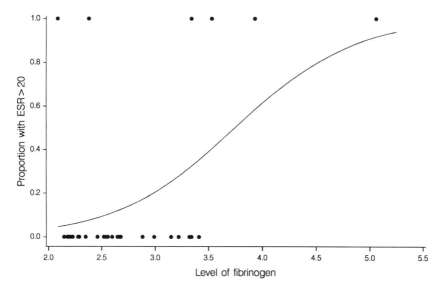

Fig. 3.4 Relationship between the proportion of patients with an ESR reading greater than 20 mm/h and the level of plasma fibrinogen.

number of degrees of freedom, this cannot be taken to mean that the model provides a satisfactory summary of the data. The response variable here is binary and so the deviance does not provide information about the goodness of fit of the model. Other methods for assessing the adequacy of models fitted to binary data are reviewed in Chapter 5, and the analysis of this data set is further discussed in examples 5.21–5.25.

3.10 LINEAR TRENDS IN PROPORTIONS

Experimental studies often involve factors with quantitative levels. For example, in an animal experiment, a treatment factor may be one whose levels are the amounts of a particular amino acid in the diet of the animal; in a clinical trial, different concentrations of a particular drug may form the levels of a quantitative factor. In circumstances such as these, it is usually important to investigate the relationship of the response to changing levels of the factor, for example, to determine if any such relationship is linear.

A similar situation arises in observational studies, where information is collected on continuous explanatory variables, such as age, height and systolic blood pressure, for each person in the study. Rather than assume that the response is linearly dependent upon such variables at the outset, a factor may be defined for each continuous variable, whose levels correspond to consecutive intervals of that variable. For example, the levels of a factor corresponding to the continuous variable age would index a series of consecutive age groups. The manner in which the logistic transform of the response probability varies over the levels of the resulting factor will then be of interest. If this relationship is linear, there is a justification for using the corresponding explanatory variable as a continuous variable in modelling the binary response data.

The approach used to investigate whether there is a linear trend across the levels of a quantitative factor applies equally to quantitative factors in experimental and observational studies and is illustrated in the following example.

Example 3.7 Toxicity of cypermethrin to moths
The larvae of the tobacco budworm, *Heliothis virescens*, are responsible for much damage to cotton crops in the United States, and Central and Southern America. As a result of intensive cropping practices and the misuse of pesticides, particularly synthetic pyrethroids, the insect has become an important crop pest. Many studies on the resistance of the larvae to pyrethroids have been conducted, but the object of an experiment described by Holloway (1989) was to examine levels of resistance in the adult moth to the pyrethroid *trans*-cypermethrin. In the experiment, batches of pyrethroid-resistant moths of each sex were exposed to a range of doses of cypermethrin two days after emergence from pupation. The number of moths which were either knocked down (movement of the moth is uncoordinated) or dead (the moth is unable to move and does not respond to a poke from a blunt instrument) was recorded 72 h after treatment. The data obtained are given in Table 3.5.

We will first analyse these data as a factorial experiment involving two factors, namely sex with two levels and dose of cypermethrin with six levels. These factors are labelled as sex and dose, respectively. On fitting linear logistic models to the

Table 3.5 Mortality of the tobacco budworm 72 h after exposure to cypermethrin

Sex of moth	Dose (in μg) of cypermethrin	Number affected out of 20
Male	1.0	1
	2.0	4
	4.0	9
	8.0	13
	16.0	18
	32.0	20
Female	1.0	0
	2.0	2
	4.0	6
	8.0	10
	16.0	12
	32.0	16

Table 3.6 Analysis of deviance on fitting linear logistic models to the data in Table 3.5

Terms fitted in model	Deviance	d.f.
Sex	118.80	10
Dose	15.15	6
Sex + dose	5.01	5

proportion of insects affected by the toxin, the analysis of deviance in Table 3.6 is obtained. From this table, the residual deviance after fitting sex and dose is not significantly large relative to percentage points of the χ^2-distribution on 5 d.f., showing that a sex $\times$ dose interaction is not needed in the model. Next, the deviance due to fitting sex, adjusted for dose, is 10.14 on 1 d.f., which is highly significant. Also, the deviance due to fitting dose, allowing for the effect of sex, is 113.79 on 5 d.f., the significance of which is beyond doubt. We may therefore conclude that the proportion of moths affected by the toxin depends on both the sex of the moth and on the dose of the chemical that is applied.

The factor dose is a quantitative factor, whose levels are increasing concentrations of cypermethrin. In order to investigate the form of the relationship between the response and the levels of dose, a continuous variable D is defined, whose values are the dose levels of 1, 2, 4, 8, 16 and 32 µg of cypermethrin. When this variable is included in a linear logistic model, it represents the linear effect of dose. On fitting a linear logistic model that includes sex and the variable D, the deviance is 27.97 on 9 d.f., and so the deviance due to D, adjusted for sex, is $118.80 - 27.97 = 90.83$ on 1 d.f. Although this is highly significant, before we can conclude that there is a linear relationship between the logistic transform of the response probability and the dose of cypermethrin, the extent of any non-linearity in the relationship must be determined. This is reflected in the difference between the deviance for the model that contains sex and D, and the model that contains sex and dose, which is $90.83 - 5.01 = 85.82$ on 4 d.f. Because this is highly significant, we conclude that the relationship between the logistic transform of the response probability and dose is not linear. The next step could be to successively examine whether quadratic, cubic or higher powers of D are needed in the model in addition to sex and the linear variable D. If the terms D, D^2, D^3, D^4 and D^5 are fitted in addition to sex, the effect of dose is explained by five parameters, and the model is equivalent to that which contains sex and the factor dose. The deviance for the model sex $+ D + D^2 + D^3 + D^4 + D^5$ is therefore identical to that for the model sex + dose, that is, 5.01. The result of fitting the sequence of models that include increasing powers of D is given in Table 3.7. The figures in this table show that the model that contains sex and the linear, quadratic and cubic powers of D fits as well as the model with all five powers of D. This shows that on the original dose scale, there is evidence of a cubic relationship between the logistic transformation of the response probability and dose of cypermethrin.

Because the doses of cypermethrin increase in powers of two from 1 to 32, it is

Table 3.7 Deviances on including linear, quadratic, and higher-order powers of dose in a linear logistic model fitted to the data of Table 3.5

Terms fitted in model	Deviance	d.f.
Sex	118.80	10
Sex + D	27.97	9
Sex + D + D^2	14.84	8
Sex + D + D^2 + D^3	8.18	7
Sex + D + D^2 + D^3 + D^4	5.52	6
Sex + D + D^2 + D^3 + D^4 + D^5	5.01	5

more natural to use the logarithm of the dose in the linear logistic modelling. If the variable L is taken to be $\log_2(\text{dose})$, then L takes the values $0, 1, \ldots, 5$, but logarithms to any other base could have been used. The deviance on including L in the model, together with sex, is 6.76 on 9 d.f. The deviance on fitting sex $+ L + L^2 + L^3 + L^4 + L^5$ is again the same as that for sex + dose, namely 5.01 on 5 d.f., and so the deviance due to deviation from linearity is $6.76 - 5.01 = 1.75$ on 4 d.f., which is certainly not significant. A linear relationship between the logistic transform of the probability of a moth being affected and the logarithm of the amount of cypermethrin to which the insect is exposed is therefore a reasonable model. The analysis of deviance is summarized in Table 3.8. This analysis has shown that there is a linear relationship between the logistic transform of the response probability and dose on the logarithmic scale. An alternative approach to the analysis of the data in Table 3.5 is to compare the form of the dose–response relationship between the two sexes. This approach is described in the next section.

Table 3.8 Analysis of deviance to examine the extent of linearity in the relationship between the logit of the response probability and $\log_2(\text{dose})$ for the data of Table 3.5

Source of variation	Deviance	d.f.
Sex adjusted for dose	10.14	10
Dose adjusted for sex	113.79	5
Linear trend	112.04	1
Residual (non-linearity)	1.75	4

3.11 COMPARING STIMULUS–RESPONSE RELATIONSHIPS

Many problems lead to a number of different groups of data, where in each group there is a relationship between the response probability and certain stimulus variables. Differences in the form of this relationship between the groups will then be of interest. For instance, in Example 1.8 on the role of women in society, the relationship between the probability of an individual agreeing with the controversial statement is modelled as a function of the number of years of education. There are two groups of data, from male and female respondents, and one objective in analysing these data is to study the extent to which this relationship differs between the sexes.

When there are a number of groups of data and a single explanatory variable, x, four different models for the linear dependence of the logistic transformation of the response probability on x can be envisaged. The most general model, model (1), specifies that there are different logistic regression lines for each group of data, so that the model can be written as

$$\text{Model (1):} \quad \text{logit}(p) = \alpha_j + \beta_j x$$

Here it is more convenient to denote the intercept and slope of the logistic regression line for the jth group by α_j, β_j rather than β_{0j}, β_{1j} to avoid having to use double subscripts.

A model somewhat simpler than model (1) would have a common slope for the groups, but different intercepts, on the logit scale, so that this model, model (2), is

$$\text{Model (2):} \quad \text{logit}(p) = \alpha_j + \beta x$$

This model leads to a series of parallel regression lines with slope β. A model with the same number of parameters as model (2), model (3), would have common intercepts, but different slopes, so that this model can be expressed as

$$\text{Model (3):} \quad \text{logit}(p) = \alpha + \beta_j x$$

Under this model, the lines are concurrent at a response probability of α when $x = 0$. In any particular problem, one would not usually be interested in both model (2) and model (3). For the majority of applications, it is more appropriate to investigate whether the logistic regression lines are parallel rather than whether they are concurrent.

The most simple model is model (4) for which

$$\text{Model (4):} \quad \text{logit}(p) = \alpha + \beta x$$

In this model, the relationship between the response probability and x is the same for each of the different groups of data.

These models for comparing relationships between different groups combine a factor that is used to index the groups with an explanatory variable. Terms such as α_j and $\beta_j x$ are fitted through the use of indicator variables as discussed in

Section 3.2.1. Fitted models can then be compared on the basis of their deviance. Let $D(1)$ to $D(4)$ denote the deviance on fitting the four models described above and suppose that their respective degrees of freedom are v_1 to v_4. In order to test if the different logistic regression lines can be taken to be parallel, models (1) and (2) would be compared using the result that $D(2) - D(1)$ has a χ^2-distribution on $v_2 - v_1$ degrees of freedom, when the β_j are equal. If the observed value of the chi-squared statistic is significantly large, the two models are not comparable. If they are comparable, one could then go on to test if they are actually coincident using the result that $D(4) - D(2)$ has a χ^2-distribution on $v_4 - v_2$ degrees of freedom, again when the hypothesis of constant intercepts is true. In this sequential procedure, model (3) could be substituted for model (2) if it was desired to test for concurrency followed by coincidence.

Example 3.8 Women's role in society
The data from Example 1.8 are modelled by fitting a linear logistic regression on the number of years of education, to the probability that an individual agrees with the statement that women should take care of the homes and leave running the country up to men. The observed proportion of women in agreement with the statement who have received two years of education is 0/0. This observation is totally uninformative and so is omitted from the modelling process. At the outset, two separate logistic regression lines are fitted to the data, one for each sex. The model used is

$$\text{logit}(p) = \alpha_j + \beta_j \text{ years}, \quad j = 1, 2$$

where $j = 1$ refers to the data from the male respondents and $j = 2$ to that of the females. GLIM output from fitting this model follows. Note that a constant term is omitted from the model so that the estimated intercepts can be interpreted directly. In this output, the two levels of the factor labelled **SEX** correspond to the males and females, respectively.

scaled deviance = 57.103 at cycle 4
d.f. = 37

	estimate	s.e.	parameter
1	2.098	0.2355	SEX(1)
2	3.003	0.2724	SEX(2)
3	−0.2340	0.02019	SEX(1).YEAR
4	−0.3154	0.02365	SEX(2).YEAR

The deviance on fitting the separate lines model is 57.10 on 37 d.f. Although this deviance is rather large relative to its number of degrees of freedom, it is of interest to compare this logistic model to one that has a common coefficient for years, but different intercepts. The appropriate GLIM output is given below.

scaled deviance = 64.007 at cycle 4
d.f. = 38

	estimate	s.e.	parameter
1	2.509	0.1839	SEX(1)
2	2.498	0.1828	SEX(2)
3	− 0.2706	0.01541	YEAR

The increase in deviance on constraining the slopes of the fitted logistic regression lines to be equal is $64.01 - 57.10 = 6.91$ on 1 d.f. This increase is highly significant ($P = 0.009$) and so we conclude that two separate logistic regression lines are needed. Since the deviance for this model is uncomfortably high, the next step is to determine whether there is any evidence of lack of fit. If a logistic regression line is fitted to the data from each of the two sexes taken separately, the individual contributions to the overall deviance can be obtained. For males, the deviance is found to be 18.95 on 19 d.f., while for females it is 38.16 on 18 d.f. A straight line logistic model fits the male data very well, but appears not to be such a good model for the females. This can be studied more closely by plotting the data for each sex on a logit scale. In several cases, the observed proportion of individuals in agreement with the statement is zero or one and so the **empirical logit**, $\log \{(y_i + 0.5)/(n_i - y_i + 0.5)\}$ is used in constructing the graphs of Fig. 3.5. This is simply the logistic transform of an observed proportion y_i/n_i, adjusted so that finite values are obtained when y_i is equal to either zero or n_i. The fitted linear logistic regression lines have been superimposed on the graphs in Fig. 3.5. This figure shows that the data do not deviate systematically from the fitted model in either case. However, this can be examined in more detail by fitting the quadratic logistic regression model

$$\text{logit}\,(p) = \alpha + \beta \text{ years} + \gamma \text{ years}^2$$

to the data for each sex. On including the quadratic term years^2 in the model for each sex, the deviance is reduced by a non-significant amount in both cases. The reason for the large deviance for the female data is due in part to the presence of several empirical logits that are quite distant from the fitted line, particularly the first and the last two. The proportions to which they correspond, namely 4/6, 1/3 and 2/6 are based on relatively few individuals and so it is difficult to come to any firm conclusions about the extent to which they are discordant. In addition to there being observations with small binomial denominators, the fitted probabilities under the assumed model are also very small for some groups, particularly, for females with 18, 19 or 20 years of education. This suggests that the chi-squared approximation to the distribution of the deviance may not be very good and the residual deviance is not a reliable measure of lack of fit of the model.

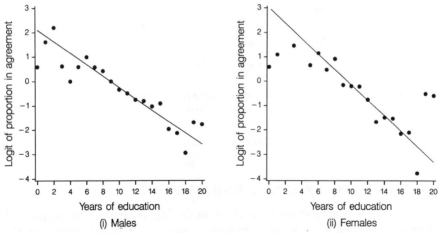

Fig. 3.5 Plot of the empirical logits of the proportion of individuals in agreement with the controversial statement, against the number of years of education, for (i) males and (ii) females.

We may therefore conclude that the model which describes the relationship between the proportion of individuals in agreement with the controversial statement and years of education by two different logistic regression lines adequately summarizes the observed data. From GLIM output given previously, the equations

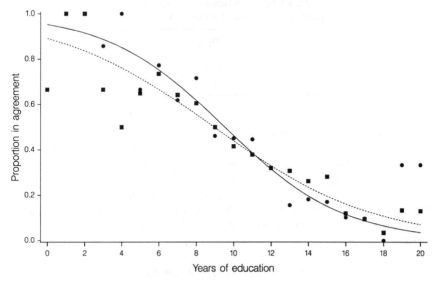

Fig. 3.6 Proportion of individuals in agreement with the controversial statement, against the number of years of education, and the fitted probabilities for males (···■···) and females (—●—).

of the fitted models are

Males: $\text{logit}(p) = 2.098 - 0.234$ years

Females: $\text{logit}(p) = 3.003 - 0.315$ years

The observed proportions are plotted in Fig. 3.6, together with the fitted curves. In summary, the probability that a male with little formal education agrees with the statement is less than the corresponding probability for a female, but for females, this probability declines more sharply as the number of years of education increase.

3.12 NON-CONVERGENCE AND OVERFITTING

When using a computer package to fit linear logistic models to binary and binomial response data, warning messages about the non-convergence of the iterative process used to produce the parameter estimates are sometimes encountered. The most likely cause of this phenomenon is that the model is an exact fit to certain binary observations, or to observed proportions of either zero or one.

As an illustration, suppose that the proportion of individuals with a certain characteristic, at the two levels of a factor A, are as shown in Table 3.9. Let p_i be

Table 3.9 Data on the proportion of individuals with a particular characteristic at each of two levels of a factor A

Level of A	Proportion
1	0/4
2	5/6

the true success probability at the ith level of A, $i = 1, 2$. In order to compare these two probabilities, the linear logistic model where $\text{logit}(p_i) = \beta_0$, a constant term, would be compared with the model where $\text{logit}(p_i) = \beta_0 + \alpha_i$. The second of these two models is overparameterized, in that only two of the three unknown parameters in the linear part of the model can be estimated from the two observations. This model will therefore be a perfect fit to the two observed proportions, and the fitted probability at the first level of A, $\hat{p}_1$, will be zero. Now, under the second model,

$$\text{logit}(\hat{p}_1) = \log(\hat{p}_1/(1 - \hat{p}_1)) = \hat{\beta}_0 + \hat{\alpha}_1$$

and so as $\hat{p}_1$ tends to zero, $\hat{\beta}_0 + \hat{\alpha}_1$ will tend to $-\infty$. In using GLIM for this modelling, the parameter α_1 is set equal to zero to avoid the overparameterization, which means that $\hat{\beta}_0$ must tend to $-\infty$. The computer software will detect this and ultimately report non-convergence of the fitting procedure.

On using GLIM to fit the model logit $(p_i) = \beta_0 + \alpha_i$ to the data in Table 3.9, the parameter estimates after ten cycles of the iterative procedure, are as follows.

	estimate	s.e.	parameter
1	-12.37	146.9	1
2	13.97	146.9	A(2)

From these estimates, logit $(\hat{p}_1) = -12.37$, so that $\hat{p}_1$ is practically equal to zero. The fitted probability at the second level of A, $\hat{p}_2$, will be finite, and since logit $(\hat{p}_2) = \hat{\beta}_0 + \hat{\alpha}_2$, $\hat{\alpha}_2$ will be forced towards $+\infty$ to compensate for the large negative value of $\hat{\beta}_0$. From the estimates given, logit $(\hat{p}_2) = -12.37 + 13.97 = 1.60$, and so $\hat{p}_2 = 0.83$, which is the same as the observed proportion of 5/6.

Models that reproduce all or part of an observed data set are not particularly helpful, and such models are said to be **overfitted**. These models contain more unknown parameters than can be justified by the data. When overfitting occurs, some or all of the parameter estimates will be crucially dependent on the observed data, in that slight changes to the data could result in substantial changes to the model-based parameter estimates. This in turn means that these parameters are imprecisely estimated, and so they will have unusually large standard errors. This feature is exhibited by the standard errors of the parameter estimates in the preceding illustration.

Problems of non-convergence and overfitting generally occur when the observed data are sparse, that is when there are some combinations of explanatory variables with very few binary observations. The chi-squared approximation to the distribution of the deviance will not then be reliable, which means that a large residual deviance, relative to the number of degrees of freedom, may not necessarily indicate that the fitted model is unsatisfactory. In these circumstances, any attempt to include additional terms in the model to bring the residual deviance down to a more acceptable level could result in an overfitted model.

When analysing data from observational studies, or highly structured data from an experimental study, it can be difficult to spot if too complicated a model is being entertained, before attempting to fit it. However, if convergence problems are encountered, and the standard errors of some of the parameter estimates become unusually large, this strongly suggests that the model being fitted is too complex.

3.13 A FURTHER EXAMPLE ON MODEL SELECTION

In this section, the application of logistic regression modelling to the analysis of a more highly structured data set that those considered elsewhere is discussed. The data to be used for this illustration are from the observational study on nodal involvement in patients presenting with prostatic cancer, given as Example 1.9 of

Chapter 1. In the data base there are observations on six variables for each of 53 patients; the variables were as follows.

Age: Age of patient in years

Acid: Level of serum acid phosphatase

X-ray: Result of an X-ray examination (0 = negative, 1 = positive)

Size: Size of tumour (0 = small, 1 = large)

Grade: Grade of tumour (0 = less serious, 1 = more serious)

Nodes: Nodal involvement (0 = no involvement, 1 = involvement)

The object of the study was to construct a model that could be used to predict the value of the binary response variable, nodes, on the basis of the five explanatory variables age, acid, X-ray, size and grade. In modelling the binary response probability, each of the categorical variables X-ray, size and grade can be treated as either a two-level factor or as a variate. In fitting one of these terms as a factor using GLIM, a single indicator variable would be constructed which takes the value zero for the first level of the factor and unity for the second. This indicator variable corresponds exactly with the manner in which the observed levels of these factors were recorded, and so each of these terms is fitted as a variate.

Without considering interactions or powers of the explanatory variables, there are 32 possible linear logistic models that could be fitted to this data set. In this example, it is feasible to fit all 32 of these in order not to miss anything. If there were many more explanatory variables, some selection strategy would be necessary. In such circumstances, determination of the most appropriate subset from a number of explanatory variables is best accomplished using a subjective approach, based on underlying knowledge of the variables and their likely inter-relationships. Because of their arbitrary nature, analogues of the automatic variable selection routines, forward selection and backward elimination used in multiple regression analysis should be avoided.

An initial examination of the data reveals that the distribution of the values of acid over the 53 patients is rather skew. This suggests that log (acid) should be taken to be the explanatory variable in modelling variation in the binary response variable. The deviance on fitting a linear logistic regression on acid is 67.12 on 51 d.f. while that for a regression on log (acid) is 64.81. This confirms that log (acid) on its own is a better predictor of nodal involvement than acid. Looking at the values of acid for patients with and without nodal involvement, it is seen that the spread of values is rather similar, apart from an unusually high value for patient number 24. When this observation is omitted, the deviances for a logistic regression on acid and log (acid) are 60.45 and 59.76, respectively. The evidence for transforming this variable is now not as strong. However, in the absence of any medical reason for excluding the data for this patient from the data base, this individual is retained and log (acid) is used as an explanatory variable in the modelling process.

The influence of particular observations on the form of the fitted model is discussed later in section 5.9.

The deviance, and corresponding number of degrees of freedom, for each of 32 possible logistic regression models fitted to these data is given in Table 3.10.

The variable which, on its own, has the greatest predictive power is X-ray, since this is the variable which leads to the biggest drop in deviance from that on fitting a constant term alone. However, the deviance for this single-variable model is

Table 3.10 The result of fitting 32 linear logistic regression models to the data on nodal involvement in prostatic cancer patients

Terms fitted in model	Deviance	d.f.
Constant	70.25	52
Age	69.16	51
log (acid)	64.81	51
X-ray	59.00	51
Size	62.55	51
Grade	66.20	51
Age + log (acid)	63.65	50
Age + X-ray	57.66	50
Age + size	61.43	50
Age + grade	65.24	50
log (acid) + X-ray	55.27	50
log (acid) + size	56.48	50
log (acid) + grade	59.55	50
X-ray + size	53.35	50
X-ray + grade	56.70	50
Size + grade	61.30	50
Age + log (acid) + X-ray	53.78	49
Age + log (acid) + size	55.22	49
Age + log (acid) + grade	58.52	49
Age + X-ray + size	52.08	49
Age + X-ray + grade	55.49	49
Age + size + grade	60.28	49
log (acid) + X-ray + size	48.99	49
log (acid) + X-ray + grade	55.03	49
log (acid) + size + grade	54.51	49
X-ray + size + grade	52.78	49
Age + log (acid) + X-ray + size	47.68	48
Age + log (acid) + X-ray + grade	50.79	48
Age + log (acid) + size + grade	53.38	48
Age + X-ray + size + grade	51.57	48
log (acid) + X-ray + size + grade	47.78	48
Age + log (acid) + X-ray + size + grade	46.56	47

considerably larger than that for models with more than one term. The smallest deviance is naturally that obtained on including all five variables in the model, but this deviance is not much smaller than the deviance for either of two four-variable models, [age + log (acid) + X-ray + size] and [log (acid) + X-ray + size + grade], and one three-variable model [log (acid) + X-ray + size]. All three of these models have the variables log (acid), X-ray and size in common. The addition of either age or grade to this basic model reduces the deviance by only a very small amount.

On including quadratic terms in age and log (acid) into the model with all five explanatory variables, the deviance is again reduced by only a very small amount. There is therefore no reason to include higher-order powers of log (acid) and age in the logistic regression model.

The next step is to see if any interaction terms need to be incorporated in the model. To examine this, each of the ten terms formed from products of the five explanatory variables is added, in turn, to the model that contains the five variables. The reduction in deviance on including each of these product terms, from its original value of 46.56, is then calculated. Two interactions reduce the deviance by relatively large amounts: size × grade reduces it by 8.13 on 1 d.f., while grade × log (acid) reduces it by 4.04 on 1 d.f. No other interactions reduce the deviance by more than 1.05. To see if both of these interactions are needed, each is fitted after the other. The reduction in deviance due to size × grade, adjusted for grade × log (acid), is 7.18 on 1 d.f. and that for grade × log (acid), adjusted for size × grade is 3.09 on 1 d.f. Both of these reductions in deviance are significant at the 10% level, at least, and so both interaction terms are retained in the model. To check that no other two-factor interaction term is needed, the eight remaining interactions are added to the model with the five basic variables and the two important interaction terms. The deviance is reduced only by 5.17 on 8 d.f., confirming that no other two-factor interactions need be included. There is also no evidence of a three-factor interaction between size, grade and log (acid).

Finally, the model is examined to see if any terms not involved in an interaction can be omitted. When age is omitted from the model, the deviance increases by 0.95 on 1 d.f., whereas when X-ray is omitted, the deviance increases by 5.77 on 1 d.f. The variable age may therefore be dropped. After omitting age, both interactions still need to be included in the model. In particular, the increase in deviance on omitting size × grade from the model that contains X-ray, size, grade, log (acid), size × grade and grade × log (acid) is 6.26, while that on omitting grade × log (acid) is 4.17.

The parameter estimates under the final model, obtained from GLIM output, are given below, in which **SXG** is the parameter associated with size × grade, and **GXLA** is that associated with grade × log (acid).

	estimate	s.e.	parameter
1	−2.553	1.035	1
2	2.340	1.077	XRAY
3	3.138	1.167	SIZE
4	9.958	4.597	GRAD

5	1.708	1.418	LACD
6	−5.647	2.411	SXG
7	10.42	6.543	GXLA

From these estimates, the fitted models for individuals with each of the two grades of tumour are

Grade 0:   logit $(\hat{p}) = -2.55 + 2.34$ X-ray $+ 3.14$ size $+ 1.71$ log (acid)

Grade 1:   logit $(\hat{p}) = 7.40 + 2.34$ X-ray $- 2.51$ size $+ 12.13$ log (acid)

where $\hat{p}$ is the fitted probability of nodal involvement and the variables X-ray and size are themselves binary.

This model indicates that on the logit scale, there is a linear relationship between the probability of nodal involvement and log (acid). This relationship has a different slope for each of the two grades of tumour, and different intercepts depending on the size and grade of the tumour and also on the X-ray result. The fitted logistic regression lines are shown in Fig. 3.7. In qualitative terms, the probability of nodal involvement rises much more sharply with increasing serum acid phosphatase levels for individuals with a more serious tumour (grade = 1) than for those with a less serious one (grade = 0). There is a higher probability of nodal involvement in patients with a less serious tumour when the tumour size is large (size = 1)

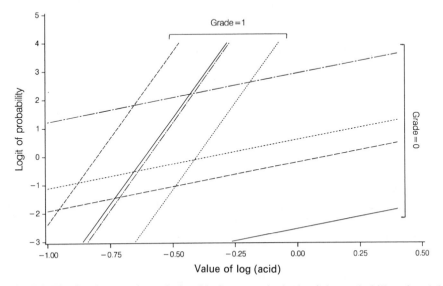

Fig. 3.7 The fitted regression relationship between the logit of the probability of nodal involvement and log (acid) for individuals with X-ray = 0, size = 0 (——); X-ray = 0, size = 1 (······); X-ray = 1, size = 0 (-------), and X-ray = 1, size = 1 (—·—).

compared to when it is small (size = 0). However, for patients with a more serious tumour, the reverse seems to be the case; those with a small tumour have a higher probability of nodal involvement than those with a large tumour. A positive X-ray result suggests a higher probability of nodal involvement for patients with tumours of either grade.

The next stage is to validate the model that has now been identified. An important ingredient of this validation process will be discussions with the clinical investigator on the composition of the model to determine if it is sensible on medical grounds. In addition, a statistical analysis can be carried out to determine if there are individual patients in the data base for whom the model does not satisfactorily predict nodal involvement, and to identify whether there are particular patients that have an undue influence on the structure of the model. If data from additional patients become available, the model based on the original 53 individuals can be further tested by seeing how well it predicts nodal involvement in these new patients.

3.14 PREDICTING A BINARY RESPONSE PROBABILITY

A common goal in linear logistic modelling is to predict the value of a binary response variable. The predicted response probability may subsequently form the basis for allocating an individual according to one of two groups. Examples include the prediction of the disease state of an individual on the basis of prognostic factors, determining whether or not an individual is mentally sub-normal from the results of a battery of psychological tests, and assessing the chance that an industrial component is faulty from a number of metallurgical analyses. In situations such as these, the first step is to collect binary response data, and corresponding values of the explanatory variables, for a particular set of individuals. A linear logistic model is fitted to the observed data and this is then used to predict the response probability for a new individual for which the values of the explanatory variables in the model are available.

To assist in the interpretation of the predicted value of the response probability for a new individual, $\hat{p}_0$, a confidence interval for the corresponding true response probability, p_0, will be required. This is best obtained by constructing a confidence interval for logit (p_0) and then transforming the resulting limits to give an interval estimate for p_0 itself. Suppose that the fitted response probability for the new individual is such that logit $(\hat{p}_0) = \hat{\eta}_0$, where $\hat{\eta}_0 = \sum \hat{\beta}_j x_{j0}$ is the linear predictor and x_{j0} is the value of the jth explanatory variable for the new individual, $j = 1, 2, ..., k$. The variance of this linear predictor is calculated from the variances and covariances of the estimates of the β-parameters using the standard result

$$\text{Var}(\hat{\eta}_0) = \sum_{j=1}^{k} x_{j0}^2 \, \text{Var}(\hat{\beta}_j) + 2 \sum_{j=1}^{k} \sum_{h=1}^{j} x_{h0} x_{j0} \, \text{Cov}(\hat{\beta}_h, \hat{\beta}_j) \tag{3.8}$$

However, it can be obtained directly using many statistical packages by adding the values of the explanatory variables for the new observation, together with an arbitrary value of the response variable, to the basic set of data and giving this observation zero weight; see section 9.3.1 for fuller details.

An approximate $100(1 - \alpha)\%$ confidence interval for logit (p_0) is $\hat{\eta}_0 \pm z_{\alpha/2}\sqrt{\text{Var}(\hat{\eta}_0)}$, where $z_{\alpha/2}$ is the upper $\alpha/2$ point of the standard normal distribution, leading to an approximate confidence interval for p_0 itself.

To assign an individual to one of two groups, say, group 1 and group 2, on the basis of their predicted response probability, a threshold value π_0 has to be identified, where π_0 is such that the individual is assigned to group 1 if $\hat{p}_0 < \pi_0$ and to group 2 if $\hat{p}_0 \geqslant \pi_0$. While it may often be reasonable to take $\pi_0 = 0.5$, it is sometimes the case that the two groups should not be regarded symmetrically; a value of π_0 other than 0.5 may then be appropriate. The choice of a suitable value of π_0 can be made on the basis of the original data, by identifying the value that minimizes either the overall proportion of misclassifications, or by compromising between the minimization of the two misclassification probabilities, namely, the probability of declaring an individual to be in group 1 when they should be in group 2 and *vice versa*. This is illustrated in the following example.

Example 3.9 Prostatic cancer
In the analysis of the data concerning nodal involvement in prostatic cancer patients discussed in section 3.13, a model that could be used to predict the probability of nodal involvement was identified. Suppose that a future patient presents with the following values of the explanatory variables in that model: X-ray = 0, size = 1, grade = 1, acid = 0.7. Using the model identified in section 3.13, the predicted probability of nodal involvement for this patient, $\hat{p}_0$, is such that logit $(\hat{p}_0) = 0.570$, so that $\hat{p}_0 = 0.64$.

The approximate variance of logit $(\hat{p}_0)$ can be found from the estimated variances and covariances of the $\hat{\beta}$s using equation (3.8). These quantities can be obtained from most statistical packages in the form of a **variance–covariance matrix** whose diagonal elements are the variances of the parameter estimates and whose off-diagonal elements are the covariances between pairs of estimates. The variance–covariance matrix of the seven parameter estimates in this example, obtained using GLIM, is

Constant	1.072						
X-ray	−0.560	1.160					
Size	−0.848	0.494	1.361				
log (acid)	0.403	0.168	0.378	2.012			
Grade	−1.443	1.329	1.175	−0.291	21.132		
Size × grade	1.102	−1.022	−1.586	−0.455	−8.258	5.813	
Grade × log (acid)	−0.982	1.033	0.133	−1.838	27.765	−7.852	42.807
	Constant	X-ray	Size	log (acid)	Grade	Size × grade	Grade × log (acid)

From this matrix, $\text{Var}(\hat{\beta}_0) = 1.072$, $\text{Var}(\hat{\beta}_1) = 1.160, \ldots$, $\text{Var}(\hat{\beta}_7) = 42.807$, and

$\mathrm{Cov}\,(\hat{\beta}_0, \hat{\beta}_1) = -0.560$, $\mathrm{Cov}\,(\hat{\beta}_0, \hat{\beta}_2) = -0.848$, ..., $\mathrm{Cov}\,(\hat{\beta}_5, \hat{\beta}_6) = -7.852$, where $\hat{\beta}_0$ is the estimated constant term and $\hat{\beta}_1, \hat{\beta}_2, ..., \hat{\beta}_6$ are the coefficients of X-ray, size, log (acid), grade, size × grade and grade × log (acid), respectively, in the fitted model. On substituting for these in equation (3.8) it is found that $\mathrm{Var}\,[\mathrm{logit}\,(\hat{p}_0)] \approx 1.274$. A 95% confidence interval for $\mathrm{logit}\,(\hat{p}_0)$ is therefore $0.570 \pm 1.96\,\sqrt{(1.274)}$, that is, the interval from -1.642 to 2.783. The corresponding confidence interval for the probability that this new individual has nodal involvement ranges from 0.06 to 0.84. This wide interval indicates that the predicted probability is imprecisely estimated. Had the value of X-ray for this individual been 1 rather than 0, the predicted probability of nodal involvement would have been 0.95 with a 95% confidence interval ranging from 0.52 to 1.00. This interval is somewhat narrower and it is correspondingly easier to classify this individual.

To study possible classification rules for future patients in more detail, the fitted probabilities for the 53 patients in the original data set can be compared with the observation on nodal involvement. The proportion of patients who, on the basis of a range of threshold values, would be classified as (i) having nodal involvement when they did not (false positive), and (ii) not having nodal involvement when they did (false negative) can then be computed. These values are summarized in Table 3.11. In this example, it is important not to miss anyone who may have nodal

Table 3.11 Probabilities of misclassifying prostatic cancer patients

Threshold value	Proportion of false positives		Proportion of false negatives		Total proportion misclassified	
0.25	0.321	(9/28)	0.040	(1/25)	0.189	(10/53)
0.35	0.250	(6/24)	0.069	(2/29)	0.151	(8/53)
0.45	0.227	(5/22)	0.097	(3/31)	0.151	(8/53)
0.50	0.210	(4/19)	0.147	(5/34)	0.170	(9/53)
0.55	0.176	(3/17)	0.167	(6/36)	0.170	(9/53)
0.65	0.167	(2/12)	0.244	(10/41)	0.226	(12/53)

involvement and so it is appropriate to pay rather more attention to the chance of a false negative than to that of a false positive. On these grounds, it seems reasonable to take a threshold value of 0.35, so that a new patient with a predicted probability less than 0.35 will be classified as not having nodal involvement.

In this analysis, the same patients have been used both to determine the classification rule and to judge its performance, a procedure which leads to biased estimates of the misclassification probabilities. More reliable estimates of these probabilities can be obtained by determining the allocation rule using data from all but one of the patients, and to then use this rule to classify the omitted patient. However, this **cross-validation** approach has the disadvantage of being computationally intensive. One further point to be borne in mind is that just as particular observations in the basic data set may have an undue influence on the structure of the fitted model, so might they affect predicted probabilities and the classification of future individuals. This matter will be considered further in Chapter 5.

FURTHER READING

Comprehensive introductions to the theory and practice of linear regression modelling are given by Draper and Smith (1981), and Montgomery and Peck (1982). Methods of estimation are described in many text books on statistics, such as Cox and Hinkley (1974), and the different approaches are reviewed and contrasted in Barnett (1982).

The logistic regression model was first suggested by Berkson (1944), who showed how the model could be fitted using iteratively weighted least squares. Comparisons between the logistic and probit transformations are made by Chambers and Cox (1967), and Berkson (1951) explains why he prefers logits to probits. The equivalence of iterative weighted least squares and the method of maximum likelihood is discussed by Thisted (1988; Section 4.5.6). The computation of maximum likelihood estimates using the method of scoring is due to Fisher (1925), but a more readable summary of the method is included in Everitt (1987). The algorithm used to obtain the maximum likelihood estimates of the parameters of a linear logistic model was given in a general form by Nelder and Wedderburn (1972) and is reproduced in McCullagh and Nelder (1989) and Aitken et al. (1989).

The use of the deviance as a measure of goodness of fit was first proposed by Nelder and Wedderburn (1972) and fuller details on the adequacy of the chi-squared approximation to its null distribution are contained in McCullagh and Nelder (1989). The argument against using the deviance as a measure of goodness of fit for binary data was given by Williams (1983). The empirical logistic transform and the relationship between logistic regression and discriminant analysis are described in greater detail by Cox and Snell (1989). Krzanowski (1988) includes a clear account of methods of classification in his general text on multivariate methods.

Issues to be faced before an attempt is made to model an observed response variable are discussed by Chatfield (1988). Cox and Snell (1989; Appendix 2) provide an excellent summary of how to choose a subset of explanatory variables for use in logistic regression modelling.

4

Bioassay and some other applications

The development of models for binary and binomial response data was originally prompted by the needs of a type of experimental investigation known as a **biological assay**, or **bioassay** for short. In a typical assay, different concentrations of a chemical compound are applied to batches of experimental animals. The number of animals in each batch that respond to the chemical is then recorded, and these values are regarded as observations on a binomial response variable. In this application area, models based on the logistic and probit transformations can be motivated through the notion of a tolerance distribution. This is described in Section 4.1. Some specific aspects of bioassay are then discussed in Sections 4.2–4.4, and the extension from linear to non-linear logistic modelling is considered in Section 4.5. A transformation of the binary response probability which is not as widely used as the logistic or probit transformations is the complementary log–log transformation. Nevertheless, models based on this transformation do arise naturally in a number of areas of application, some of which are described in Section 4.6.

4.1 THE TOLERANCE DISTRIBUTION

Consider an insecticidal trial in which a particular insecticide is applied at known concentrations to batches of insects. If a low dose of the insecticide is administered to a given batch, none of the insects in that batch may die, whereas if a high dose is given, they may all die. Whether or not a particular insect dies when exposed to a given dose of the insecticide depends on the **tolerance** of that individual to the compound. Those insects with a low tolerance will be more likely to die from exposure to a given dose than those with a high tolerance.

Over the population of insects, from which those used in the trial have been drawn, there will be a distribution of tolerance levels. Those insects with a tolerance less than d_i will die when dose d_i is administered. Let U be a random variable associated with the tolerance distribution and let u be the tolerance of a particular individual. The corresponding probability density function of U is denoted $f(u)$, and the probability of death when exposed to dose d_i is then given by

$$p_i = P(U \leqslant d_i) = \int_{-\infty}^{d_i} f(u) \, du$$

Now suppose that the tolerances are normally distributed, so that U has a normal distribution with mean μ and variance σ^2. Then,

$$f(u) = \frac{1}{\sigma\sqrt{(2\pi)}} \exp\left\{-\frac{1}{2}\left(\frac{u-\mu}{\sigma}\right)^2\right\}, \quad -\infty < u < \infty$$

and

$$p_i = \frac{1}{\sigma\sqrt{(2\pi)}} \int_{-\infty}^{d_i} \exp\left\{-\frac{1}{2}\left(\frac{u-\mu}{\sigma}\right)^2\right\} du = \Phi\left(\frac{d_i-\mu}{\sigma}\right)$$

where Φ is the standard normal distribution function. Now let $\beta_0 = -\mu/\sigma$ and $\beta_1 = 1/\sigma$ so that $p_i = \Phi(\beta_0 + \beta_1 d_i)$, or probit$(p_i) = \Phi^{-1}(p_i) = \beta_0 + \beta_1 d_i$. This **probit regression model** for the relationship between the probability of death, p_i, and dose of insecticide, d_i, arises as a direct result of the assumption of a normal tolerance distribution.

A probability distribution that is very similar to the normal distribution is the **logistic distribution**. The probability density function of a random variable U that has a logistic distribution is

$$f(u) = \frac{\exp\{(u-\mu)/\tau\}}{\tau[1 + \exp\{(u-\mu)/\tau\}]^2}, \quad -\infty < u < \infty$$

where $-\infty < \mu < \infty$, $\tau > 0$. The mean and variance of this distribution are μ and $\pi^2\tau^2/3$, respectively. If the tolerances are taken to have this logistic distribution,

$$p_i = \int_{-\infty}^{d_i} \frac{\exp\{(u-\mu)/\tau\}}{\tau[1 + \exp\{(u-\mu)/\tau\}]^2} \, du = \frac{\exp\{(d_i-\mu)/\tau\}}{1 + \exp\{(d_i-\mu)/\tau\}}$$

Writing $\beta_0 = -\mu/\tau$ and $\beta_1 = 1/\tau$, we find that

$$p_i = \frac{\exp(\beta_0 + \beta_1 d_i)}{1 + \exp(\beta_0 + \beta_1 d_i)}$$

whence $\text{logit}(p_i) = \beta_0 + \beta_1 d_i$. The assumption of a logistic tolerance distribution therefore leads to a linear logistic regression model for p_i.

If an assumed underlying logistic tolerance distribution is constructed to have the same mean and variance as a corresponding normal distribution, an approximate relationship between the logit and probit of p can be found. We have seen that $\text{logit}(p_i) = (d_i - \mu)/\tau$, while $\text{probit}(p_i) = (d_i - \mu)/\sigma$. Equating the variances of the two underlying tolerance distributions gives $\sigma^2 = \pi^2\tau^2/3$ and so the logit of p_i will be approximately equal to $\pi/\sqrt{3}$ multiplied by the probit of p_i. The similarity of the two transformed values of p after equating the variances of the underlying normal and logistic tolerance distributions can be seen from the table of Appendix A; multiplying a probit by $\pi/\sqrt{3}$ gives a value very similar to the corresponding logit.

Since the tolerance must always be positive, and there will usually be some individuals with a very high tolerance, the distribution of tolerances is unlikely to be symmetric. The logarithms of the tolerances are rather more likely to be symmetrically distributed and so it is customary to work with the logarithm of the concentration or dose of the compound being assayed, rather than the concentration itself. It does not really matter what base of logarithms is used for this, and in this book logarithms to the base e will be used unless indicated otherwise. Using the logarithm of concentration as an explanatory variable also means that a zero concentration of the compound, as used in a control batch for example, corresponds to a log concentration of $-\infty$, leading to a zero probability of death under a logistic or probit regression model. Methods for allowing for natural mortality in a batch of insects are discussed in Section 4.4.

On some occasions, a logistic or probit regression model may fit better when the logarithm of the concentration is used, whereas on other occasions it may not be necessary or desirable to transform the explanatory variable. In the absence of any underlying scientific reason for using transformed values of an explanatory variable in the model, the choice between alternative models will rest on statistical grounds alone and the model which best fits the available data would be adopted.

Example 4.1 Anti-pneumococcus serum
A typical example of a bioassay, which concerned the relationship between the proportion of mice dying from pneumonia and the dose of a protective serum, was given in Example 1.6. Let y_i be the number of mice which die out of n_i exposed to the ith dose, $i = 1, 2, \ldots, 5$. A plot of the empirical logits, $\log\{(y_i + 0.5)/(n_i - y_i + 0.5)\}$, against the explanatory variables dose and $\log(\text{dose})$ is given in Fig. 4.1. This figure suggests that the relationship between the logistic transform of the response probability, $\text{logit}(p)$, and $\log(\text{dose})$ is more linear than that between $\text{logit}(p)$ and dose. However, for illustrative purposes, an analysis of deviance will be used to determine whether it is preferable to use $\log(\text{dose})$ as the explanatory variable in the model. In addition, the goodness of fit of a linear logistic model will be compared with that of a probit model.

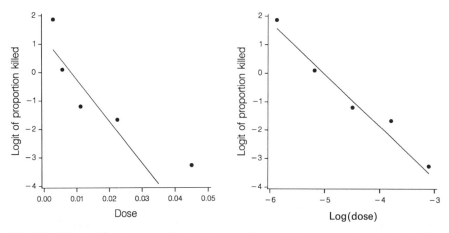

Fig. 4.1 Relationship between the logistic transform of the proportion of mice killed and (i) dose, (ii) log (dose) for the data from Example 1.6.

The GLIM output given below was obtained on fitting a logistic regression model to the observed data, using the logarithm of dose (**LDOS**) as the explanatory variable.

$$\text{scaled deviance} = \textbf{2.8089 at cycle 4}$$
$$\textbf{d.f.} = \textbf{3}$$

	estimate	s.e.	parameter
1	−9.189	1.255	1
2	−1.830	0.2545	LDOS

The deviance on fitting the model is 2.81 on 3 d.f. Since the deviance is so close to the number of degrees of freedom on which it is based, the model would appear to fit the data very well.

For comparison, if dose rather than log (dose) is used in the logistic regression model, the deviance is 15.90 on 3 d.f. This model is certainly not satisfactory, the deviance being significantly high at the 0.5% level of significance. The fit of the logistic regression model to the data using (i) dose and (ii) log (dose) as the explanatory variable, respectively, is also shown in Fig. 4.1. There is an obvious lack of fit in the regression model that uses dose rather than log (dose), and so we conclude that log (dose) should be used as the explanatory variable in the model.

On fitting a probit regression model using log (dose), the deviance is 3.19, but if dose is used, the deviance increases to 19.09. Since the deviance on fitting a logistic regression model with log (dose) as the explanatory variable is slightly smaller than the deviance for the corresponding probit model, 2.81 as opposed to 3.19, one might consider that the linear logistic model is the better fit, but there is very little in it.

Informal comparisons between a model based on log (dose) with one based on dose and between a probit and a logistic model can be made using the deviance. However, a formal significance test cannot be used in either situation. This is because the difference between two deviances has only an approximate χ^2-distribution when the models are nested, that is, when one model contains a subset of terms in the other. In comparing models with log (dose) and dose or a probit model with a logistic, this is not the case.

4.2 ESTIMATING AN EFFECTIVE DOSE

When a response probability is modelled as a function of a single explanatory variable x, it is often of interest to estimate the value of x that corresponds to a specified value of the response probability. For example, in bioassay, one is often interested in the concentration of a chemical that is expected to produce a response in 50% of the individuals exposed to it. This dose is termed the **median effective dose** and is commonly referred to as the *ED50* value. When the response being studied is death, this quantity is renamed the **median lethal dose** or *LD50* value. The concentration that is expected to produce a response in other proportions is expressed in a similar manner. For example, the dose which causes 90% of individuals to respond is the *ED90* value, and so on. Quantities such as these summarize the potency of the chemical that is being assayed and may subsequently form the basis of comparisons between different compounds.

Suppose that a linear logistic model is used to describe the relationship between the probability of a response, p, and the dose, d, in a bioassay. Then

$$\text{logit}(p) = \log\left(\frac{p}{1-p}\right) = \beta_0 + \beta_1 d$$

The dose for which $p = 0.5$ is the *ED50* value and since $\text{logit}(0.5) = \log 1 = 0$, the *ED50* value satisfies the equation

$$\beta_0 + \beta_1 ED50 = 0$$

so that $ED50 = -\beta_0/\beta_1$. After fitting the linear logistic model, estimates $\hat{\beta}_0$, $\hat{\beta}_1$ of the unknown parameters are obtained and so the *ED50* value is estimated using

$$\hat{ED50} = -\frac{\hat{\beta}_0}{\hat{\beta}_1}$$

To obtain the *ED90* value, $p = 0.9$ is used in the linear logistic model to give

$$\log\left(\frac{0.9}{0.1}\right) = \beta_0 + \beta_1 ED90$$

so that the *ED90* value is $(2.1972 - \beta_0)/\beta_1$, which can be estimated by $\hat{ED90} = (2.1972 - \hat{\beta}_0)/\hat{\beta}_1$.

If log (dose), rather than dose, is used as the explanatory variable in the model, these formulae will need some modification. When the model

$$\text{logit}(p) = \beta_0 + \beta_1 \log d$$

is fitted, the *ED50* value is such that

$$\beta_0 + \beta_1 \log(ED50) = 0$$

so that now $ED50 = \exp(-\beta_0/\beta_1)$, which is estimated by $E\hat{D}50 = \exp(-\hat{\beta}_0/\hat{\beta}_1)$. Similarly, the *ED90* value would now be estimated by

$$E\hat{D}90 = \exp\left\{\frac{2.1972 - \hat{\beta}_0}{\hat{\beta}_1}\right\}$$

When logarithms to other bases are used in the model, that base must be substituted for e in these expressions. In computer-based analyses, it is essential to know which base of logarithms (10 or e) has been used in order that further calculations based on the output can be carried out correctly.

Estimates of the *ED50* and *ED90* values can similarly be obtained under a probit regression model. For example, when log (dose) is used as the explanatory variable, the model is

$$\text{probit}(p) = \Phi^{-1}(p) = \beta_0 + \beta_1 \log d$$

On setting $p = 0.5$, probit $(0.5) = 0$ and, as for the logistic model, the *ED50* value will be estimated by $E\hat{D}50 = \exp(-\hat{\beta}_0/\hat{\beta}_1)$, where $\hat{\beta}_0$, $\hat{\beta}_1$ are the parameter estimates under the probit model. When $p = 0.9$, probit $(0.9) = 1.2816$ and so the *ED90* value is now estimated by $\exp[(1.2816 - \hat{\beta}_0)/\hat{\beta}_1]$.

Having obtained an estimate of a quantity such as the *ED50*, it is generally desirable to provide a standard error or confidence interval for it so that the precision of the estimate can be judged. Because the arguments used to obtain the standard errors of functions of parameter estimates and corresponding confidence intervals are important and not straightforward, they are presented here in some detail. The approximate standard error of the estimated *ED50* value is first derived in section 4.2.1, and approximate confidence intervals for the true *ED50* value, based on this standard error, are given. More accurate confidence intervals for the true *ED50* value can be constructed using Fieller's theorem, an approach which is described in section 4.2.2.

4.2.1 Approximate standard error of an estimated effective dose

Observing that $E\hat{D}50$ is a function of two parameter estimates, say, $g(\hat{\beta}_0, \hat{\beta}_1)$, a standard result for the approximate variance of the function can be used to obtain the standard error of the estimate. This result states that the variance of $g(\hat{\beta}_0, \hat{\beta}_1)$

is approximately

$$\left(\frac{\partial g}{\partial \hat{\beta}_0}\right)^2 \text{Var}\,(\hat{\beta}_0) + \left(\frac{\partial g}{\partial \hat{\beta}_1}\right)^2 \text{Var}\,(\hat{\beta}_1) + 2\left(\frac{\partial g}{\partial \hat{\beta}_0}\frac{\partial g}{\partial \hat{\beta}_1}\right) \text{Cov}\,(\hat{\beta}_0, \hat{\beta}_1) \qquad (4.1)$$

When the model includes dose as the explanatory variable, $g\,(\hat{\beta}_0, \hat{\beta}_1)$ is taken to be $-\hat{\beta}_0/\hat{\beta}_1$. Writing $\hat{\rho} = \hat{\beta}_0/\hat{\beta}_1$, $v_{00} = \text{Var}\,(\hat{\beta}_0)$, $v_{11} = \text{Var}\,(\hat{\beta}_1)$ and $v_{01} = \text{Cov}\,(\hat{\beta}_0, \hat{\beta}_1)$, it is found using expression (4.1) that

$$\text{Var}\,(E\hat{D}50) \approx \frac{v_{00} - 2\hat{\rho}v_{01} + \hat{\rho}^2 v_{11}}{\hat{\beta}_1^2}$$

In practice, the variances and covariance in this expression will not be known exactly and approximate values $\hat{v}_{00}$, $\hat{v}_{11}$ and $\hat{v}_{01}$ will be used in their place. The standard error of the estimated $ED50$ value, s.e. $(E\hat{D}50)$, will then be given by

$$\text{s.e.}(E\hat{D}50) \approx \left\{\frac{\hat{v}_{00} - 2\hat{\rho}\hat{v}_{01} + \hat{\rho}^2 \hat{v}_{11}}{\hat{\beta}_1^2}\right\}^{1/2} \qquad (4.2)$$

The standard error of an estimate obtained in this way can be used to obtain an approximate confidence interval for the corresponding parameter. For example, a 95% confidence interval for the $ED50$ value could be obtained using $E\hat{D}50 \pm 1.96$ s.e.$(E\hat{D}50)$. It can happen that the lower limit of a confidence interval calculated in this manner is negative. Although this could simply be replaced by zero, one way of avoiding this difficulty altogether is to work with the logarithm of the $ED50$ value. The standard error of $\log (E\hat{D}50)$ is obtained using the standard result

$$\text{s.e.}\,\{\log(E\hat{D}50)\} \approx (E\hat{D}50)^{-1}\,\text{s.e.}\,(E\hat{D}50)$$

An approximate 95% confidence interval for the $\log (ED50)$ value is $\log (E\hat{D}50) \pm 1.96$ s.e. $\{\log (E\hat{D}50)\}$, and a confidence interval for the $ED50$ value follows from exponentiating these two limits. Although this procedure ensures that the lower limit of the confidence interval is non-negative, the procedure cannot be generally recommended. This is because the estimated $\log (ED50)$ value is less likely to be symmetrically distributed than the $ED50$ value itself, and hence, the assumption of a normal distribution, used in constructing the confidence interval, is less likely to be reasonable.

When the explanatory variable used in the logistic model is $\log (\text{dose})$, the $ED50$ value is estimated by $\exp(-\hat{\beta}_0/\hat{\beta}_1)$ and, from relation (4.2), the standard error of $\log (E\hat{D}50)$ is given by

$$\text{s.e.}\,\{\log (E\hat{D}50)\} \approx \left\{\frac{\hat{v}_{00} - 2\hat{\rho}\hat{v}_{01} + \hat{\rho}^2 \hat{v}_{11}}{\hat{\beta}_1^2}\right\}^{1/2} \qquad (4.3)$$

The standard error of $E\hat{D}50$ itself can be found using

$$\text{s.e.}\,(E\hat{D}50) \approx E\hat{D}50\ \text{s.e.}\,\{\log\,(E\hat{D}50)\} \tag{4.4}$$

When log (dose) is used as the explanatory variable, it is better to derive a confidence interval for the $ED50$ value from exponentiating the limits for the logarithm of the $ED50$ value, obtained using relation (4.3), rather than use the symmetric interval based on relation (4.4).

A similar procedure can be used to obtain the standard error of other quantities of interest, such as the estimated $ED90$ value, and corresponding confidence intervals.

4.2.2 Confidence intervals for the effective dose using Fieller's theorem

Fieller's theorem is a general result that enables confidence intervals for the ratio of two normally distributed random variables to be obtained. This result will first be given in fairly general terms before applying it to the construction of a confidence interval for the $ED50$ value.

Suppose that $\rho = \beta_0/\beta_1$ where β_0, β_1 are estimated by $\hat{\beta}_0$, $\hat{\beta}_1$ and these estimates are assumed to be normally distributed with means β_0, β_1, variances v_{00}, v_{11} and covariance v_{01}. Consider the function $\psi = \hat{\beta}_0 - \rho\hat{\beta}_1$. Then, $E(\psi) = \beta_0 - \rho\beta_1 = 0$, since $\hat{\beta}_0$, $\hat{\beta}_1$ are unbiased estimates of β_0 and β_1, respectively, and the variance of ψ is given by

$$V = \text{Var}\,(\psi) = v_{00} + \rho^2 v_{11} - 2\rho v_{01} \tag{4.5}$$

Since $\hat{\beta}_0$, $\hat{\beta}_1$ are assumed to be normally distributed, ψ will also be normally distributed and

$$\frac{\hat{\beta}_0 - \rho\,\hat{\beta}_1}{\sqrt{(V)}}$$

has a standard normal distribution. Consequently, if $z_{\alpha/2}$ is the upper $\alpha/2$ point of the standard normal distribution, a $100(1-\alpha)\%$ confidence interval for ρ is the set of values for which

$$|\hat{\beta}_0 - \rho\,\hat{\beta}_1| \leqslant z_{\alpha/2}\sqrt{(V)}$$

Squaring both sides and taking the equality gives

$$\hat{\beta}_0^2 + \rho^2\hat{\beta}_1^2 - 2\rho\hat{\beta}_0\hat{\beta}_1 - z_{\alpha/2}^2 V = 0$$

and after substituting for V from equation (4.5) and some rearrangement, the following quadratic equation in ρ is obtained:

$$(\hat{\beta}_1^2 - z_{\alpha/2}^2 v_{11})\,\rho^2 + (2v_{01}z_{\alpha/2}^2 - 2\hat{\beta}_0\hat{\beta}_1)\,\rho + \hat{\beta}_0^2 - v_{00}z_{\alpha/2}^2 = 0 \tag{4.6}$$

The two roots of this quadratic equation will constitute the confidence limits for ρ. This is Fieller's result.

To use this result to obtain a confidence interval for $ED50 = -\beta_0/\beta_1$, write $-ED50$ for ρ in equation (4.6). Also, since the variances and covariance of the parameter estimates obtained on fitting a linear logistic model are only large-sample approximations, the approximate variances and covariance, $\hat{v}_{00}$, $\hat{v}_{11}$ and $\hat{v}_{01}$ have to be used in place of v_{00}, v_{11}, and v_{01}. The resulting quadratic equation in $ED50$ is

$$(\hat{\beta}_1^2 - z_{\alpha/2}^2 \hat{v}_{11}) \, ED50^2 - (2\hat{v}_{01} z_{\alpha/2}^2 - 2\hat{\beta}_0\hat{\beta}_1) \, ED50 + \hat{\beta}_0^2 - \hat{v}_{00} z_{\alpha/2}^2 = 0$$

and solving this equation using the standard formula gives

$$ED50 = \frac{-\left(\hat{\rho} - g\dfrac{\hat{v}_{01}}{\hat{v}_{11}}\right) \pm \dfrac{z_{\alpha/2}}{\hat{\beta}_1}\left\{\hat{v}_{00} - 2\hat{\rho}\hat{v}_{01} + \hat{\rho}^2\hat{v}_{11} - g\left(\hat{v}_{00} - \dfrac{\hat{v}_{01}^2}{\hat{v}_{11}}\right)\right\}^{1/2}}{1 - g} \qquad (4.7)$$

for the $100(1-\alpha)\%$ confidence limits for the true $ED50$ value, where $\hat{\rho} = \hat{\beta}_0/\hat{\beta}_1$ and $g = z_{\alpha/2}^2 \hat{v}_{11}/\hat{\beta}_1^2$. When there is a strong dose–response relationship, $\hat{\beta}_1$ will be highly significantly different from zero and $\hat{\beta}_1/\sqrt{(\hat{v}_{11})}$ will be much greater than $z_{\alpha/2}$. In these circumstances, g will be small; the more significant the relationship, the more negligible g becomes. When g is taken to be zero in expression (4.7), the limits for the $ED50$ value are exactly those based on the approximate standard error of the $ED50$ value given in relation (4.2).

When $\log(\text{dose})$ is being used as the explanatory variable, a confidence interval for the $ED50$ value can be found by first obtaining confidence limits for $\log(ED50) = -\beta_0/\beta_1$ using Fieller's theorem, and then exponentiating the resulting limits to give an interval estimate for the $ED50$ value itself.

Example 4.2 Anti-pneumococcus serum
Consider again the data on the extent to which a certain serum protects mice against the development of pneumonia, first given as Example 1.6. In Example 4.1, it was found that the most appropriate model for these data was a linear logistic model for the probability of death, with the logarithm of dose of serum as the explanatory variable. The fitted model is

$$\text{logit}(p) = -9.19 - 1.83 \log(\text{dose})$$

and so the $ED50$ value is estimated from

$$\widehat{ED50} = \exp(-9.19/1.83) = 0.0066 \text{ cc}$$

It is always a good idea to compare an estimated *ED50* value with the observed data to see if the estimate looks right. In this case it does, since at a dose of 0.0056 cc, the observed proportion of deaths is 21/40 – very near that required for the *ED50*.

The variance–covariance matrix of the parameter estimates, obtained using GLIM, is

$$\begin{array}{c} \text{Constant} \\ \log(\text{dose}) \end{array} \begin{bmatrix} 1.5753 & \\ 0.3158 & 0.0648 \end{bmatrix}$$

$$\qquad\qquad \text{Constant} \quad \log(\text{dose})$$

from which $\hat{v}_{00} = 1.575$, $\hat{v}_{11} = 0.065$ and $\hat{v}_{01} = 0.316$. To obtain an approximate standard error for $E\hat{D}50$, note that $\hat{\rho} = 9.19/1.83 = 5.021$. Substitution into relation (4.3) gives s.e.$\{\log(E\hat{D}50)\} = 0.1056$ and so, using relation (4.4), s.e.$(E\hat{D}50) = E\hat{D}50$ s.e.$\{\log(E\hat{D}50)\} = 0.0007$. An approximate 95% confidence interval for the $\log(ED50)$ value is $-5.021 \pm 1.96 \times 0.1056$, that is the interval from -5.228 to -4.820. A 95% confidence interval for the *ED50* value itself is therefore $(e^{-5.228}, e^{-4.820})$, that is $(0.0054, 0.0081)$.

A more exact confidence interval can be obtained from Fieller's result. The value of g required in the computation of confidence limits for the $\log(ED50)$ value is $(1.96^2 \times 0.0648)/(-1.830^2) = 0.074$. Substitution into expression (4.7) gives $(-5.250, -4.817)$ as a 95% confidence interval for the $\log(ED50)$ value. Exponentiating these limits, a 95% confidence limit for the *ED50* value itself is $(0.0052, 0.0081)$. This interval is practically identical to that formed by ignoring g altogether.

4.3 RELATIVE POTENCY

In bioassay, there is often a need to compare models fitted to different sets of data. To fix ideas, suppose that some newly developed compound (N) is to be compared with a standard (S) under similar experimental conditions. For each compound, the proportion of individuals responding at a series of escalating doses is recorded. In order to compare the two compounds, models with two different logistic regression lines, parallel lines and a common line are fitted and compared. If the dose–response relationship is different for each compound, the difference between the effect of the two compounds is not constant and there is no simple summary of their relative effectiveness. If the dose–response relationship is the same for each compound, the two compounds are equally effective. When the dose–response relationship for the two compounds is parallel on the logit scale, the relative potency of the two compounds can be assessed. Indeed, an assay designed to compare the potency of two stimuli is often referred to as a **parallel line assay**.

Suppose that the new compound is expected to give a consistently higher response than the standard, and that log (dose) is the explanatory variable, so that the equations of the parallel logistic regression lines are

Compound N: $\text{logit}(p) = \alpha_N + \beta \log(\text{dose})$

Compound S: $\text{logit}(p) = \alpha_S + \beta \log(\text{dose})$

where $\alpha_N > \alpha_S$. The relative potency of the two compounds is defined to be the ratio of equally effective doses. To derive an expression for the relative potency, suppose that dose d_N of the new drug and dose d_S of the standard produce the same response, say, λ, on the logit scale so that these two doses are equally potent. The potency of the new compound relative to the standard is then $r_{NS} = d_S/d_N$, that is, the new compound is r_{NS} times more effective than the standard. The situation is illustrated in Fig. 4.2. When $\text{logit}(p) = \lambda$, $\alpha_N + \beta \log(d_N) = \alpha_S + \beta \log(d_S)$ and so the potency of N relative to S is

$$r_{NS} = \frac{d_S}{d_N} = \exp\left\{\frac{\alpha_N - \alpha_S}{\beta}\right\}$$

Note that $\log(r_{NS}) = \log(d_S) - \log(d_N)$ is the horizontal separation of the parallel regression lines in Fig. 4.2, and that $r_{NS} = ED50_S/ED50_N$, is the ratio of the $ED50$ values for the standard and new compound, respectively. After adjudging that a parallel line model on the logit scale fits the observed dose–response data, the relative potency is estimated from the parameter estimates using $\hat{r}_{NS} = \exp\{(\hat{\alpha}_N - \hat{\alpha}_S)/\hat{\beta}\} = \hat{ED50}_S/\hat{ED50}_N$.

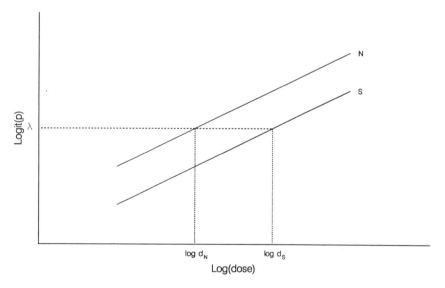

Fig. 4.2 Logistic regression lines with equality effective doses d_S and d_N.

A confidence interval for the relative potency can be found using Fieller's theorem. Following a similar procedure to that in Section 4.2, a confidence interval for $\rho = (\alpha_N - \alpha_S)/\beta$ is first obtained by considering the function $\psi = \hat{\alpha}_N - \hat{\alpha}_S - \rho\hat{\beta}$. The expected value of ψ is zero, but its variance is given by

$$V = \text{Var}(\hat{\alpha}_N) + \text{Var}(\hat{\alpha}_S) + \rho^2\text{Var}(\hat{\beta}) - 2\text{Cov}(\hat{\alpha}_N, \hat{\alpha}_S)$$
$$- 2\rho\text{Cov}(\hat{\alpha}_N, \hat{\beta}) - 2\rho\text{Cov}(\hat{\alpha}_S, \hat{\beta})$$

As before,

$$\frac{\hat{\alpha}_N - \hat{\alpha}_S - \rho\hat{\beta}}{\sqrt{(V)}}$$

has a standard normal distribution, and a $100(1 - \alpha)\%$ confidence interval for the logarithm of the relative potency is the set of values of ρ for which

$$|\hat{\alpha}_N - \hat{\alpha}_S - \rho\hat{\beta}| \leqslant z_{\alpha/2}\sqrt{(V)}$$

Squaring both sides and substituting for V leads to a quadratic expression in ρ. Finally, the roots of this equation are exponentiated to give a confidence interval for the relative potency itself.

Example 4.3 Toxicity of insecticides to flour beetles
In an insecticidal trial reported by Hewlett and Plackett (1950), flour beetles, *Tribolium castaneum*, were sprayed with one of three different insecticides in solution in Shell oil P31. The three insecticides used were dichlorodiphenyltrichloroethane (DDT) at 2.0% w/v, γ-benzene hexachloride (γ-BHC) used at 1.5% w/v and a mixture of the two. In the experiment, batches of about fifty insects were exposed to varying deposits of spray, measured in units of mg/10 cm^2. The resulting data on the proportion of insects killed after a period of six days are given in Table 4.1.

In modelling these data, the logarithm of the amount of deposit of insecticide will be used as the explanatory variable in a linear logistic model. The first step is to look at the extent to which there are differences between the insecticides in terms of the relationship between the probability of death and the amount of deposit.

Table 4.1 Toxicity to the flour beetle of sprays of DDT, γ-BHC and the two together

Insecticide	Deposit of insecticide					
	2.00	2.64	3.48	4.59	6.06	8.00
DDT	3/50	5/49	19/47	19/50	24/49	35/50
γ-BHC	2/50	14/49	20/50	27/50	41/50	40/50
DDT + γ-BHC	28/50	37/50	46/50	48/50	48/50	50/50

Logistic regression models are fitted, where the systematic part of the model has (i) different intercepts and slopes, (ii) different intercepts and common slope, and (iii) common intercept and slope. The deviance on fitting each of these three models is given below.

(i) Separate lines: Deviance = 17.89 on 12 d.f.

(ii) Parallel lines: Deviance = 21.28 on 14 d.f.

(iii) Common line: Deviance = 246.83 on 16 d.f.

The increase in deviance on constraining the fitted logistic regression lines to be parallel is $21.28 - 17.89 = 3.39$ on 2 d.f. which is not significant ($P = 0.184$). We therefore conclude that regression lines with a common slope fit the data as well as three separate lines. This means that the pattern of response to increasing doses is similar for the three insecticides. The increase in deviance on constraining the parallel lines to have similar intercepts is highly significant ($P < 0.001$) and so there are clear differences between the insecticides. Moreover, since the deviance for the parallel line model is not unacceptably high, this model can be used as a basis for comparing the effectiveness of the three insecticides.

GLIM output from fitting parallel logistic regression lines is reproduced below, in which the factor **CHEM** refers to the chemical to which the insects were exposed, and **LDEP** is the logarithm of the deposit of insecticide.

scaled deviance = 21.282 at cycle 4
d.f. = 14

	estimate	s.e.	parameter
1	−4.555	0.3611	CHEM(1)
2	−3.842	0.3327	CHEM(2)
3	−1.425	0.2851	CHEM(3)
4	2.696	0.2157	LDEP

The fitted linear logistic models for the three insecticides are:

DDT: logit $(p) = -4.555 + 2.696 \log(\text{deposit})$

γ-BHC: logit $(p) = -3.842 + 2.696 \log(\text{deposit})$

DDT + γ-BHC: logit $(p) = -1.425 + 2.696 \log(\text{deposit})$

In Fig. 4.3, the empirical logits are plotted against $\log(\text{deposit})$ and the fitted parallel logistic regression lines are superimposed.

The *ED50* values for the three insecticides are readily obtained from the parameter estimates under the fitted model, and are $5.417 \, \text{mg}/10 \, \text{cm}^2$ for DDT, $4.158 \, \text{mg}/10 \, \text{cm}^2$ for γ-BHC and $1.696 \, \text{mg}/10 \, \text{cm}^2$ for the mixture of DDT and γ-BHC. The relative potency of the mixture of compounds to the two primary

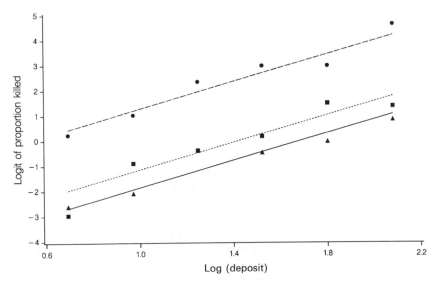

Fig. 4.3 Plot of the empirical logit of the proportion of insects killed against Log (deposit) for DDT (—▲—), γ-BHC ($\cdots$■$\cdots$) and DDT + γ-BHC (---●---).

insecticides can now be estimated from appropriate ratios of the estimated *ED50* values. Thus, the relative potency of a mixture of DDT and γ-BHC to DDT alone is $5.417/1.696 = 3.19$, while the potency of the mixture relative to γ-BHC is $4.158/1.696 = 2.45$. Thus, the mixture of DDT and γ-BHC is more than three times as potent as DDT alone and just under two and a half times more potent than γ-BHC alone. It is also clear that of the two basic toxins, γ-BHC is the more potent to the flour beetles.

4.4 NATURAL RESPONSE

In many experiments it is found that the response of an experimental unit is not solely due to the stimulus to which that unit has been exposed. For example, in an insecticidal trial some insects may die during the experimental period even when they have not been exposed to an insecticide, or insects that have been so exposed may die from a cause unrelated to the toxin. These insects are said to have a natural response or natural mortality which will generally need to be taken account of in the analysis of the mortality data. It is rarely possible to conduct a post-mortem on the individual insects used in such a trial to determine which of them died as a result of exposure to the insecticide, and so the natural response will have to be estimated from a control group or from the actual dose–response data.

Suppose that in an insecticidal trial, a proportion π respond in the absence of the insecticide. Suppose also that the expected proportion responding to the insecticide alone at a given dose, d_i, is p_i, $i = 1, 2, \ldots, n$. The proportion p_i is not directly observable because of the natural response π, so suppose that the expected overall proportion of insects which respond at dose d_i is p_i^*. This quantity is made up of the proportion π of insects which respond naturally and a proportion p_i of the remaining proportion $(1 - \pi)$ which respond to the insecticide. Hence, p_i^*, the observable response probability, is given by

$$p_i^* = \pi + (1 - \pi)p_i \tag{4.8}$$

a result sometimes known as Abbott's formula. The value π is the minimum observable response probability and could be estimated from a control batch of insects. Since a control group contains no information about the potency of a compound, it should be regarded separately from the other dose–response data. Incidentally, the adoption of this procedure means that no difficulties will then be caused in using the logarithm of dose as the explanatory variable. If the control group consists of n_0 insects, of which y_0 die over the experimental period, π can be estimated by $\pi_0 = y_0/n_0$. The number of insects used in the control group must be large in order that π_0 has a sufficient degree of precision. In circumstances where inherent variability leads to the observed response probability at low doses of the toxin being lower than π_0, π_0 is likely to be an overestimate of the natural mortality and will need to be adjusted accordingly. In this way, a reasonable estimate of π can be found.

A simple way of adjusting the analysis to take account of a non-zero estimate of the natural response, π_0, is to argue that the effective number of insects in a batch of size n_i that are susceptible to the ith dose of the insecticide, d_i, is $n_i^* = n_i(1 - \pi_0)$, and that if y_i respond at this dose, the number responding to the insecticide is effectively $y_i^* = y_i - n_i\pi_0$. The adjusted proportions y_i^*/n_i^* are then used as a basis for modelling the relationship between logit (p_i^*) and d_i. Unfortunately, this simple procedure cannot be justified on theoretical grounds, because the y_i^* can no longer be taken to have a binomial distribution. However, when π_0 is small and has been estimated quite precisely, the procedure is not seriously deficient. A more satisfactory approach is described below.

In modelling the observed proportions y_i/n_i, the y_i can be assumed to have a $B(n_i, p_i^*)$ distribution, where the logistic transformation of $p_i = (p_i^* - \pi)/(1 - \pi)$ is taken to be linearly related to the dose d_i, that is, logit $(p_i) = \beta_0 + \beta_1 d_i$. The model for p_i^* is therefore such that

$$p_i^* = \pi_0 + (1 - \pi_0)\left\{\frac{\exp(\beta_0 + \beta_1 d_i)}{1 + \exp(\beta_0 + \beta_1 d_i)}\right\} \tag{4.9}$$

This model can be fitted to the observed data by maximizing the likelihood function for the original data,

$$L(\beta) = \prod_i \binom{n_i}{y_i}(p_i^*)^{y_i}(1 - p_i^*)^{n_i - y_i} \tag{4.10}$$

where p_i^* is given in equation (4.9), in terms of β_0 and β_1. Alternatively, the model described by equation (4.9) can be expressed in the form

$$\log\left\{\frac{p_i^* - \pi_0}{1 - p_i^*}\right\} = \beta_0 + \beta_1 d_i \qquad (4.11)$$

This is not a linear logistic model for p_i^*, but the model can be fitted directly using specialist software, or computer packages that allow the user to specify their own choice of link function, such as GLIM. See Chapter 9 for further information on this.

When information about the natural response is scanty or non-existent, the most appropriate procedure is to estimate π, in addition to β_0 and β_1, from the dose–response data. Then, an appropriate model for p_i^* is

$$\log\left\{\frac{p_i^* - \pi}{1 - p_i^*}\right\} = \beta_0 + \beta_1 d_i \qquad (4.12)$$

where now π, β_0 and β_1 need to be estimated simultaneously. The model described by equation (4.12) can be fitted using any package that enables the previous model (equation (4.11)) to be fitted. The procedure is to fit the model with a range of values for π and then adopt that value of π, and the corresponding values of β_0 and β_1, for which the deviance is minimized. This procedure may be embedded in an algorithm that searches for the value of π that minimizes the deviance. Further details can be found in section 9.3.2.

This discussion has been motivated by consideration of a single explanatory variable d_i. These models can of course be generalised to those where the linear systematic component of the model contains more than one explanatory variable, factors, and combinations of factors and explanatory variables. In particular, the same procedure could be used to allow an adjustment to be made for natural mortality in a parallel line assay, as illustrated in Example 4.4 in the sequel.

In a bioassay, some of the individuals exposed to a certain stimulus may be immune to that stimulus. Such individuals will not respond to the stimulus, no matter how high the level of that stimulus is, a characteristic termed **natural immunity**. The model for p_i^* given by equation (4.8) can be generalised to allow for natural immunity in the experimental units by supposing that the maximum response probability is some value $\omega(<1)$. The observable response probability now becomes $p_i^* = \pi + (1 - \pi - \omega)p_i$. The parameters π and ω can both be estimated directly from the data, or simultaneously with the β-parameters, along the same lines as above.

Example 4.4 Toxicity of insecticides to flour beetles
The parallel line assay described in Example 4.3 actually included a control group of 200 beetles that were not sprayed with any of the three insecticides. After six days, twenty of these had died, suggesting a natural mortality rate of 10%.

The model for the observable response probability in equation (4.11), assuming a mortality rate of $\pi_0 = 0.1$, was fitted using GLIM in conjunction with the method described in Section 9.3.2. The deviance under this model is 26.39 on 14 d.f., and the equations of the fitted parallel regression lines are:

DDT:	logit $(p) =$	-5.634 + 3.113 log (deposit)
γ-BHC	logit $(p) =$	-4.724 + 3.113 log (deposit)
DDT + γ-BHC:	logit $(p) =$	-1.996 + 3.113 log (deposit)

From these equations, estimates of the *ED50* values can be obtained and relative potencies can be estimated. They turn out to be 6.109 mg/10 cm² for DDT, 4.561 mg/10 cm² for γ-BHC and 1.899 mg/10 cm² for the mixture of DDT and γ-BHC. These values are somewhat greater than those reported in Example 4.3.

In attempting to estimate the natural mortality, π, from the dose–response data, along with the two parameters in the linear component of the model, using the method described in Section 9.3.2, it was found that the deviance is minimized when $\hat{\pi}$ is zero. This suggests that there is not enough information available in the original data to justify an estimated control mortality greater than zero. The deviance when $\hat{\pi} = 0.0$ is 21.28 and the fitted models are as given in Example 4.3.

4.5 NON-LINEAR LOGISTIC REGRESSION MODELS

When the empirical logistic transform of an observed binomial response probability is plotted against a given explanatory variable, it may be that the relationship between the two is not a linear one. A more complicated model will then need to be identified. If the purpose of fitting a logistic regression model is to estimate a dose that produces an extreme response, such as the *ED90* or even the *ED99* value, it is particularly important to adopt a model that is in close agreement with the observed data at extreme values of the dose.

A model for the relationship between the true response probability, p, and an explanatory variable, x, that might be adopted on pragmatic grounds is a **polynomial logistic regression model**, whose general form is

$$\text{logit}(p) = \beta_0 + \beta_1 x + \beta_2 x^2 + \cdots + \beta_k x^k$$

This model for logit (p) is still linear in the β-parameters, and this makes it much easier to fit than other models. Morgan (1985) found that the cubic polynomial $(k = 3)$ was a good fit to a number of data sets, at extreme dose levels. He also investigated the three parameter model

$$\text{logit}(p) = \beta_1(x - \mu) + \beta_2(x - \mu)^3 \tag{4.13}$$

for which μ is the *ED50* value, and found that there was little to choose between this model and the cubic polynomial. One way of fitting the model given by equation (4.13), using packages such as GLIM, is to estimate the parameters β_1 and

β_2 for a grid of values of μ and to take the set of estimates that minimize the deviance. One disadvantage of this approach is that it does not enable the variance of $\hat{\mu}$, nor covariance terms involving $\hat{\mu}$, to be estimated. These would be needed if, for example, the standard error of an estimated *ED90* value was required.

Example 4.5 Mortality of confused flour beetles
The aim of an experiment originally reported by Strand (1930) and quoted by Bliss (1935) was to assess the response of the confused flour beetle, *Tribolium confusum*, to gaseous carbon disulphide (CS_2). In the experiment, prescribed volumes of liquid carbon disulphide were added to flasks in which a tubular cloth cage containing a batch of about thirty beetles was suspended. Duplicate batches of beetles were used for each concentration of CS_2. At the end of a five-hour period, the proportion killed was recorded and the actual concentration of gaseous CS_2 in the flask, measured in mg/l, was determined by a volumetric analysis. The mortality data are given in Table 4.2.

Table 4.2 The number of beetles killed, y, out of n exposed to different concentrations of gaseous carbon disulphide

Concentration of CS_2	Replicate 1		Replicate 2	
	y	n	y	n
49.06	2	29	4	30
52.99	7	30	6	30
56.91	9	28	9	34
60.84	14	27	14	29
64.76	23	30	29	33
68.69	29	31	24	28
72.61	29	30	32	32
76.54	29	29	31	31

In a number of articles that refer to these data, the responses from the first two concentrations are omitted because of apparent non-linearity. Bliss himself remarks that

... in comparison with the remaining observations, the two lowest concentrations gave an exceptionally high kill. Over the remaining concentrations, the plotted values seemed to form a moderately straight line, so that the data were handled as two separate sets, only the results at 56.91 mg of CS_2 per litre being included in both sets.

However, there does not appear to be any biological motivation for this and so here they are retained in the data set.

Combining the data from the two replicates and plotting the empirical logit of the observed proportions against concentration gives the graph shown in Fig. 4.4.

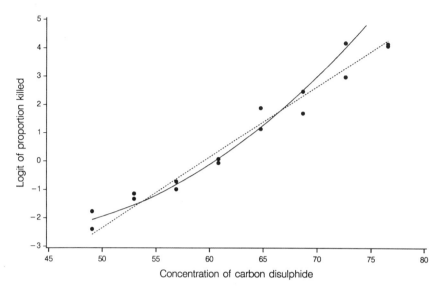

Fig. 4.4 Plot of the logistic transform of the proportion of beetles killed against concentration of CS_2 with the fitted linear ($\cdots\cdots$) and quadratic (——) logistic regression lines.

The apparent non-linear relationship between the logistic transform of the proportion of beetles killed and the concentration, x, is not removed by working with the logarithm of the concentration, and so the concentration itself will be used as the explanatory variable. On fitting the model logit $(p_i) = \beta_0 + \beta_1 x_i$, where p_i is the probability that a beetle dies when exposed to CS_2 at a concentration of x_i mg/l, the deviance is 12.51 on 14 d.f. Since the mean deviance for this model is less than unity, on the face of it this model seems acceptable. The equation of the fitted model is logit $(\hat{p}) = -14.81 + 0.249\ x$, and when this is plotted with the data in Fig. 4.4, it appears that the model does not fit the observed data.

On including a quadratic term in x, the deviance is reduced to 7.93 on 13 d.f. This reduction of 4.58 on 1 d.f. is highly significant, showing that the quadratic logistic regression model fits the data better than the linear one. When a cubic term x^3 is incorporated in the model, the deviance is only reduced further by 0.045 on 1 d.f. and the quality of the fit is not much improved. The fitted quadratic logistic regression model, where

$$\text{logit}\ (\hat{p}) = 7.968 - 0.517\ x + 0.00637\ x^2$$

is also plotted in Fig. 4.4. Although this plot suggests that the quadratic logistic regression model does not fit the observed data at a CS_2 concentration of 76.54 mg/l, this impression is misleading. The observed proportions of beetles in the two batches that are affected at this concentration are both equal to one, but use of the empirical logistic transformation of these values has enabled them to be included on the graph. The fitted response probability at this concentration is actually 0.997, which is very close to the corresponding observed values. Figure 4.5 shows that the observed proportions of beetles affected by the different concentrations of the gas are well fitted by the quadratic logistic regression model.

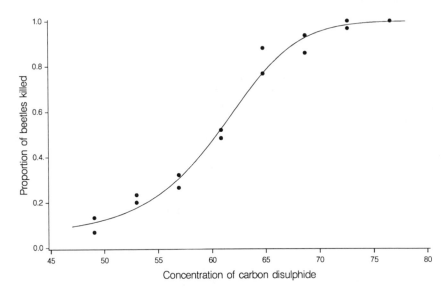

Fig. 4.5 Relationship between the proportion of beetles killed and the concentration of CS_2, with the fitted quadratic logistic regression model.

To estimate the *ED90* value on the basis of the quadratic logistic regression model, we have

$$\text{logit}(0.9) = \log(9) = 7.968 - 0.517\,ED90 + 0.00637\,ED90^2$$

from which the following quadratic equation is obtained:

$$0.00637\,ED90^2 - 0.517\,ED90 + 5.771 = 0$$

The two roots of this equation are 67.80 and 13.36, from which 67.80 mg/l is the estimated concentration of gaseous CS_2 that produces a response in 90% of beetles exposed to it.

Polynomial logistic regression models are often difficult to justify in terms of the context of the subject area, and so due care must be taken in using them. They are particularly susceptible to rather strange behaviour outside the range of dose levels used in an experiment. One alternative to using them is to model the logistic transform of a response probability by a non-linear function of unknown parameters. For example,

$$\text{logit}(p) = \beta_0 \, e^{\beta_1 x}$$

models an exponential increase or decrease in the logistic transformation of the response probability, while

$$\text{logit}(p) = \beta_0 + \beta_1 \, e^{\beta_2 x}$$

models an asymptotic regression on the logit scale. In principle, these models can be fitted by the method of maximum likelihood, but they are typically much more difficult to fit than polynomial models and there is no generally available computer software that can be used to fit them.

Another way of improving the fit of a model for binary data is to use a transformation of the response probability that is more flexible than the logistic. Models based on generalizations of the logistic transformation are considered in Section 5.3.

4.6 APPLICATIONS OF THE COMPLEMENTARY LOG–LOG MODEL

A transformation of the probability scale that is sometimes useful in modelling binomial response data is the complementary log–log transformation where $\log[-\log(1 - p)]$ is modelled as a linear function of explanatory variables. This transformation was described in Section 3.5.3. In the context of a bioassay, the complementary log–log model can be derived by supposing that the tolerances of individuals have an extreme value distribution known as the **Gumbel distribution** with probability density function

$$f(u) = \frac{1}{\kappa} e^{(u-\alpha)/\kappa} \exp[-e^{(u-\alpha)/\kappa}], \quad -\infty < u < \infty$$

where $-\infty < \alpha < \infty$ and $\kappa > 0$ are unknown parameters. Unlike the normal and logistic distributions, on which the probit and logistic transformations are based, this is an asymmetric tolerance distribution with a mode at α and a mean that is greater than α. The variance of the distribution is $1.645\kappa^2$. The probability of a response when exposed to dose d_i is

$$p_i = \int_{-\infty}^{d_i} f(u) \, du = 1 - \exp[-e^{(d_i-\alpha)/\kappa}]$$

and so $\log[-\log(1 - p_i)] = \beta_0 + \beta_1 d_i$, where $\beta_0 = -\alpha/\kappa$ and $\beta_1 = 1/\kappa$ are unknown parameters. The complementary log–log function again transforms a probability in the range $(0, 1)$ to a value in $(-\infty, \infty)$ but this function is not symmetric about $p = 0.5$; in fact, when $p = 0.5$, the transformed value is -0.3665.

A Gumbel tolerance distribution is also encountered in problems concerning the breaking strength of materials. The weakest link theory, according to which a component breaks when the applied force is larger than the tolerance of the weakest section or link, suggests that a distribution of minima, such as the Gumbel distribution, should be used to model the breaking strength. This leads to a complementary log–log model for the probability of the component failing. Some other contexts that lead directly to the complementary log–log model are described in more detail in the following sub-sections.

4.6.1 A dilution assay

When the number of infective organisms in a solution cannot be measured directly, a dilution assay may be used to estimate the number of organisms present. In this type of assay, the solution containing the organism is progressively diluted and samples from each dilution are applied to a number of plates that contain a growth medium. After a period of incubation, the proportion of plates on which growth is observed is recorded. Suppose that the original solution is diluted in powers of 2, leading to solutions that are of half strength, quarter strength, and so on. Let N be the number of organisms per unit volume in the original solution so that the corresponding number in a solution diluted by a factor of $1/2^i$ is expected to be $N/2^i$, for $i = 0, 1, 2, \ldots, m$. Under random sampling, the number of organisms per unit volume of diluent that will be deposited on a plate can be assumed to have a Poisson distribution with mean μ_i, where μ_i is the expected number of organisms in a unit volume of the ith dilution. Consequently, $\mu_i = N/2^i$, so that $\log \mu_i = \log N - i \log 2$. Under the Poisson assumption, the chance that a plate at the ith dilution contains no organisms is $e^{-\mu_i}$ and so, if p_i is the probability that growth occurs on a plate at this dilution, $p_i = 1 - e^{-\mu_i}$. Hence, $\mu_i = -\log(1 - p_i)$, so that $\log \mu_i = \log[-\log(1 - p_i)]$, and

$$\log[-\log(1 - p_i)] = \log N - i \log 2$$

The data to which this model is fitted consists of the number of plates, y_i, on which growth has occurred, out of n_i plates inoculated with the ith dilution of the original solution. Writing $\beta_0 = \log N$ and $\beta_1 = -\log 2$, the model for p_i becomes

$$\log[-\log(1 - p_i)] = \beta_0 + \beta_1 i$$

where i is an explanatory variable and β_1 is a known regression coefficient. The quantity of direct interest here is N which can be estimated by $\hat{N} = \exp(\hat{\beta}_0)$, where $\hat{\beta}_0$ is the maximum likelihood estimate of β_0 in the above model. A $100(1 - \alpha)\%$ confidence interval for N can be obtained by first calculating a confidence interval for $\log N$ from $\hat{\beta}_0 \pm z_{\alpha/2}$ s.e. $(\hat{\beta}_0)$, where $z_{\alpha/2}$ is the upper $100\alpha/2\%$ point of the standard normal distribution. The resulting confidence limits are then exponentiated to give the required confidence interval for N itself.

4.6.2 Analysis of grouped survival data

In the analysis of survival data, interest centres on the length of time measured from some time origin to an event that is generally termed failure. The corresponding time to that event is referred to as the survival time, although the failure of interest will not necessarily be death. In some situations the actual time of failure will be unknown, although the failure is known to have occurred during a particular interval of time. Data of this form are known as **grouped** or **interval censored** survival data. One possible method for analysing such data is described below in the context of an example from medicine.

In the management of patients who have been cured of ulcers, carcinomas or other recurrent conditions, the patients are usually provided with medication to maintain their recovery. These patients are subsequently examined at frequent intervals of time in order to detect whether a relapse has occurred. In a clinical trial to compare two alternate maintenance therapies, a standard and a new therapy, say, patients admitted to the trial are randomised to receive one of the two treatments. The progress of these patients is then monitored by screening them at regular intervals. In addition, some patients may experience symptoms of a relapse, and be subsequently diagnosed as having had a relapse at a time other than one of the scheduled screening times. Suppose that the patients in such a trial are followed to time t, at which point the last screening test is carried out. Information on whether or not a relapse was detected at any time up to and including the last screen is then recorded.

Let $p(t; \mathbf{x})$ be the probability of a relapse up to time t for a patient with explanatory variables $\mathbf{x}$. This set of explanatory variables should include an indicator variable, say, x_1, which takes the value zero if the individual is receiving the standard treatment and unity if the new treatment is being used. Suppose that the hazard or risk of a relapse at time t for the ith patient, $i = 1, 2, \ldots, n$, with explanatory variables $\mathbf{x}_i$, is given by

$$h(t; \mathbf{x}_i) = e^{\eta_i} h_0(t) \qquad (4.14)$$

where $\eta_i = \beta_1 x_{1i} + \beta_2 x_{2i} + \cdots + \beta_k x_{ki}$ is a linear combination of the explanatory variables for the ith individual and $h_0(t)$ is a **baseline hazard function**. This is simply the hazard of a relapse for an individual for whom the k explanatory variables in the model each take the value zero. If $h_0(t)$ is not specified, this is the **Cox proportional hazards regression model** for the analysis of survival data, introduced by Cox (1972).

Now let $S(t; \mathbf{x}_i)$ be the probability that the ith individual relapses after time t, so that

$$S(t; \mathbf{x}_i) = 1 - p(t; \mathbf{x}_i)$$

The function $S(t; \mathbf{x}_i)$ is called the **survivor function**, and is related to the hazard function by the equation

$$S(t; \mathbf{x}_i) = \exp\left\{-\int_0^t h(u; \mathbf{x}_i)\,du\right\}$$

Using equation (4.14), it then follows that

$$S(t; \mathbf{x}_i) = [S_0(t)]^{\exp(\eta_i)}$$

where $S_0(t)$ is the survivor function for an individual for whom all the explanatory variables are zero. The probability of a relapse up to time t under this model is therefore given by

$$p(t; \mathbf{x}_i) = 1 - [S_0(t)]^{\exp(\eta_i)} = 1 - \exp\left(-\exp\{\eta_i + \log[-\log S_0(t)]\}\right)$$

and consequently,

$$\log[-\log\{1 - p(t; \mathbf{x}_i)\}] = \eta_i + \log[-\log S_0(t)]$$

This is a complementary log–log model for the relapse probability, which can be fitted to the binary response variable that takes the value zero for those individuals in the study who have not experienced a relapse before time t, and unity otherwise. On fitting the model

$$\log\{-\log[1 - p(t; \mathbf{x}_i)]\} = \beta_0 + \beta_1 x_{1i} + \beta_2 x_{2i} + \cdots + \beta_k x_{ki}$$

the parameter estimate for the constant $\hat{\beta}_0$ is an estimate of $\log[-\log S_0(t)]$. The ratio of the hazards for the new treatment relative to the standard is

$$\frac{h(t; x_1 = 1)}{h(t; x_1 = 0)} = \exp(\beta_1)$$

which can be estimated by $\exp(\hat{\beta}_1)$, where $\hat{\beta}_1$ is the parameter estimate corresponding to the indicator variable for the treatment, x_1. Values of the hazard ratio less than unity suggest that the risk of a relapse at time t is smaller under the new treatment than under the standard. A confidence interval for the hazard ratio is readily obtained from the standard error of $\hat{\beta}_1$ as in Section 4.6.1.

This method of estimating the hazard ratio from grouped survival data is not particularly efficient, since data on the times of relapse are not utilized. Alternative approaches have been reviewed by Whitehead (1989), who also gives a numerical example.

4.6.3 Serological testing

When an individual has been exposed to a disease, antibodies are produced which, for many diseases, remain in the body long after the individual has recovered from the disease. The presence of antibodies can be detected using a serological test and an individual who has such antibodies is termed **seropositive**. In areas where the disease under study is endemic, it might be assumed that the risk of infection is constant. The age distribution of seropositivity in a population can then give information about that populations experience of the disease, in particular, the annual infection rate.

Let λ be the probability of infection in a small interval of time, say, one day. Then, the probability that an individual is never infected in an n-day period is $(1 - \lambda)^n$, which can be approximated by $e^{-n\lambda}$. Now suppose that age is measured in years and let $\mu = 365\lambda$ be the mean number of infections per year. The probability that an individual aged A years is not infected is then $e^{-\mu A}$ so that $1 - e^{-\mu A}$ is the probability that an individual of age A years has been infected. If it is now assumed that the random variable associated with the number of infections occurring in one year has a Poisson distribution with mean μ, the annual infection rate is $h = P[R \geqslant 1] = 1 - e^{-\mu}$. This is a quantity of direct epidemiological interest.

If p_i is the probability of infection in an individual in the ith age group, for which the mean age is A_i, then p_i is given by

$$p_i = 1 - e^{-\mu A_i} \tag{4.15}$$

and this is the expected proportion of the population that is seropositive. Data from a serological survey consist of the number of seropositive individuals y_i out of n_i in the ith age group, where y_i has a binomial distribution with parameters n_i and p_i. From equation (4.15),

$$\log[-\log(1 - p_i)] = \log \mu + \log A_i = \beta_0 + x_i$$

where $\beta_0 = \log \mu$ and the explanatory variable $x_i = \log A_i$ has a known coefficient of one.

Example 4.6 Serological testing for malaria
Draper, Voller and Carpenter (1972) discuss the interpretation of serological data on malaria, collected from a number of areas in Africa and South America. In a serologic survey carried out in 1971 in two area of Amozonas, Brazil, the indirect fluorescent antibody test was used to detect the presence of antibodies to the malarial parasite *Plasmodium vivax* in the villagers. The data reproduced in Table 4.3 refer to the proportion of individuals in each of seven age groups who were found to be seropositive. Also given in the table is the mid-point of each age range; that for the final age group has been estimated from one of the figures in the original paper. GLIM output from fitting the model $\log[-\log(1 - p_i)] = \beta_0 + \log A_i$, where p_i is the expected proportion of individuals of age A_i who are seropositive, and A_i is taken to be the mid-point of the age range, is given after Table 4.3.

Table 4.3 Seropositivity rates for villagers in Amozonas, Brazil in 1971

Age group	Mid-point of age range in years	Proportion seropositive
0–11 months	0.5	3/10
1–2 years	1.5	1/10
2–4 years	3.0	5/29
5–9 years	7.0	39/69
10–14 years	12.0	31/51
15–19 years	17.0	8/15
⩾ 20 years	30.0	91/108

scaled deviance = 20.253 at cycle 3
d.f. = 6

	estimate	s.e.	parameter
1	− 2.607	0.08550	1

In the terminology of generalized linear modelling, implicit in the package GLIM, a term with a known regression coefficient is known as an **offset**, so in this example, the term $\log A_i$ is declared to be an offset. The quantity of direct interest in this output is the estimate of $\beta_0 = \log \mu$ which is $\hat{\beta}_0 = -2.607$. From this, $\hat{\mu} = e^{-2.607} = 0.074$, and so the infection rate h is estimated by $\hat{h} = 1 - e^{-\hat{\mu}} = 0.071$. The annual infection rate for this type of malaria in Amozonas is therefore estimated to be about 71 infections per 1000 individuals.

The standard error of the estimated annual infection rate can be determined from the standard error of $\hat{\beta}_0 = \log \hat{\mu}$, given in the GLIM output. An approximation to the variance of a function $g(\hat{\vartheta})$ of some parameter estimate $\hat{\vartheta}$ is obtained from

$$\text{Var}\,[g(\hat{\vartheta})] \approx \left\{ \frac{\partial g(\hat{\vartheta})}{\partial \hat{\vartheta}} \right\}^2 \text{Var}\,(\hat{\vartheta}) \qquad (4.16)$$

a result that is a special case of that given in expression (4.1). In this example, the variance of $\hat{h}$ is required, where

$$\hat{h} = 1 - \exp[-e^{\hat{\beta}_0}]$$

Using relation (4.16) to obtain the approximate variance of $\hat{h}$, the standard error of $\hat{h}$ is given by

$$\text{s.e.}\,(\hat{h}) \approx e^{\hat{\beta}_0} \exp\,[-e^{\hat{\beta}_0}]\,\text{s.e.}\,(\hat{\beta}_0)$$

Since $\hat{\beta}_0 = -2.607$ and s.e. $(\hat{\beta}_0) = 0.085$, we find that s.e. $(\hat{h}) \approx 0.0059$. An approximate confidence interval for the true annual infection rate is therefore $0.071 \pm 1.96 \times 0.0059$, that is the interval from 0.060 to 0.083. There is therefore a 95% chance that the interval from 60 to 83 infections per 1000 individuals includes the true annual infection rate.

The deviance on fitting the complementary log–log model essentially measures the extent to which the data support the hypothesis that there is a constant infection rate. In this case, the deviance of 20.25 on 6 d.f. is rather large, significant at the 1% level in fact, and so there is evidence to doubt the hypothesis of a constant infection rate. A more complex model would have a different infection rate for the individuals in each age group, so that some indication of time trend in the infection rate could be detected. One model for this situation would be

$$\log\left[-\log\left(1 - p_i\right)\right] = \log \mu_i + \log A_i$$

and the infection rate for the individuals in the ith age group is estimated by $\hat{h}_i = 1 - \exp\left(-\hat{\mu}_i\right)$. When this model is fitted, using the levels of the factor **AGE** to index the seven age groups, the following GLIM output is obtained.

scaled deviance = 0.00000 at cycle 5
d.f. = 0

	estimate	s.e.	parameter
1	−0.3378	0.5804	AGE(1)
2	−2.656	1.000	AGE(2)
3	−2.763	0.4479	AGE(3)
4	−2.129	0.1648	AGE(4)
5	−2.551	0.1862	AGE(5)
6	−3.105	0.3622	AGE(6)
7	−2.787	0.1204	AGE(7)

The deviance for this model is zero because a full model is being fitted. In fitting the model, a constant has been omitted so that the parameter estimates associated with fitting the term $\log \mu_i$ in the model can be interpreted directly. From the estimated values, $\log \hat{\mu}_i$, the estimated annual number of infections per 1000 individuals in the seven groups are 510, 68, 61, 11, 75, 44 and 60, respectively. There is no obvious pattern here, but it is clear that the estimated rate for the first age group, corresponding to that experienced over the twelve months prior to the survey, is far greater than that further back in time. In addition, the rate for the 5–9 age group is somewhat lower than at other periods. Perhaps some sort of malaria eradication programme was in operation at that time.

FURTHER READING

An account of the method of probit analysis and its application to the analysis of binomial response data from bioassay is given by Finney (1971). Finney summarises the history of the probit method and gives much detail on the estimation of the median effective dose, relative potency and natural mortality. Fieller's theorem was first given in Fieller (1940), but was presented more clearly in Fieller (1954). Models for the analysis of data from assays involving mixtures of drugs, such as the assay described in Example 4.3, were given by Hewlett and Plackett (1964), but a more recent article is that of Giltinan, Capizzi and Malani (1988).

Non-linear logistic regression models have been presented by a number of authors. For example, Copenhaver and Mielke (1977) proposed a three-parameter model, called the **quantit model**, which is discussed further by Morgan (1983). Other generalizations of the logistic and probit models have been described by Prentice (1976), Aranda-Ordaz (1981), Guerrero and Johnson (1982) and Stukel (1988).

The complementary log-log transformation and its use in the analysis of dilution assays was described by Fisher (1922). Details on the use of the model in reliability theory are contained in Crowder *et al.* (1991). Kalbfleisch and Prentice (1980), Lawless (1982), and Cox and Oakes (1984) provide comprehensive accounts of the analysis of survival data, while Whitehead (1989) explores a number of different approaches to the analysis of grouped survival data. The application of the complementary log-log model to serological testing was given in an appendix to Draper, Voller and Carpenter (1972) provided by Carpenter.

5
Model checking

Once a model has been fitted to the observed values of a binary or binomial response variable, it is essential to check that the fitted model is actually valid. Indeed, a thorough examination of the extent to which the fitted model provides an appropriate description of the observed data is a vital aspect of the modelling process.

There are a number of ways in which a fitted model may be inadequate. The most important of these is that the linear systematic component of the model may be incorrectly specified; for example, the model may not include explanatory variables that really should be in the model, or perhaps the values taken by one or more of the explanatory variables need to be transformed. The transformation of the response probability used may not be correct; for example, it may be that a logistic transformation of a binary response probability has been used when it would have been more appropriate to use the complementary log–log transformation. The data may contain particular observations, termed **outliers**, that are not well fitted by the model, or observations, termed **influential values**, that have an undue impact on the conclusions to be drawn from the analysis. Finally, the assumption that the observed response data come from a particular probability distribution, for example the binomial distribution, may not be valid.

The techniques used to examine the adequacy of a fitted model are known collectively as **diagnostics**. These techniques may be based on formal statistical tests, but more frequently involve a less formal evaluation of tables of the values of certain statistics or a graphical representation of these values. Some of these statistics are based on differences between the fitted values under a model and the binomial observations to which that model has been fitted, while others summarize the effect of deleting one or more observations from the data set on certain aspects of the fit.

In this chapter, methods for exploring the adequacy of the systematic part of the model and the transformation of the response probability, and diagnostic methods for detecting outliers and influential values are reviewed and illustrated in the context of modelling binomial data. The manner in which the validity of the binomial assumption can be tested will also be discussed, although a review of alternative models will be deferred until the next chapter.

Throughout this chapter, the diagnostics will be presented for linear logistic modelling. Similar diagnostics can be constructed for models based on other link functions, such as the probit and complementary log–log transformations, although some of the algebraic expressions will then be different. A number of diagnostic techniques proposed for binomial data are uninformative when used with binary data. Accordingly, in presenting diagnostic methods, a distinction needs to be drawn between methods that can be used for binomial data, in which the binomial denominators exceed one, and those methods that can be used when the data are binary. Sections 5.1–5.6 describe diagnostic procedures for binomial data while methods for binary data are discussed in Section 5.7. A summary of the role of the different diagnostic methods is given in Section 5.8. This section also includes some guidance on which methods are most likely to be useful on a routine basis.

Much of the algebraic detail underpinning the diagnostic methods that are described in this chapter will be omitted. The final section of this chapter contains full details of the source of the results quoted, and so these references should be consulted for a fuller appreciation of the theoretical background.

5.1 DEFINITION OF RESIDUALS

Measures of agreement between an observation on a response variable and the corresponding fitted value are known as **residuals**. These quantities, and summary statistics derived from them, can provide much information about the adequacy of the fitted model. For convenience, different definitions of residuals for binomial data are presented in this section, although the manner in which they are actually used in model checking will be described in subsequent sections.

Suppose that a linear logistic model is fitted to n binomial observations of the form y_i/n_i, $i = 1, 2, \ldots, n$, and that the corresponding fitted value of y_i is $\hat{y}_i = n_i \hat{p}_i$. The ith **raw residual** is then the difference $y_i - \hat{y}_i$, and provides information about how well the model fits each particular observation. Because each of the observations, y_i, may be based on different numbers of binary responses, n_i, and because they will all have different success probabilities, p_i, these raw residuals are difficult to interpret. In particular, a large difference between y_i and the corresponding fitted value $\hat{y}_i$ will be less important when y_i has a low precision compared to when y_i has a high precision. The precision of y_i is reflected in its standard error, given by s.e. $(y_i) = \sqrt{\{n_i \hat{p}_i (1 - \hat{p}_i)\}}$; a low precision corresponds to a high standard error. The raw residuals can be made more comparable by dividing them by s.e.(y_i), giving

$$X_i = \frac{y_i - n_i \hat{p}_i}{\sqrt{\{n_i \hat{p}_i (1 - \hat{p}_i)\}}} \tag{5.1}$$

These residuals are known as **Pearson residuals**, since the sum of their squares is $X^2 = \sum X_i^2$ which is Pearson's X^2-statistic, first introduced in Chapter 2. These residuals therefore measure the contribution that each observation makes to a statistic that is a summary measure of the goodness of fit of the fitted linear logistic model. For this reason, they are intuitively appealing as measures of model adequacy.

The standardization used in the construction of the Pearson residuals does not yield residuals that have even approximate unit variance, since no allowance has been made for the inherent variation in the fitted values $\hat{y}_i$. A better procedure is to divide the raw residuals by their standard error, s.e. $(y_i - \hat{y}_i)$. This standard error is quite complicated to derive, but it is found to be given by

$$\text{s.e.} (y_i - \hat{y}_i) = \sqrt{\{\hat{v}_i(1 - h_i)\}}$$

where $\hat{v}_i = n_i \hat{p}_i(1 - \hat{p}_i)$ and h_i is the ith diagonal element of the $n \times n$ matrix $H = W^{1/2}X(X'WX)^{-1}X'W^{1/2}$. In this expression for H, W is the $n \times n$ diagonal matrix of weights used in fitting the model, and X is the $n \times p$ design matrix, where p is the number of unknown parameters in the model. The notation X' denotes the transpose of the matrix X. For the linear logistic model, the ith diagonal element of W is $n_i \hat{p}_i(1 - \hat{p}_i)$, the estimated variance of y_i, but it will be different for other link functions. The first column of the design matrix X will be a column of ones when the model includes a constant term, while the remaining columns are the values of the explanatory variables (or indicator variables) in the model. Fortunately, many statistical packages give the values h_i, or quantities from which h_i can easily be found; see Section 9.3.3. The resulting standardized residuals are

$$r_{P_i} = \frac{y_i - n_i \hat{p}_i}{\sqrt{\{\hat{v}_i(1 - h_i)\}}} \tag{5.2}$$

These residuals are simply the Pearson residuals, X_i, divided by $\sqrt{(1 - h_i)}$, and are therefore known as **standardized Pearson residuals**.

Another type of residual can be constructed from the deviance that is obtained after fitting a linear logistic model to binomial data, given by

$$D = 2\sum_i \left\{ y_i \log\left(\frac{y_i}{\hat{y}_i}\right) + (n_i - y_i)\log\left(\frac{n_i - y_i}{n_i - \hat{y}_i}\right) \right\}$$

The signed square root of the contribution of the ith observation to this overall deviance is

$$d_i = \text{sgn}(y_i - \hat{y}_i)\left[2y_i \log\left(\frac{y_i}{\hat{y}_i}\right) + 2(n_i - y_i)\log\left(\frac{n_i - y_i}{n_i - \hat{y}_i}\right) \right]^{1/2}$$

where $\text{sgn}(y_i - \hat{y}_i)$ is the function that makes d_i positive when $y_i \geqslant \hat{y}_i$ and negative when $y_i < \hat{y}_i$. The quantity d_i is known as a **deviance residual** and is such that the overall deviance is $D = \sum d_i^2$. The deviance residuals can also be standardized to have approximate unit variance by dividing by $\sqrt{(1 - h_i)}$ to give **standardized**

deviance residuals, defined by

$$r_{Di} = \frac{d_i}{\sqrt{(1 - h_i)}} \qquad (5.3)$$

Yet another way of deriving a residual is to compare the deviance obtained on fitting a linear logistic model to the complete set of n binomial observations, with the deviance obtained when the same model is fitted to the $n - 1$ observations, excluding the ith, for $i = 1, 2, \ldots, n$. This gives rise to a quantity that measures the change in the deviance when each observation in turn is excluded from the data set. Exact values of these statistics could be obtained by fitting the model to all n observations, and to the n different data sets resulting from omitting each observation in turn. This computationally intensive procedure can be avoided by using the result that the change in deviance on omitting the ith observation from the fit is well approximated by

$$h_i r_{Pi}^2 + (1 - h_i) r_{Di}^2$$

where r_{Pi} and r_{Di} are the standardized Pearson and deviance residuals given by equations (5.2) and (5.3), respectively. Since r_{Pi}, r_{Di} and h_i are all obtained from fitting the model to the original set of n observations, this approximation neatly avoids having to refit the model a further n times. The signed square root of these values,

$$r_{Li} = \text{sgn}\,(y_i - \hat{y}_i)\sqrt{\{h_i r_{Pi}^2 + (1 - h_i) r_{Di}^2\}} \qquad (5.4)$$

are known as **likelihood residuals** for a reason that will become apparent in Section 5.4. Note that r_{Li}^2 is a weighted combination of r_{Di}^2 and r_{Pi}^2. Moreover, the values of h_i will usually be small, and so the values of r_{Li} will be similar to those of r_{Di}. To see why the diagonal elements of H are small, first note that H is **symmetric**, that is, $H = H'$, and has the property that $H = H^2$, in other words it is **idempotent**. Because of this,

$$h_i = \sum_{j=1}^{n} h_{ij}^2 = h_i^2 + \sum_{j \neq i} h_{ij}^2$$

where h_{ij} is the (i, j)th element of the matrix H and $h_i \equiv h_{ii}$. It then follows that $h_i > h_i^2$ and so the values of h_i must all lie between zero and one. Furthermore, the sum of the diagonal elements of H is equal to its rank, which is p, the number of unknown parameters being fitted in the model. Hence, the average value of h_i is p/n, where n is the number of observations, and so the values of h_i will usually be small.

An argument similar to that used in the previous paragraph suggests that a residual might be constructed from the change in the value of the X^2-statistic on deleting the ith observation. It turns out that this difference can be approximated by r_{Pi}^2, the square of the standardised Pearson residual, further justifying the relevance of these values in assessing model adequacy.

In analysing normally distributed data, the deviance residual would be further standardized by division by s, the square root of the residual mean square after fitting a model. Then, the standardized deviance residuals, standardized Pearson residuals and the likelihood residuals all lead to the same quantity, the standardized residual $(y_i - \hat{y}_i)/[s\sqrt{(1 - h_i)}]$, where now h_i is the ith diagonal element of the matrix $H = X(X'X)^{-1}X'$. This matrix is such that $\hat{\mathbf{y}} = H\mathbf{y}$, so that it is the matrix that 'puts the hat on the $\mathbf{y}$'. It is therefore widely referred to as the **hat matrix**. In addition, because h_i measures the effect that y_i has on the determination of $\hat{y}_i$, h_i is generally known as the **leverage**. Although the relationship $\hat{\mathbf{y}} = H\mathbf{y}$ does not hold exactly for binomial data, h_i can still be used as an approximate measure of the effect of a binomial observation on its corresponding fitted value, as discussed later in section 5.5.

The interpretation of a set of residuals is greatly helped if their distribution, under the assumption that the fitted model is correct, is known. The exact distribution of all the residuals defined in this section is intractable. This suggests that if we can find some function of the binomial observations, $A(y_i)$, that is approximately normally distributed, an appropriate residual, standardised to have approximate unit variance would be

$$r_{Ai} = \frac{A(y_i) - A(\hat{y}_i)}{\text{s.e.}\{A(y_i) - A(\hat{y}_i)\}} \tag{5.5}$$

This method of constructing residuals was first proposed by Anscombe (1953), and so they have come to be known as **Anscombe residuals**. The appropriate function A for binomial data, and the standard error in the denominator of equation (5.5), are quite complicated expressions. However, they are not too difficult to compute, as shown in section 9.3.4.

To illustrate the similarities and differences between these four different types of residual, their values have been calculated for two data sets: the data from Example 1.6 on the susceptibility of mice to pneumonia, and that from Example 4.5 on the mortality of confused flour beetles.

Example 5.1 Anti-pneumococcus serum
In modelling the data from Example 1.6 on the extent to which a particular serum protects mice from pneumonia, a linear logistic model for the relationship between the probability of a mouse dying from pneumonia and the logarithm of the dose of serum, was found to be satisfactory. This analysis was given in Example 4.1. Table 5.1 gives the number of mice, y_i, dying from pneumonia, out of 40, for each of the five doses, together with the corresponding fitted number, $\hat{y}_i$, the values of h_i, the Pearson and deviance residuals, X_i, d_i, the standardized Pearson and deviance residuals, r_{Pi}, r_{Di}, the likelihood residuals, r_{Li}, and the Anscombe residuals, r_{Ai}. See Appendix C.1 for the GLIM macro used to compute the Anscombe

Table 5.1 Fitted values and residuals after fitting a linear logistic model to the data from Example 1.6

Dose of serum	y_i	$\hat{y}_i$	h_i	X_i	d_i	r_{Pi}	r_{Di}	r_{Li}	r_{Ai}
0.0028	35	33.08	0.577	0.801	0.834	1.231	1.283	1.254	1.284
0.0056	21	22.95	0.410	−0.624	−0.621	−0.812	−0.808	−0.810	−0.808
0.0112	9	10.99	0.360	−0.704	−0.719	−0.880	−0.898	−0.892	−0.898
0.0225	6	3.82	0.385	1.171	1.090	1.493	1.390	1.430	1.392
0.0450	1	1.15	0.269	−0.146	−0.150	−0.171	−0.175	−0.174	−0.175

residuals and Chapter 9 for details on how the quantities in Table 5.1 can be obtained using different software packages.

Table 5.1 shows that the standardised deviance residuals and the Anscombe residuals are very similar, and that they do not differ greatly from the standardized Pearson and the likelihood residuals. The exact values of the likelihood residuals were also calculated by omitting each observation from the fit in turn and taking the signed square root of the difference between the resulting deviance and that for the model fitted to all five observations. These values were found to be 1.244, −0.814, −0.885, 1.461 and −0.174, respectively. The similarity between these exact values and the approximate values r_{Li} in Table 5.1 suggests that the approximation on which r_{Li} is based is a good one in this case.

Example 5.2 Mortality of confused flour beetles
In Example 4.5, data were given on the proportion of confused flour beetles killed on exposure to different concentrations of gaseous carbon disulphide. Although a quadratic logistic regression model for the relationship between the probability that a beetle is killed and the concentration of carbon disulphide appeared to fit the data, for illustrative purposes the fitted values and residuals on fitting a linear logistic regression line are given in Table 5.2. The standardized deviance, likelihood and Anscombe residuals are again very similar, but the standardised Pearson residuals do differ somewhat from the others. The greatest differences between the four types of residual occur when the residuals are relatively large and the fitted probabilities are close to zero or one. The most striking aspect of Table 5.2 is that the three largest standardized Pearson residuals do not rank the observations in the same order as the other three types of residual. For example, on the basis of r_{Pi}, it is the second observation which deviates most from its corresponding fitted value, whereas it is the fourteenth observation that is the most discrepant on the basis of r_{Di}, r_{Li} and r_{Ai}.

In this section, a number of residuals have been defined and so one must enquire whether some of them have more to recommend them than others. The empirical evidence presented in Examples 5.1 and 5.2, together with more detailed numerical studies by Williams (1984) and Pierce and Schafer (1986), suggest that the

Table 5.2 Fitted values and four types of residual on fitting a logistic regression line to the data from Example 4.5

Concentration of CS_2	y_i	n_i	$\hat{y}_i$	r_{Pi}	r_{Di}	r_{Li}	r_{Ai}
49.06	2	29	2.035	−0.027	−0.027	−0.027	−0.027
49.06	4	30	2.105	1.461	1.311	1.333	1.315
52.99	7	30	5.019	1.061	1.013	1.021	1.015
52.99	6	30	5.019	0.525	0.513	0.515	0.513
56.91	9	28	9.742	−0.317	−0.319	−0.319	−0.319
56.91	9	34	11.830	−1.116	−1.141	−1.137	−1.142
60.84	14	27	15.846	−0.767	−0.762	−0.762	−0.762
60.84	14	29	17.020	−1.216	−1.206	−1.207	−1.208
64.76	23	30	23.714	−0.345	−0.340	−0.341	−0.340
64.76	29	33	26.086	1.352	1.441	1.428	1.444
68.69	29	31	28.193	0.542	0.569	0.565	0.569
68.69	24	28	25.465	−1.028	−0.956	−0.965	−0.957
72.61	29	30	28.916	0.086	0.087	0.087	0.087
72.61	32	32	30.844	1.154	1.618	1.577	1.719
76.54	29	29	28.597	0.658	0.927	0.914	0.984
76.54	31	31	30.570	0.682	0.960	0.946	1.019

standardized deviance residuals, the likelihood residuals and the Anscombe residuals will usually be very similar to each other. In addition, all three are likely to perform similarly in terms of the ranking of extreme observations. The numerical studies also indicate that all three of these residuals are reasonably well approximated by a standard normal distribution when the binomial denominators are not too small. For this reason, the values of these residuals will generally lie between −2 and 2, when the fitted model is satisfactory. Because the Anscombe residuals are much more difficult to compute than any of the other types of residual, numerical integration being required, there is no great advantage in using them. The standardized Pearson residuals are not as closely approximated by a normal distribution and furthermore, may not rank extreme observations appropriately. In summary then, either the standardized deviance residuals, r_{Di}, or the likelihood residuals, r_{Li}, may be used routinely in model checking.

5.2 CHECKING THE FORM OF THE LINEAR PREDICTOR

The residuals obtained after fitting a linear logistic model to an observed set of data form the basis of a large number of diagnostic techniques for assessing model adequacy. In this section, we see how residuals, and other quantities derived from them, can be used to examine whether the linear component of the model is appropriate. The emphasis will on graphical methods designed to detect particular forms of misspecification, rather than formal tests of statistical significance.

Information about the adequacy of a fitted model can sometimes be obtained directly from the tabulated values of the residuals. For example, in the case of data from designed experiments, where the binomial observations are cross-classified with respect to one or more factors, a display of the residuals, cross-classified by these factors, can be illuminating. The occurrence of relatively large residuals, or patterns in the residuals across parts of the display, indicates that there are particular combinations of factor levels that are not well fitted by the model. In situations where the observations are located in identifiable spatial positions, such as in agricultural experiments where treatments are allocated to plots of land, a display of the residuals according to their locations can provide useful information about the model. For example, consider an agricultural experiment designed to compare a number of fertiliser treatments in terms of the yield of a particular crop. The occurrence of positive residuals in a series of adjacent plots of land, and clusters of negative residuals over other areas may indicate that the experimental area is not homogeneous in terms of fertility.

Example 5.3 Germination of lupin seed
In an experiment carried out by research staff of the Royal Botanic Gardens at Kew, seeds of *Lupinus polyphyllus* were germinated after storage for a period of five days under various conditions. Batches of seeds were stored at each of three constant temperatures (21 °C, 42 °C and 62 °C) and at each of three moisture levels (low, medium and high). For 100 seeds from each combination of levels of storage temperature and moisture level, a germination test was carried out at two temperatures (11 °C and 21 °C). The data in Table 5.3 give the number of seeds that germinated out of 100 for each combination of the levels of the three factors.

Table 5.3 Number of seeds germinating, out of 100, at two different temperatures after being stored at different temperatures and different moisture levels

Germination temperature	Moisture level	Storage temperature		
		21 °C	42 °C	62 °C
11 °C	low	98	96	62
11 °C	medium	94	79	3
11 °C	high	92	41	1
21 °C	low	94	93	65
21 °C	medium	94	71	2
21 °C	high	91	30	1

A linear logistic model that contains the main effects of storage temperature, moisture level and germination temperature, and the interaction between storage temperature and moisture level, appears to fit the data, the deviance for this model being 4.43 on 8 d.f. The standardized deviance residuals under this model are presented in Table 5.4. To enhance the visual impact of the display, the residuals have been expressed correct to just one decimal place. This display shows that the model fits least well to the data obtained from seeds stored at a temperature of 62°C and a low moisture level. Table 5.4 also shows that the residuals for a storage temperature of 42°C are all positive when the germination temperature is 11°C and negative when this temperature is 21°C. This means that the germination probabilities under the fitted model tend to be underestimates at the lower germination temperature and overestimates at the higher temperature, suggesting that a term that represents the interaction between storage temperature and germination temperature should be included in the model. However, when this term is added to the model, the deviance is reduced by 2.54 on 1 d.f., which is not significant at the 10% level. The evidence for including this interaction term is therefore slight.

Table 5.4 Standardized deviance residuals after fitting a linear logistic model to the data on the germination of lupin seed in Table 5.3

Germination temperature	Moisture level	Storage temperature 21°C	42°C	62°C
11°C	low	1.2	0.5	−1.6
11°C	medium	−0.5	0.5	0.2
11°C	high	−0.3	0.8	−0.2
21°C	low	−1.0	−0.5	1.6
21°C	medium	0.5	−0.5	−0.2
21°C	high	0.3	−0.8	0.2

5.2.1 Plots of residuals

Although tables of residuals can be useful in circumstances such as those just described, graphical summaries of the residuals are often more informative, particularly when the data are from observational studies. The simplest graphical display is a plot of the residuals against the corresponding observation number, or index. This is known as an **index plot**. Although this plot is particularly suited to the detection of outliers, that is, observations that have unusually large residuals, a systematic pattern in the plot indicates that the model is not correct.

A graph which can be more informative than an index plot is a plot of the residuals against the values of the linear predictor, $\hat{\eta}_i = \sum \hat{\beta}_j x_{ji}$. Again, the occurrence of a systematic pattern in the plot suggests that the model is incorrect in some way. For example, additional explanatory variables, or transformed values, such as squares or cross-products, of variables that have already been included in the model may be needed. Outliers will also be identifiable as observations that have unusually large residuals.

A plot of residuals against particular explanatory variables in the model can also be used to identify if any particular variable needs to be transformed. A trend in a plot of residuals against potential explanatory variables not included in the model indicates that the response does depend on that variable, suggesting that it should be included.

5.2.2 Half-normal plots of residuals

All of the graphs described above can be constructed from either the standardized deviance residuals or the likelihood residuals, whichever is more convenient. Since the distribution of either of these two types of residual can be approximated by a standard normal distribution, when the fitted model is correct, a normal probability plot of the residuals might also be expected to be useful in model checking. The idea behind this plot is that if the residuals are arranged in ascending order and plotted against an approximation to their expected values derived under the assumption that they are normally distributed, an approximate straight line will indicate that the residuals can be assumed to have a normal distribution. Denoting the ordered values of n standardized deviance or likelihood residuals by $r_{(i)}$, their expected values can be approximated by $\Phi^{-1}\{(i - \frac{3}{8})/(n + \frac{1}{4})\}$, the probit of $\{(i - \frac{3}{8})/(n + \frac{1}{4})\}$; see Blom (1958). A normal probability plot of the residuals is then a plot of $r_{(i)}$ against $\Phi^{-1}\{(i - \frac{3}{8})/(n + \frac{1}{4})\}$.

Although this plot was originally designed to examine whether the residuals can be taken to be normally distributed, in linear logistic modelling it is more useful for diagnosing model inadequacy or revealing the presence of outliers. Such features are shown up more effectively in a half-normal plot, in which the absolute values of the residuals are arranged in ascending order and plotted against $\Phi^{-1}\{(i + n - \frac{1}{8})/(2n + \frac{1}{2})\}$. Outliers will appear at the top right of the plot as points that are separated from the others, while systematic departure from a straight line could indicate that the model is unsatisfactory. However, even when the fitted model is correct, the residuals used in constructing a half-normal plot will not be uncorrelated and may not be approximately normally distributed. Because of this, a half-normal plot of the residuals will not necessarily give a straight line when the fitted model is in fact correct.

The interpretation of a half-normal plot of the standardized deviance or likelihood residuals is very much helped by constructing a **simulated envelope** for the plot, proposed by Atkinson (1981). This envelope is such that if the fitted model is correct, the plotted points are all likely to fall within the boundaries of the envelope.

A simulated envelope for a half-normal plot of the standardized deviance residuals is constructed in the following way. For each of the n observations, nineteen additional observations are simulated from a binomial distribution with parameters n_i and $\hat{p}_i$, where n_i is the binomial denominator and $\hat{p}_i$ is the estimated response probability for the ith observation in the original data set, $i = 1, 2, \ldots, n$. The model fitted to the original data is then fitted to each of these nineteen sets of simulated values of the binomial response variable, where the explanatory variables in the model keep their original values. From each of these fits, the absolute values of the standardized deviance residuals, $|r_{Di}|$, are obtained. These values are then ordered to give the values $|r_D|_{(i)}$, where $|r_D|_{(1)} < |r_D|_{(2)} < \cdots < |r_D|_{(n)}$. The mean, minimum and maximum of the values of $|r_D|_{(i)}$ over the nineteen simulated data sets, for $i = 1, 2, \ldots, n$, are then computed and plotted on the half-normal plot of the standardized deviance residuals from the original data set. An algorithm for generating pseudo-random numbers from the binomial distribution is described in section 9.3.5 and a GLIM macro for constructing a half-normal plot of the standardized deviance residuals is given in Appendix C.2.

By using nineteen simulations, there is a chance of 1 in 20, or 5%, that the largest absolute residual from the original data set lies outside the simulated envelope, when the fitted model is correct. This result can be used to assess whether the observation that yields the largest residual can be regarded as an outlier. However, the main advantage of the simulation envelope is that the half-normal plot can be interpreted without having to make assumptions about the distribution of the residuals. In particular, deviation of the points from the mean of the simulated values, or the occurrence of points near to or outside the simulated envelope, indicates that the fitted model is not appropriate.

Example 5.4 Mortality of confused flour beetles
Residuals on fitting a logistic regression line to the data from Example 4.5 were given in Table 5.2. The model for the fitted response probabilities, $\hat{p}_i$, is logit $(\hat{p}_i) = \hat{\eta}_i$, where $\hat{\eta}_i$ is the linear predictor given by

$$\hat{\eta}_i = -14.81 + 0.249\, x_i$$

and x_i is the concentration of CS_2 to which the ith batch of beetles is exposed, $i = 1, 2, \ldots, 16$.

A plot of the standardized deviance residuals against the values of the linear predictor, given in Fig. 5.1, shows that the fitted model is inappropriate; the

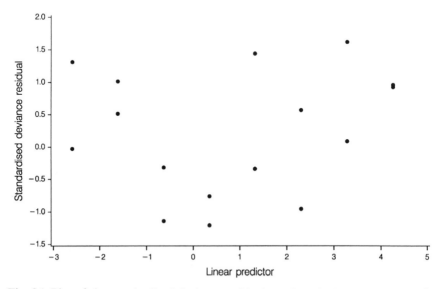

Fig. 5.1 Plot of the standardized deviance residuals against the linear predictor after fitting a logistic regression line to the data from Example 4.5.

curvature in this plot suggests that the model might be improved by the addition of a quadratic term in x_i. When the fitted model includes a single explanatory variable, x, no additional information is provided by a plot of the standardized deviance residuals against the values of x, since they are linearly related to the values of the linear predictor. A half-normal plot of the sixteen standardized deviance residuals is obtained by plotting their ordered absolute values against $\Phi^{-1}\{(i + 15.875)/32.5\}$. The graph obtained is shown in Fig. 5.2. Although the plotted points all lie within the simulated envelope, they do deviate systematically from the line drawn through the means of the simulated values, suggesting that the fitted model is inappropriate.

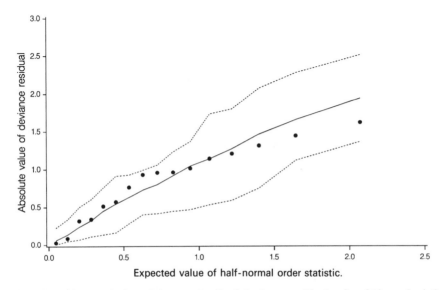

Fig. 5.2 Half-normal plot of the standardized deviance residuals after fitting a logistic regression line to the data from Example 4.5. Also plotted are the simulated envelopes (······), and the means of the simulated values (——).

When x_i^2 is added to the logistic regression model, the resulting plot of the r_{Di} against the values of the linear predictor, $\hat{\eta}_i$, given as Fig. 5.3, shows much less pattern than Fig. 5.1. A half-normal plot of the standardized deviance residuals for the quadratic logistic regression model is shown in Fig. 5.4. This plot shows that the residuals under the quadratic logistic regression model are smaller than would have been expected if the data were binomially distributed, suggesting that the model is overfitted.

So far, the diagnostic methods described in this section are relatively unsophisticated and are designed to provide a general indication of whether a fitted linear logistic model is acceptable. During the last few years, a number of more

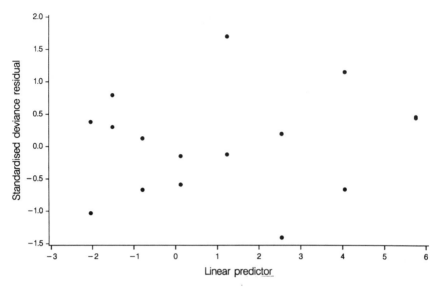

Fig. 5.3 Plot of the standardized deviance residuals against the values of the linear predictor after fitting a quadratic logistic regression model to the data from Example 4.5.

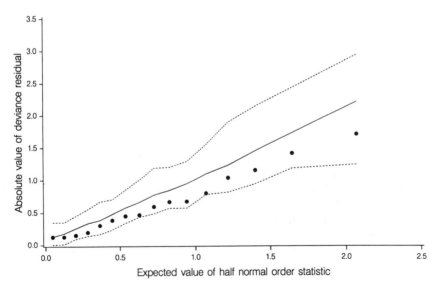

Fig. 5.4 Half-normal plot of the standardized deviance residuals after fitting a quadratic logistic regression model to the data from Example 4.5. Also plotted are the simulated envelopes (·····), and the means of the simulated values (——).

sophisticated graphical methods have been proposed which focus on particular aspects of the structure of the linear component of the model. These methods, which are referred to as **added variable plots**, **partial residual plots** and **constructed variable plots** are described in the following sections. They are all analogues of graphical methods used in the assessment of model adequacy in linear regression models for continuous response data.

5.2.3 The added variable plot

The added variable plot, described by Wang (1985), provides a graphical means of identifying whether or not a particular explanatory variable should be included in a linear logistic model, in the presence of other explanatory variables. The variable that is the candidate for inclusion in the model may be a new explanatory variable, or it may simply be a higher power of one that is currently included in the model.

Suppose that a linear logistic model with k explanatory variables is fitted to n binomial observations. Let X_i denote the ith Pearson residual, defined by equation (5.1), and let H be the hat matrix and W the weight matrix under this fitted model. Now suppose that u is a further explanatory variable measured on each observation, and it is desired to explore whether u should be added to the original model. The added variable plot for the new variable u is a plot of X_i against the ith element of the vector $(I - H)W^{1/2}\mathbf{u}$, where I is the $n \times n$ identity matrix and $\mathbf{u}$ is the vector of values of u for the n observations.

It can be shown that the ith element of $(I - H)W^{1/2}\mathbf{u}$ is simply $(u_i - \hat{u}_i)\sqrt{(w_i)}$, where $w_i = n_i \hat{p}_i (1 - \hat{p}_i)$, $\hat{p}_i$ is the ith fitted value under the model fitted to the k explanatory variables excluding u, u_i is the value of the new explanatory variable for the ith observation, and $\hat{u}_i$ is the ith fitted value from a weighted least squares regression of u on the k explanatory variables in the model, using weights w_i. The quantities $(u_i - \hat{u}_i)\sqrt{(w_i)}$ are known as **added variable residuals** or **u-residuals**. If there is no particular pattern in the added variable plot, then u is not an important variable. On the other hand, if the plot has a linear trend, one would conclude that the variable u is needed in the presence of the other variables. The scatter in the plot indicates the strength of the relationship between $\text{logit}(p_i)$ and u_i, in the presence of the other k variables.

Example 5.5 Mortality of confused flour beetles
As an illustration, the added variable plot will be used to indicate whether a quadratic term x^2 is needed in addition to x in modelling the data from Example 4.5. Here, the added variable residuals, or x^2-residuals, have been calculated from $(u_i - \hat{u}_i)\sqrt{(w_i)}$, where $u_i = x_i^2$; $\hat{u}_i$ is obtained from regressing x_i^2 on x_i with weights $n_i \hat{p}_i (1 - \hat{p}_i)$. Alternatively, computer software for matrix algebra could be used to obtain the added variable residuals from $(I - H)W^{1/2}\mathbf{u}$. The Pearson residuals, X_i, and the x^2-residuals are given in Table 5.5; the added variable plot is given in Fig. 5.5. The linear trend in this plot indicates that the x^2-term may be needed in the logistic regression model.

Table 5.5 Pearson residuals and the x^2-residuals for the added variable plot for x^2

Unit	X_i	x^2-residual
1	−0.025	140.66
2	0.969	50.00
3	−0.294	−56.38
4	−0.721	−98.42
5	−0.321	−52.94
6	0.505	34.91
7	0.082	100.28
8	0.639	129.49
9	1.355	143.06
10	0.480	50.00
11	−1.019	−62.13
12	−1.139	−102.00
13	1.246	−55.53
14	−0.965	33.18
15	1.095	103.56
16	0.661	133.76

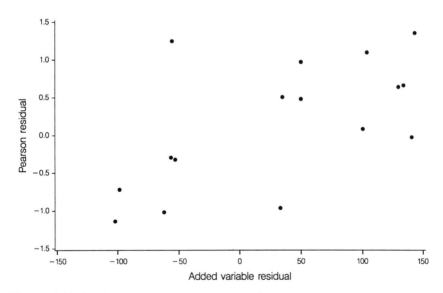

Fig. 5.5 Added variable plot for the inclusion of x^2 in a linear logistic regression model fitted to the data from Example 4.5.

5.2.4 The partial residual plot

The partial residual plot was originally proposed as a method for identifying whether a particular explanatory variable should be transformed to give a non-linear term. The application of this plot to linear logistic modelling was described by Landwehr, Pregibon and Shoemaker (1984).

Suppose that a linear logistic model with k explanatory variables $x_1, x_2, \ldots, x_k$ is fitted to n proportions, y_i/n_i, $i = 1, 2, \ldots, n$, and let $\hat{p}_i$ be the ith fitted response probability under this model. A partial residual plot for the jth explanatory variable, $j = 1, 2, \ldots, k$, is a plot of the **partial residuals**

$$\frac{y_i - n_i\hat{p}_i}{n_i\hat{p}_i(1 - \hat{p}_i)} + \hat{\beta}_j x_{ji} \tag{5.6}$$

against x_{ji}, where $\hat{\beta}_j$ is the coefficient of the jth explanatory variable, x_j, in the model. Note the absence of a square root sign in the denominator of the first term in expression (5.6). If the resulting plot is an approximate straight line, the variable x_j does not need to be transformed. On the other hand, a non-linear plot indicates the need for a transformation, and the pattern observed provides a rough indication of the transformation required. For example, a cubic curve would suggest that a term in x_j^3 is needed.

Example 5.6 Mortality of confused flour beetles
On fitting a linear logistic regression line to the data from Example 4.5, it is found that

$$\text{logit}(\hat{p}_i) = -14.81 + 0.249x_i$$

where x_i is the concentration of carbon disulphide to which the ith batch of beetles is exposed, $i = 1, 2, \ldots, 16$. The partial residuals for x are obtained from $[(y_i - n_i\hat{p}_i)/n_i\hat{p}_i(1 - \hat{p}_i)] + 0.249x_i$; a plot of these against x_i is given in Fig. 5.6. There is a definite curvature in this plot, which indicates that a non-linear term in x is needed. The parabolic shape suggests that a term in x^2 may well improve the fit of the model. After adding such a term, the fitted model becomes

$$\text{logit}(\hat{p}_i) = 7.968 - 0.517x_i + 0.00637x_i^2$$

A partial residual plot for x^2 in this model can now be used to see if the introduction of a quadratic term has improved the fit. The partial residual plot for the variable x^2, given in Fig. 5.7, is a plot of $[(y_i - n_i\hat{p}_i)/n_i\hat{p}_i(1 - \hat{p}_i)] + 0.00637x_i^2$ against x_i^2, where now $\hat{p}_i$ is the ith fitted value under the model that includes both x_i and x_i^2. This plot is quite straight and so the addition of the quadratic term has indeed improved the adequacy of the fitted model.

Although the partial residual plot provides an indication of whether a non-linear

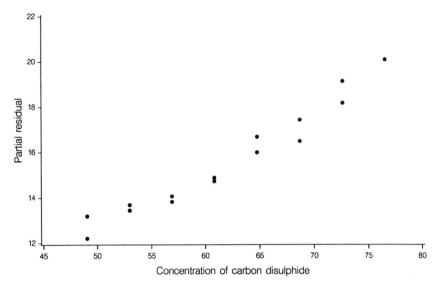

Fig. 5.6 Partial residual plot for x in a linear logistic regression model fitted to the data from Example 4.5.

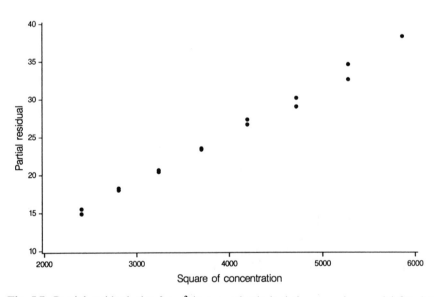

Fig. 5.7. Partial residual plot for x^2 in a quadratic logistic regression model fitted to the data from Example 4.5.

transformation of an explanatory variable should be included in the model, there are occasions when the plot can be misleading. For example, if the fitted probabilities are all close to zero or one, the plot can fail to detect non-linearity in an explanatory variable. An alternative to this plot is the constructed variable plot which is specifically designed to show non-linear dependence between the logistic transform of a response probability and an explanatory variable.

5.2.5 The constructed variable plot

Suppose that an explanatory variable x is included in the model and it is desired to investigate if x should be replaced by some power of x. Let

$$
x^{(\lambda)} = \begin{cases} \dfrac{x^{\lambda} - 1}{\lambda}, & \lambda \neq 0 \\[2mm] \log x, & \lambda = 0 \end{cases}
$$

so that $x^{(\lambda)}$ is the power transformation considered by Box and Cox (1964). When $\lambda = 1$, $x^{(\lambda)} = x - 1$, and this corresponds to no transformation of x. Also, $\lambda = 0.5$ corresponds to the square-root transformation, $\lambda = -1$ to the reciprocal transformation, and as $\lambda \to 0$, $x^{(\lambda)} \to \log x$, so that $x^{(\lambda)}$ is a continuous function of λ.

To identify whether λ is different from unity for the jth explanatory variable in the model, x_j, $1 \leqslant j \leqslant k$, the **constructed variable** z_j is defined, where $z_j = \hat{\beta}_j x_j \log x_j$ and $\hat{\beta}_j$ is the coefficient of x_j in the fitted model. The **constructed variable plot**, due to Wang (1987), is an added variable plot for the constructed variable z_j. That is, the Pearson residuals X_i are plotted against the elements of $(I - H)W^{1/2}z_j$, where z_j is the vector of values of the constructed variable z_j for the n observations. The ith element of $(I - H)W^{1/2}z_j$ is $(z_{ji} - \hat{z}_{ji})\sqrt{(w_i)}$, where $w_i = n_i\hat{p}_i(1 - \hat{p}_i)$, $\hat{p}_i$ is the ith fitted probability under the model fitted to the original set of explanatory variables, $z_{ji} = \hat{\beta}_j x_{ji} \log x_{ji}$, and $\hat{z}_{ji}$ is the ith fitted value from a least squares regression of z_j on the k explanatory variables with weights w_i. The values of $(z_{ji} - \hat{z}_{ji})\sqrt{(w_i)}$ are called **constructed variable residuals**.

A linear trend in the constructed variable plot suggests that a non-linear term in x is needed. The slope of the line in the plot can be used to indicate the value of the transformation parameter λ, since the estimated slope is equal to $\hat{\lambda} - 1$, where $\hat{\lambda}$ is the estimated value of λ. In practice, the resulting estimate will usually be rounded off to the nearest half. After transforming a variable, a constructed variable plot for the transformed variable can again be used to identify whether the transformation has been successful.

Example 5.7 Mortality of confused flour beetles
Consider again the data from Example 4.5 on the mortality of confused flour beetles. The constructed-variable plot for the explanatory variable x is a plot of the Pearson residuals, given in Table 5.5, against the constructed variable residuals, obtained from $(z_i - \hat{z}_i)\sqrt{(w_i)}$. In this expression, z_i is the ith value of the constructed variable, given by $z_i = \hat{\beta} x_i \log x_i$, $\hat{\beta}$ is the coefficient of x in the fitted model, that is

0.249, $w_i = n_i \hat{p}_i(1 - \hat{p}_i)$, and $\hat{p}_i$ is the ith estimated response probability from the fitted logistic regression model. The resulting plot is given in Fig. 5.8. This plot is very similar to the added-variable plot in Fig. 5.5, and again confirms that a non-linear term in x is required in the model. The slope of a straight line fitted to the plotted points in Fig. 5.8 is 0.75, so that the power, λ, to which x should be raised is estimated to be 1.75, which is not too far from 2. Again, the plot indicates that a quadratic term in x is needed in the model.

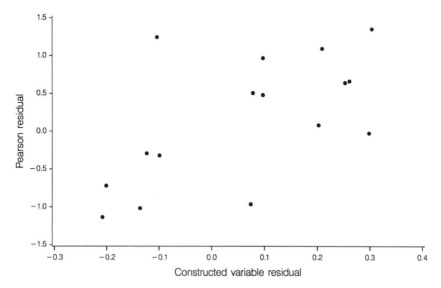

Fig. 5.8 Constructed variable plot for x in a linear logistic regression model fitted to the data from Example 4.5.

5.3 CHECKING THE ADEQUACY OF THE LINK FUNCTION

In modelling binary and binomial data, a particular function of the response probability, known as the link function, is related to a linear combination of the explanatory variables in the model. The logistic transformation is the one that is most commonly used, but rather than adopt this choice of link function uncritically, one should consider whether a different transformation leads to a more simple model or to a model that fits the data better. For example, a linear logistic model with a given set of explanatory variables may not be such a good fit to an observed set of data as, say, a complementary log–log model. Alternatively, the linear component of, say, a complementary log–log model, may require fewer terms than the corresponding component of a logistic model. Thus determination of a suitable choice of link function is bound up with the determination of the structure of the linear part of the model and so any study of the adequacy of a given link function has to be made on the basis of a fixed set of explanatory variables. In this section, a procedure is described that can be used to examine whether a particular transformation of the true response probability, such as the logistic transformation, is satisfactory. This procedure can also be used to provide an indication of the transformation that is most appropriate for the observed data.

Suppose that the link function used in modelling an observed set of binomial data depends on some parameter, different values of which lead to different link functions. If α_0 is the value of this parameter that is being used in modelling an observed set of binomial data, the link function for the ith response probability, $i = 1, 2, \ldots, n$, can be denoted by $g(p_i; \alpha_0)$, so that

$$g(p_i; \alpha_0) = \eta_i \qquad (5.7)$$

where η_i is the linear component of the model for the ith observation. Also suppose that the correct, though unknown, link function is actually $g(p_i; \alpha)$. The family of link functions proposed by Aranda-Ordaz (1981), where

$$g(p_i; \alpha) = \log \left\{ \frac{(1 - p_i)^{-\alpha} - 1}{\alpha} \right\} \qquad (5.8)$$

is particularly useful in modelling binary and binomial data. When $\alpha = 1$, $g(p_i; \alpha) = \log\{p_i/(1 - p_i)\}$, the logistic transform of p_i. As $\alpha \to 0$, $[(1 - p_i)^{-\alpha} - 1]/\alpha \to \log(1 - p_i)^{-1}$, and so then $g(p_i; \alpha) = \log[-\log(1 - p_i)]$, the complementary log–log link function. In many cases, the hypothesized link function, $g(p_i; \alpha_0)$ will be the logistic function, for which $\alpha_0 = 1$.

The function $g(p_i; \alpha)$ can be approximated by the first two terms in a Taylor series expansion of the function about α_0, that is

$$g(p_i; \alpha) \approx g(p_i; \alpha_0) + (\alpha - \alpha_0) \left. \frac{\partial g(p_i; \alpha)}{\partial \alpha} \right|_{\alpha_0}$$

The correct model, $g(p_i; \alpha) = \eta_i$, can therefore be approximated by the model

$$g(p_i; \alpha_0) = \eta_i + \gamma z_i \qquad (5.9)$$

where $\gamma = \alpha_0 - \alpha$ and $z_i = \partial g(p_i; \alpha)/\partial\alpha|_{\alpha_0}$. This model uses the hypothesized link function and includes an additional explanatory variable z which is a further example of a constructed variable.

Before the model in equation (5.9) can be fitted, the values of z must be determined. These will depend on p_i, which are unknown, and so estimates $\hat{p}_i$ are used, where $\hat{p}_i$ is the fitted response probability for the ith observation, obtained from fitting the model in equation (5.7) where the initial choice of link function is used.

For the link function given in equation (5.8), the constructed variable is

$$z_i = \frac{\log(1 - \hat{p}_i)}{(1 - \hat{p}_i)^\alpha - 1} - \alpha^{-1}$$

and in the particular case where $\alpha = 1$, that is when the logistic transformation is the hypothesized link function, this reduces to

$$z_i = -[1 + \hat{p}_i^{-1}\log(1 - \hat{p}_i)] \tag{5.10}$$

If $\gamma = 0$, then $\alpha = \alpha_0$ and the original link function is suitable. Consequently, a test of the hypothesis that $\gamma = 0$ in equation (5.9) provides a test of the adequacy of the initial choice of link function. This hypothesis can be tested by looking at the reduction in deviance on adding z into the model. If this reduction in deviance is large when compared with percentage points of the χ^2-distribution on one degree of freedom, one would conclude that the original link function is unsatisfactory. This procedure is referred to as a **goodness of link test**.

The estimated value of γ under the extended model described by equation (5.9) is $\hat{\gamma} = \alpha_0 - \hat{\alpha}$, and so $\hat{\alpha} = \alpha_0 - \hat{\gamma}$ is an estimate of the unknown parameter in the link function. This gives information about which link function may be suitable in modelling the observed data, although in practice its main use is in distinguishing between the logistic and complementary log–log transformations. If a link function different from that used originally is found to be more appropriate, the structure of the linear predictor of the model will need to be re-examined to ensure that this part of the model remains satisfactory under the new link function.

Example 5.8 Mortality of confused flour beetles
We saw in Example 4.5 that a logistic regression model that included concentration, x, as a linear term did not fit the data, but that the fit was considerably improved by the addition of a quadratic term. A similar conclusion was reached in Examples 5.5–5.7. In this example, we will investigate whether the fit of a model that contains the linear term alone can be improved by changing the link function. The equation of the fitted logistic regression model is

$$\text{logit}(\hat{p}_i) = -14.81 + 0.249x_i$$

where $\hat{p}_i$ is the estimated response probability and x_i the concentration of CS_2 for the ith batch of beetles. The deviance under this model is 12.50 on 14 d.f. Here, the hypothesised link function is the logistic transformation, and to determine whether this is satisfactory, the constructed variable z is added to this model, where the ith value of z is given by equation (5.10). When z is included in the logistic regression model, the deviance becomes 7.91 on 13 d.f. The reduction in deviance of 4.59 on adding the constructed variable into the model is significant at the 5% level, suggesting that the logistic link function is not satisfactory. The coefficient of z in this model, $\hat{\gamma}$, is 1.232 and so the parameter in the general link function given by equation (5.8) is estimated by $\hat{\alpha} = 1 - \hat{\gamma} = -0.232$, which is not so different from zero. This strongly suggests that the complementary log–log model would be better than a logistic model.

On fitting the complementary log–log model to the data, using just a linear term in x, the deviance is 8.67 on 14 d.f. The reduction in deviance on adding a quadratic term to this model is only 0.368 on 1 d.f., and so there is now no reason to include it. The equation of the fitted model is

$$\log\{-\log(1 - \hat{p}_i)\} = -9.755 + 0.155x_i$$

To compare this model with the quadratic logistic regression model, the fitted values of the response probabilities under the two models, $\hat{p}_C$ and $\hat{p}_Q$, say, are plotted against each other. This graph is shown in Fig. 5.9, on which the line of

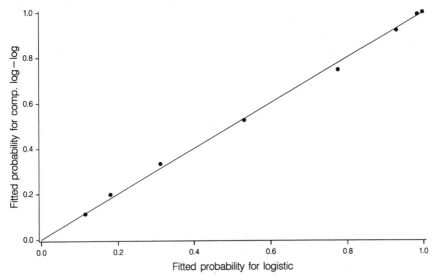

Fig. 5.9 The fitted probabilities under the complementary log–log model, $\hat{p}_C$, plotted against those for the quadratic logistic regression model, $\hat{p}_Q$.

equal fitted probabilities has been superimposed. Since there is no obvious systematic deviation about the line of equal estimated probabilities, the two models cannot be distinguished. However, the complementary log–log model has fewer terms, and might also be preferred to the quadratic logistic regression model for reasons given in section 4.5.

5.4 IDENTIFICATION OF OUTLYING OBSERVATIONS

Observations that are surprisingly distant from the remaining observations in the sample are termed outlying values, or, more briefly, **outliers**. Such values may occur as a result of measurement errors, that is errors in reading, calculating or recording a numerical value; they may be due to an execution error, in which a faulty experimental procedure has been adopted, or they may be just an extreme manifestation of natural variability.

When the data have a relatively simple structure, it is often possible to identify outliers through a visual inspection of the data. In other cases, such as when the logistic transform of a probability is regressed on a single explanatory variable, a plot of the observed data may reveal the presence of an outlier. But in more complicated situations, such as when a linear logistic regression model with a number of explanatory variables is being fitted, it may not be possible to identify outliers on a visual basis. An outlier identification procedure will then be needed before the effect of any such observations can be investigated.

In Section 5.1, it was argued that a residual provides a measure of the extent to which each observation deviates from its fitted value under an assumed model. Consequently, an observation that leads to an abnormally large residual is an outlier. Outlying observations may be detected from the tabulated residuals, from an index plot or a half-normal plot of the residuals, or from a plot of the residuals against the values of the linear predictor.

Example 5.9 Toxicity of insecticides to flour beetles
In Example 4.3, three parallel logistic regression lines were fitted to the data on the proportion of insects killed as a result of exposure to different doses of three insecticides. The original data, fitted values and likelihood residuals under the fitted logistic regression model are given in Table 5.6. Five of the six likelihood residuals

Table 5.6 Fitted values and likelihood residuals for the data from Example 4.3

Observation	Treatment	Deposit	y_i	n_i	$\hat{y}_i$	r_{Li}
1	DDT	2.00	3	50	3.19	−0.120
2	DDT	2.64	5	49	6.17	−0.571
3	DDT	3.48	19	47	10.93	2.962
4	DDT	4.59	19	50	19.50	−0.166
5	DDT	6.06	24	49	28.17	−1.410
6	DDT	8.00	35	50	37.04	−0.791
7	γ-BHC	2.00	2	50	6.10	−2.242
8	γ-BHC	2.64	14	49	11.12	1.100
9	γ-BHC	3.48	20	50	19.10	0.297
10	γ-BHC	4.59	27	50	28.30	−0.424
11	γ-BHC	6.06	41	50	36.70	1.642
12	γ-BHC	8.00	40	50	42.68	−1.195
13	DDT + γ-BHC	2.00	28	50	30.46	−0.949
14	DDT + γ-BHC	2.64	37	50	38.36	−0.532
15	DDT + γ-BHC	3.48	46	50	43.70	1.142
16	DDT + γ-BHC	4.59	48	50	46.80	0.787
17	DDT + γ-BHC	6.06	48	50	48.43	−0.357
18	DDT + γ-BHC	8.00	50	50	49.25	1.252

for the batches of insects treated with DDT are negative, whereas the remaining one has a large positive value. This shows that the response data for those insects treated with DDT are not well fitted by the model. This is also shown from the plot of the fitted logistic regression lines in Fig. 4.3. The third observation also stands out from the rest as having a relatively large residual, although it is easier to compare their values from an index plot of the likelihood residuals given in Fig. 5.10. The outlying nature of the third observation is more obvious from this plot

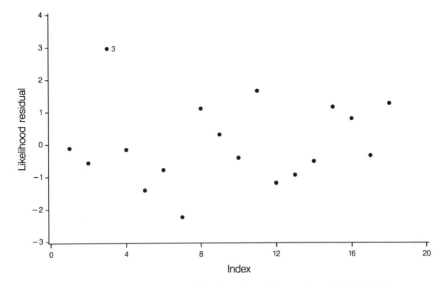

Fig. 5.10 Index plot of the likelihood residuals given in Table 5.6.

than it is from the tabulated values of the residuals. A half-normal plot of the likelihood residuals is given in Fig. 5.11. The likelihood residual corresponding to the third observation lies outside the simulated envelope, confirming that this observation is an outlier.

In some circumstances, the physical cause of an outlier may be identified. The observation can then be discarded on the grounds that it is not from the population under study and, if possible, replaced by a corrected value. For data in the form of proportions, y_i/n_i, an outlying value may be the result of one or more of the n_i component binary observations being misclassified. In this case, the binomial observation itself would not be discarded, but y_i would be adjusted to take account of the misclassification.

More often than not, there is no obvious cause of an outlier and it will be necessary to determine whether the outlier is, in probabilistic terms, too extreme to have arisen by chance under the assumed model for the data-generating process. A value (or set of values) which on the basis of an objective statistical criterion is inconsistent with the rest of the sample is said to be **discordant** and the procedure itself is called a **test of discordancy**.

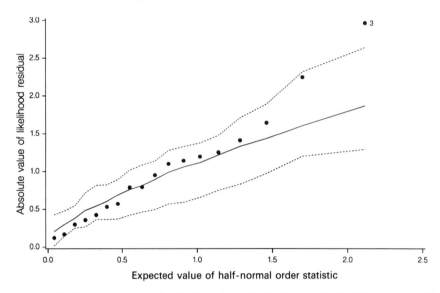

Fig. 5.11 Half-normal plot of the likelihood residuals given in Table 5.6, together with the simulated envelopes (·····), and the means of the simulated values (——).

Suppose that the presence of a potentially discordant value, the rth value, say, is modelled by taking

$$
\left.
\begin{aligned}
\text{logit}\,(p_i) &= \beta_0 + \sum_{j=1}^{k} \beta_j x_{ji}, \quad i \neq r \\[2mm]
\text{logit}\,(p_r) &= \beta_0 + \zeta + \sum_{j=1}^{k} \beta_j x_{jr}
\end{aligned}
\right\}
\tag{5.11}
$$

If ζ is equal to zero, the same model is being used for each observation and there is no discordant value. If ζ is not equal to zero, the rth observation is discordant, because logit (p_r) has been shifted by an amount ζ from what it would have been had that observation been concordant with the remaining $n - 1$ observations.

In order to test the hypothesis that $\zeta = 0$ in equations (5.11), the change in deviance that results from constraining ζ to be zero is computed. When ζ is non-zero, the model fitted to the rth observation contains a parameter whose estimated value is entirely determined by that one observation. The fitted value for this observation will then be the same as the observed value, and the deviance under this model will be the same as the deviance obtained on fitting the model with $\zeta = 0$ to the $n - 1$ observations excluding the rth. The test statistic will therefore be the difference between the deviance on fitting the model logit $(p_i) = \beta_0 + \sum_j \beta_j x_{ji}$ to the full data set and the deviance on fitting the same model to the $n - 1$ observations excluding the rth. In Section 5.1 we saw that this statistic is approximated by r_{Lr}^2 and so these values provide a natural basis for a test of discordancy. Since this discordancy test is actually a likelihood ratio test, the r_{Li} defined by equation (5.4) are called likelihood residuals.

In practice, interest will centre on the observation that has been identified as being the most extreme, that is the one that leads to the likelihood residual that is largest in absolute value. This suggests that $\tau = \max_i |r_{Li}|$ will be an appropriate test statistic for identifying a single discordant value in linear logistic modelling. Large values of τ indicate that the observation that yields the largest likelihood residual in absolute value, is discordant. An approximate test, based on the assumption that the likelihood residuals are normally distributed, is to declare the outlier to be discordant at the $100\alpha\%$ level if $\tau > \Phi^{-1}\{1 - (\alpha/2n)\}$, where n is the number of observations. Tables of the standard normal distribution function can be used to obtain percentage points of this test statistic. Alternatively, the P-value associated with the observed value of the test statistic, τ, is $2n[1 - \Phi(\tau)]$. This test will not be valid if some of the binomial denominators are near to or equal to one, since under these circumstances the likelihood residuals are not normally distributed.

Example 5.10 Toxicity of insecticides to flour beetles
In this example, the most extreme observation in the data from Example 4.3 on the toxicity of three insecticides to flour beetles is tested for discordancy. From Example 5.9, the third observation appears to be an outlier, in that it is the observation that has the largest likelihood residual in absolute value. Assuming that this observation is not known to be the result of a measurement error or an execution error, the extent to which it is discordant will be investigated. Here, $n = 18$ and so the 5% point of the null distribution of the statistic τ is $\Phi^{-1}\{1 - (0.05/36)\} = 2.99$. The observed value of τ is 2.962 and so the outlier is almost significant as a discordant value at the 5% level. The P-value associated with this discordancy test is $36[1 - \Phi(2.962)] = 0.054$, which is sufficiently small for the most extreme value to be adjudged discordant.

What is to be done about this discordant value? Its presence may suggest that the assumed underlying model for the data is incorrect and needs to be modified in order to incorporate the outlier. This may mean that the model has to be extended through the addition of further explanatory variables to allow for a non-linear relationship between the logistic transform of the response probability and log (dose) for DDT, but it could also mean that the logistic link function should be changed. If the original model is accepted, the outlier that has been identified as discordant may itself be made the subject of further scrutiny. For example, it may reflect a feature of real practical importance, such as the fact that the batch of 47 mice exposed to $3.48\,\text{mg}/10\,\text{cm}^2$ of DDT were rather more susceptible to the poison than the other batches. One further course of action is to reject the outlying value and to refit the model to the revised data set. Although this may be appropriate when the adopted model is taken to be inviolable and the outlier is deemed to be an invalid observation, this procedure should certainly not be adopted on a routine basis.

In Example 5.9, our attention was drawn to a single outlying value. How should we react to the appearance of more than one outlier? This is a difficult issue and at present there is no reliable objective procedure that can be recommended for assessing the discordancy of a group of two or more outliers on a routine basis. Instead, an informal approach should be adopted. Thus, if two or more observations give rise to residuals that are unusually large relative to the others, these observations are declared to be outliers and are made the subject of a detailed investigation to identify whether they are the result of a measurement error or an execution error.

Whenever outliers are identified, their effect on the results of the analysis can be assessed by re-analysing the data after omitting them. If essentially the same inferences are drawn from the data both with and without the outliers, one need not be too concerned about their presence. On the other hand, if the outliers do affect model-based inferences, the decision on how to regard them (to include them, to revise the model or to omit them) will be crucial. Generally speaking, this decision should not be made on statistical grounds alone; discussions with those responsible for collecting the data will be an essential ingredient of this part of the analysis.

5.5 IDENTIFICATION OF INFLUENTIAL OBSERVATIONS

An observation is said to be influential if its omission from the data set results in substantial changes to certain aspects of the fit of the linear logistic model. The last paragraph of the previous section suggested that it is more important to focus attention on outliers that are influential than those that are not. Although outliers may also be influential observations, an influential observation need not necessarily be an outlier. In particular, an influential observation that is not an outlier will occur when the observation distorts the form of the fitted model to such an extent that the observation itself has a small residual. This is illustrated in the following example.

Example 5.11 Artificial data – 1
The artificial data in Table 5.7, refer to eight binomial observations, where y_i is the number of individuals out of n_i responding at the ith value of an explanatory variable x, $i = 1, 2, ..., 8$.

On fitting a linear logistic regression model to the relationship between the success probability p_i and the value x_i for these eight observations, the deviance is 7.69 on 6 d.f. and the equation of the fitted model is

$$\text{logit}(\hat{p}_i) = -0.587 + 0.715x_i$$

The fitted values, values of the leverage, and the standardised Pearson and deviance residuals are given in Table 5.8. The likelihood residuals are included in Table 5.10, but they differ from the standardised deviance residuals in Table 5.8 by at most 0.02.

Table 5.7 Artificial data on the proportion of
individuals responding to escalating values of an
explanatory variable

Unit	x_i	y_i	n_i
1	0.1	2	10
2	0.2	2	10
3	0.3	4	10
4	0.4	7	10
5	0.5	5	10
6	0.6	6	10
7	0.7	5	10
8	1.5	5	10

Table 5.8 Fitted values, $\hat{y}_i$, values of the leverage, h_i, standardised Pearson residuals, r_{Pi}, and standardised deviance residuals, r_{Di}, for the artificial data in Table 5.7

Unit	$\hat{y}_i$	h_i	r_{Pi}	r_{Di}
1	3.740	0.262	−1.324	−1.388
2	3.909	0.208	−1.391	−1.458
3	4.081	0.168	−0.057	−0.057
4	4.255	0.141	1.894	1.891
5	4.430	0.128	0.389	0.387
6	4.607	0.131	0.948	0.947
7	4.786	0.150	0.147	0.147
8	6.193	0.812	−1.787	−1.761

Since the deviance under the fitted model is very near to its corresponding number of degrees of freedom, and there is no observation with an unusually large residual, we might conclude that the fitted model is satisfactory. A plot of the data and the fitted logistic regression line, given in Fig. 5.12, suggests otherwise! The eighth observation is unusual in that it is widely separated from the remaining observations in terms of the value of the explanatory variable, x. If this observation is excluded from the data set and the same model fitted to the seven remaining observations, the equation of the fitted model becomes

$$\text{logit}(\hat{p}_i) = -1.333 + 2.716x_i$$

The deviance for this model is 4.46 on 5 d.f., and from Fig. 5.12, we can see that this model fits the first seven observations better than the previous one. Here, the eighth observation is an influential observation in that when it is omitted, the intercept and slope of the fitted logistic regression line are considerably altered. The

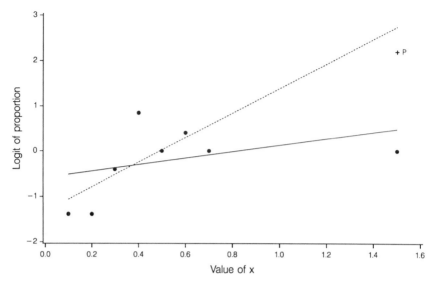

Fig. 5.12 Plot of the data from Example 5.11, together with the fitted logistic regression line with the eighth observation included in (——), and excluded from (······), the data set.

exclusion of any other observation from the fit does not materially alter the fitted model.

Example 5.11 shows that the presence of an influential observation cannot necessarily be detected from a direct examination of the residuals, and so additional diagnostic techniques are required. When an observation is distant from the remaining observations in terms of the values of its explanatory variables, as in Example 5.11, it may be influential. In general, the distance of the ith observation from the remaining observations can be measured by the leverage h_i, defined in Section 5.1. Since the hat matrix, H, depends only on the values of the response variable through their effect on the weight matrix, W, the leverage will depend mainly on the values of the explanatory variables. Observations with relatively large values of h_i are distant from the others on the basis of the values of their explanatory variables; they may be influential but will not necessarily be. Such observations are sometimes called **high leverage points**.

The values of h_i for the data from Example 5.11 were given in Table 5.8. The value of h_i for the eighth observation clearly stands out from the rest, indicating that it might be an influential value. However, if the eighth observation were actually 9/10 instead of 5/10, its position would be at the point labelled 'P' in Fig. 5.12, and its exclusion from the data set would not be expected to change the slope and intercept of the fitted logistic regression model. The values of h_i for this revised data set are 0.289, 0.228, 0.178, 0.150, 0.149, 0.177, 0.230, 0.598. The eighth observation still has a relatively large value of h_i, because it remains separated from the others in terms of its x-value, but the observation would not be declared influential.

In general terms, a value of h_i would be described as unusually large if it is greater than $2p/n$, which is twice the average value of the diagonal elements of H (see Section 5.1). An index plot of the h_i can be useful in identifying observations that are potentially influential. A plot of the absolute values of the standardized deviance or likelihood residuals against the values of h_i can also be informative, since this display combines information on outliers and potentially influential observations. Such a plot for the artificial data from Example 5.11 is given in Fig. 5.13. This figure shows that although the eighth observation is not an outlier, it does have a high leverage and so it may well be influential.

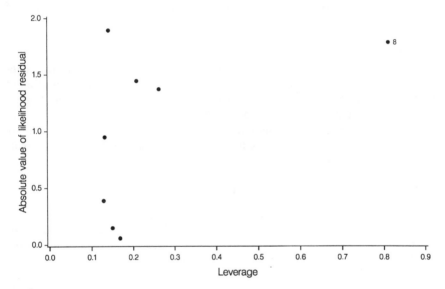

Fig. 5.13 Plot of the absolute values of the likelihood residuals against the leverage for the artificial data from Example 5.11.

In Example 5.11, an observation was described as influential because of the impact that it had on the values of the estimated parameters in the linear logistic model. Observations may be influential in other respects. For example, they may affect one or more parameter estimates of particular interest, or they may influence the comparison of two deviances to determine if a particular explanatory variable should be included in the model. In the following sub-sections, statistics are presented for examining the impact of a single observation on a number of different aspects of influence. Usable versions of these statistics will generally be based on approximations to their exact values, but algebraic details are omitted.

5.5.1 Influence of the ith observation on the overall fit of the model

The influence of a particular observation on the overall fit of a model can be assessed from the change in the value of a summary measure of goodness of fit that results from excluding the observation from the data set. We have seen that two measures of goodness of fit are in common use, the deviance and the X^2-statistic. From Section 5.1, the change in deviance on omitting the ith observation from the fit can be approximated by r_{Li}^2, the square of the ith likelihood residual given by equation (5.4). Also, the change in the value of the X^2-statistic on omitting the ith observation from the fit is approximated by r_{Pi}^2, the square of the standardised Pearson residual given by equation (5.2). Either of these two statistics can therefore be used to measure the impact of each observation on the overall fit of the model, but since the deviance is the most widely used summary measure of goodness of fit, r_{Li}^2 is commended for general use. An index plot of these values is the best way of assessing the contribution of each observation to the overall goodness of fit of the model.

Example 5.12 Artificial data – 1
The likelihood residuals on fitting a logistic regression line to the artificial data from Example 5.11, are very close to the standardized deviance residuals given in Table 5.8. From the squares of their values, no single observation has a great impact on the overall fit of the model to the data, as measured by the deviance. For example, on omitting the eighth observation, the deviance decreases from 7.691 on 6 d.f. to 4.457 on 5 d.f. This reduction of 3.234 is very similar to the approximating value of $r_{L8}^2 = 3.175$.

5.5.2 Influence of the ith observation on the set of parameter estimates

To examine how the ith observation affects the set of parameter estimates, the vector of parameter estimates for the full data set, $\hat{\beta}$, is compared with the vector of estimates, $\hat{\beta}_{(i)}$, obtained when the same model is fitted to the data excluding the ith observation. To compare $\hat{\beta}$ with $\hat{\beta}_{(i)}$, the statistic

$$D_{1i} = \frac{1}{p}(\hat{\beta} - \hat{\beta}_{(i)})' X' W X (\hat{\beta} - \hat{\beta}_{(i)})$$

can be used, where as in the definition of the hat matrix in section 5.1, X is the design matrix, W is the weight matrix and p is the number of unknown parameters in $\hat{\beta}$. In technical terms, this statistic measures the **squared distance** between $\hat{\beta}$ and $\hat{\beta}_{(i)}$. The statistic D_{1i} can also be written as

$$D_{1i} = \frac{1}{p} \sum_{j=1}^{n} \left\{ \text{logit}(\hat{p}_j) - \text{logit}(\hat{p}_{j(i)}) \right\}^2 w_i$$

where $\hat{p}_j$ is the fitted probability for the jth observation when the model is fitted to the complete data set, $\hat{p}_{j(i)}$ is the fitted probability for the jth observation when the model has been fitted to the $n - 1$ observations, excluding the ith, and w_i is the ith diagonal element of the weight matrix. This formulation shows that D_{1i} is an overall measure of the effect of deleting the ith observation on the logistic transform of the fitted probabilities. Another way of looking at the statistic D_{1i} is that it is an approximation to

$$D_{2i} = \frac{2}{p}[\log L(\hat{\beta}) - \log L(\hat{\beta}_{(i)})]$$

where L denotes the likelihood function for the n binomial observations. Since the maximized likelihood summarizes the information in the observed values of the binomial response variable about the unknown parameters in the model, this statistic is again seen to be a measure of the difference between $\hat{\beta}$ and $\hat{\beta}_{(i)}$.

The computation of $\hat{\beta}_{(i)}$ for each of the n sets of data resulting from the deletion of each of the n observations in turn can again be avoided by using an approximation to $\hat{\beta}_{(i)}$. It can be shown that

$$\hat{\beta}_{(i)} \approx \hat{\beta} - (1 - h_i)^{-1/2} \hat{v}_i^{1/2} r_{\text{P}i}(X' WX)^{-1} \mathbf{x}_i \tag{5.12}$$

where $\hat{v}_i = n_i \hat{p}_i(1 - \hat{p}_i)$, $r_{\text{P}i}$ is the ith standardized Pearson residual and $\mathbf{x}_i$ is the vector of explanatory variables for the ith observation. It is from this expression that most of the approximations to measures of influence are derived. Using this result, both D_{1i} and D_{2i} can be approximated by

$$D_i = \frac{h_i r_{\text{P}i}^2}{p(1 - h_i)} \tag{5.13}$$

and this is the most useful statistic for assessing the extent to which the set of parameter estimates is affected by the exclusion of the ith observation. Relatively large values of this statistic will indicate that the corresponding observations are influential. An index plot of the values of this statistic is again the most useful way of presenting these values.

Example 5.13 Artificial data – 1
The eight values of the statistic D_i computed from equation (5.13) using the values of h_i and $r_{\text{P}i}$ given in Table 5.8, and taking $p = 2$, are 0.311, 0.254, 0.000, 0.294, 0.011, 0.068, 0.002 and 6.851, respectively. The eighth observation is highlighted as being influential, in that omitting it from the data set has a great effect on the set of parameter estimates. The actual amount by which the estimates change was indicated in Example 5.11. This example clearly shows that the statistic D_i gives information about the adequacy of the fitted model that is additional to that contained in the residuals.

If the eighth observation were in fact 9/10, so that it now lies much closer to the logistic regression line fitted to the seven remaining observations (see Fig. 5.12), the eight values of D_i are 0.077, 0.133, 0.003, 0.292, 0.000, 0.010, 0.110, and 0.071. Now, the eighth observation is no longer influential, in spite of being separated from the other observations in terms of the value of the explanatory variable.

5.5.3 Influence of the *i*th observation on the *j*th parameter estimate

The D-statistic introduced in Section 5.5.2 measures the overall effect of the ith observation on the parameter estimates. In certain situations, such as when some of the terms in the model are of particular interest, one may want to examine the influence that each observation has on the jth parameter estimate, $\hat{\beta}_j$, where, for models that include a constant term, $j = 0, 1, \ldots, k$. Note that it is possible for an observation to exert an influence on the parameter estimates in such a way that the fitted probabilities change very little; such observations are unlikely to be identified as influential on the basis of the D-statistic.

Using relation (5.12), an approximation to the change in the value of $\hat{\beta}_j$ on excluding the ith observation from the data set, standardised by dividing by the standard error of $\hat{\beta}_j$, is

$$\Delta_i\hat{\beta}_j = \frac{(X'WX)_{j+1}^{-1}\mathbf{x}_i(y_i - \hat{y}_i)}{(1 - h_i)\,\text{s.e.}\,(\hat{\beta}_j)} \tag{5.14}$$

where $(X'WX)_{j+1}^{-1}$ is the $(j + 1)$th row of the variance–covariance matrix of the parameter estimates and $\mathbf{x}_i$ is the vector of explanatory variables for the ith observation. The statistic described by equation (5.14) is widely referred to as a **delta–beta**. Relatively large values of this statistic suggest that the corresponding observation influences the relevant parameter estimate. An index plot of the values of this statistic for selected parameters will be the most useful basis for a study of the influence of each observation on the parameter estimate.

An algorithm for computing the value of the $\Delta\hat{\beta}$-statistic is given in section 9.3.7 and a GLIM macro to compute the values of $\Delta_i\hat{\beta}_j$, for $i = 1, 2, \ldots, n$, and a given value of j, is given in Appendix C.4.

Example 5.14 Artificial data – 1
The values of the $\Delta\hat{\beta}$-statistic for the two parameters in the logistic regression model fitted to the artificial data from Example 5.11, $\Delta_i\hat{\beta}_0$ and $\Delta_i\hat{\beta}_1$, for $i = 1, 2, \ldots, 8$, are given in Table 5.9. From this table, or from a corresponding index

Table 5.9 The values of the statistics for assessing the influence of each observation on the estimates of β_0 and β_1

Unit	$\Delta_i\hat{\beta}_0$	$\Delta_i\hat{\beta}_1$
1	−0.786	0.580
2	−0.695	0.458
3	−0.024	0.013
4	0.634	−0.247
5	0.100	−0.014
6	0.174	0.056
7	0.016	0.023
8	1.869	−3.416

plot of the values of $\Delta_i\hat{\beta}_0$ and $\Delta_i\hat{\beta}_1$, it is seen that it is only the eighth observation that has a great influence on the two parameter estimates.

5.5.4 Influence of the ith observation on the comparison of two deviances

One of the aims of linear logistic modelling is to identify whether a particular explanatory variable, or a set of such variables, should be included in the model. As we saw in Chapter 3, whether or not a particular term should be included in the model can be assessed from the change in deviance on including a term in, or excluding it from, the model. Suppose that two models, model (1) and model (2), say, are to be compared, where model (1) contains a subset of the terms in model (2). Denoting the deviance under these two models by D_1 and D_2, respectively, the extent to which the two models differ is judged by comparing $D_1 - D_2$ with percentage points of a χ^2-distribution. If the difference in deviance is significantly large, model (2) is adopted in preference to model (1). It is often important to examine whether any particular observation has an undue impact on this test. For example, one would be rather cautious about including a term in the model if the resulting decrease in deviance was mainly due to one particular observation.

The influence of the ith observation on the difference between two deviances can be measured from the amount by which this difference changes when the ith observation is excluded from the data set. From Section 5.1, the change in the deviance on excluding the ith observation can be approximated by r_{Li}^2, and so an approximation to the change in the difference between the two deviances, due to excluding the ith observation, is

$$L_i = r_{Li}^2(1) - r_{Li}^2(2) \qquad (5.15)$$

where $r_{Li}^2(1)$ and $r_{Li}^2(2)$ are the squares of the ith likelihood residuals under model (1) and model (2), respectively. Hence, once the likelihood residuals have been calculated for model (1) and model (2), those observations which most influence the difference in deviance between these two models can readily be identified. Negative vales of L_i will be found when omitting the ith observation increases the difference in deviance between two models, while positive values will be found when the difference between the two deviances is reduced. The occurrence of several large negative values of the L-statistic would suggest that a small subset of the observations account for the change in deviance on adding a particular term into the linear logistic model.

Information about the extent to which each observation exerts an influence on whether or not a particular explanatory variable should be included in a linear logistic model can also be obtained from an added variable plot for that variable. This plot was described in Section 5.2.3. Any point that particularly influences the slope of a line drawn through the plotted points in an added variable plot will correspond to an observation that affects the extent to which that explanatory variable appears to be important.

Example 5.15 Artificial data – 1
For the data from Example 5.11, when the variable x is included in a logistic regression model that contains a constant term alone, the deviance decreases from 9.360 on 7 d.f. to 7.691 on 6 d.f. This reduction of 1.669 on 1 d.f. is not significant, but we might enquire as to whether a particular observation unduly influences this change in deviance. Table 5.10 gives the value of the likelihood residuals, r_{Li}, under these two models, together with the corresponding values of the L-statistic for each observation.

Table 5.10 Likelihood residuals on fitting model (1): logit $(p_i) = \beta_0$, and model (2): logit $(p_i) = \beta_0 + \beta_1 x_i$, to the data from Example 5.11, and the values of the L-statistic

Unit	Model (1) r_{Li}	Model (2) r_{Li}	L_i
1	−1.764	−1.371	1.232
2	−1.764	−1.444	1.028
3	−0.341	−0.057	0.113
4	1.706	1.891	−0.667
5	0.339	0.387	−0.035
6	1.017	0.947	0.137
7	0.339	0.147	0.093
8	0.339	−1.762	−3.060

The eighth observation certainly influences the change in deviance that results from the inclusion of x in the logistic regression model. From the value of L_8, we can deduce that the reduction in deviance due to including x will be larger when the eighth observation is omitted from the fit. The evidence for including x in the model will then be correspondingly greater. In fact, when the eighth observation is omitted, the deviance for the model with just a constant term is 9.245 on 6 d.f., and when x is included, the deviance is reduced to 4.458. This reduction of 4.787 is clearly significant, and suggests that x should now be included in the model. The reduction in deviance when the eighth observation is omitted is approximated by the difference between 1.669, the difference in deviance due to including x in the model fitted to all eight observations, and -3.060, the vlaue of L_8 from Table 5.10. This gives 4.729, and comparing it to the actual figure of 4.787, we see that the approximation is very good in this illustration.

An added variable plot provides an alternative means of identifying whether any particular observation has an undue influence on whether or not x should be included in the logistic regression model fitted to the data in Table 5.7. An added variable plot for the explanatory variable x is given in Fig. 5.14, in which the values of the Pearson residuals are plotted against the added variable residuals for x, that is the x-residuals. The slope of a line fitted to the eight points in this graph would

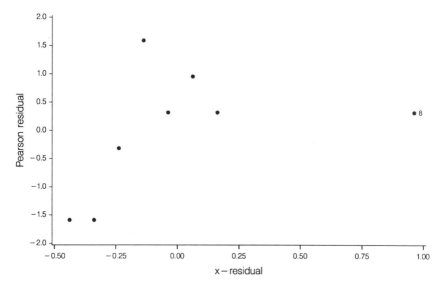

Fig. 5.14 Added variable plot for x on fitting a logistic regression line to the data from Example 5.11.

be very much greater when the eighth observation is excluded. We therefore deduce that this observation has an influence on whether or not x should be included in the model, and that if this observation were omitted, there would then be much more reason to include x in the linear logistic model.

5.5.5 Influence of the ith observation on the goodness of link test

In Section 5.3 we saw how the hypothesis that a particular link function is appropriate can be tested by looking at the change in deviance when a constructed variable is added to the model based on the assumed link function. If the resulting change in deviance is significantly large, the hypothesised link function is rejected. In general, one would be wary about changing a link function simply because it improved the fit of the model at one particular data point and so it is important to determine whether the rejection of the assumed link function is influenced by any single observation. Since this amounts to looking at the influence of each observation on the change in deviance when the constructed variable is added to the model, the technique described in Section 5.5.4 can be used to study this. To apply this technique, the likelihood residuals under the model with the hypothesised link function, and those for the model augmented by the inclusion of the constructed variable, are obtained and the values of the L-statistic given by equation (5.15) are computed. Observations with relatively large values of L_i correspond to observations which have an undue influence on the result of the goodness of link test.

Example 5.16 Mortality of confused flour beetles
In this example, we look at the influence of each observation on the result of the test to determine if the logistic link function is satisfactory for the data from Example 4.5. The likelihood residuals under the model

$$\text{logit}(p_i) = \beta_0 + \beta_1 x_i$$

are denoted by $r_{Li}(1)$, while those under the model

$$\text{logit}(p_i) = \beta_0 + \beta_1 x_i + \gamma z_i$$

where z_i is the value of the constructed variable given by equation (5.10), are denoted by $r_{Li}(2)$. The sixteen values of $L_i = r_{Li}^2(1) - r_{Li}^2(2)$ are, respectively, -0.817, 0.537, 0.096, 0.573, 0.110, 0.296, -0.516, 0.641, 1.556, 0.220, 0.776, 1.175, -0.940, -1.296, 1.329 and 0.686. The reduction in deviance on including z in the model is 4.590 and these values of L_i are approximations to the amount by which this number would change if the ith observation is omitted. The ninth observation leads to the largest positive value of L_i, and if this observation were deleted, the reduction in deviance due to including z would be approximately $4.590 - 1.556 = 3.034$. This reduction is significant at the 10% level and so even without the ninth observation, there is still some evidence that the logistic link function is inappropriate. We may therefore conclude that the result of the goodness of link test is not crucially dependent on any single observation.

5.5.6 Influence of observations on the choice of transformation of an explanatory variable

In Section 5.2.5, we saw how a constructed variable plot could be used to determine if a particular explanatory variable should be transformed before including it in a linear logistic model. This plot can also be used to identify observations that influence the choice of transformation. Since the slope of a line drawn through the points in a constructed variable plot gives information about the value of the transformation parameter λ, any point that exerts a strong influence on the slope of this line corresponds to an observation that influences the transformation suggested by the plot. One would be very cautious in using a transformation if the motivation for it was based on a very small proportion of the observed data.

Example 5.17 Mortality of confused flour beetles
A constructed variable plot for the concentration of carbon disulphide, x, in the data from Example 4.5 was given in Fig. 5.8. The plot suggested that a quadratic term in x was needed in the model. From Fig. 5.8, there does not appear to be any one observation that unduly affects the slope of the fitted line, and so we conclude that the choice of transformation is not influenced by any particular observation.

5.5.7 Influence of the *j*th observation on the fit at the *i*th point

One complication in interpreting the influence measures described in Sections 5.5.1–5.5.6 is that they relate to the influence of a single observation on the fit. The exclusion of a particular observation from the data set may lead to the fitted model changing in such a way that an observation that was previously well fitted by the model is no longer fitted so well. This phenomenon is illustrated in the following example.

Example 5.18 Artificial data – 2
The eight observations given in Table 5.11 refer to the values of a binomial

Table 5.11 A second set of artificial data on the proportions of individuals responding to different values of an explanatory variable

Unit	x_i	y_i	n_i
1	0.1	3	10
2	0.2	4	10
3	0.3	4	10
4	0.4	7	10
5	0.5	6	10
6	0.6	7	10
7	0.8	4	10
8	1.6	5	10

proportion, y_i/n_i, at a series of values of an explanatory variable x. A plot of the logistic transform of y_i/n_i against x_i is given in Fig. 5.15. On fitting a logistic regression model for the relationship between the success probability, p_i, and x_i, shown as a solid line in Fig. 5.15, the eighth observation, labelled P, is a high leverage point that will be identified as an influential observation. In contrast, the point labelled Q is neither an outlier nor an influential value. However, if P is omitted from the data set and the same model fitted to the seven remaining observations, the fitted model, which is shown as a dotted line in Fig. 5.15, is not such a good fit to the seventh observation, Q. The removal of P has had a greater effect on the fit of the model at the point Q than at any of the other points, suggesting that the fit of the model can only be improved by the deletion of observations P and Q.

A statistic that can be used to identify circumstances such as those depicted in Example 5.18 is the difference in the contribution of the *i*th observation to the deviance with and without the *j*th observation, for $j = 1, 2, \ldots, n$. A large value of this statistic would indicate that the *j*th observation was seriously affecting the fit of the model to the *i*th observation. Again, this change in deviance can be approximated by a statistic that depends only on quantities that can be obtained

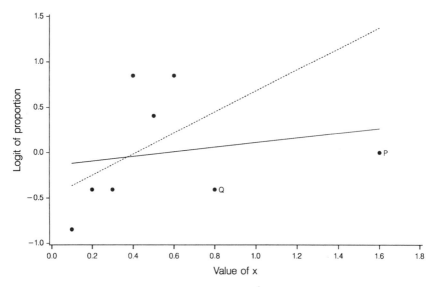

Fig. 5.15 Plot of the data from Example 5.18, together with the fitted logistic regression line when the eighth observation is included in (———), and excluded from (·······), the data set.

from fitting the model to the full data set, namely,

$$\Delta_i D_j = \frac{2X_j h_{ij} X_j}{1 - h_j} + \frac{X_j^2 h_{ij}^2}{(1 - h_j)^2} \tag{5.16}$$

where X_i, X_j are the Pearson residuals for the ith and jth observations and h_{ij} is the (i, j)th element of the hat matrix H. A relatively large value of this statistic for the ith observation, when the jth is omitted indicates that the fit of the model at the ith point is affected by deletion of the jth from the data set. In addition, a positive value of $\Delta_i D_j$ implies that the fit at the ith observation deteriorates when the jth observation is removed, while a negative value implies that the fit at the ith observation is improved. An index plot of $\Delta_i D_j$ for a subset of observations that are to be omitted from the data set provides an informative display. Normally, the values of this statistic would only be computed for a limited number of observations, such as those that have previously been identified as influential on the basis of the D-statistic.

An algorithm that can be used to obtain the values h_{ij} is given in section 9.3.8 and a GLIM macro that uses this procedure to compute the values of the ΔD-statistic, for a given value of j, is given in Appendix C.5.

Example 5.19 Artificial data – 2
For the data from Example 5.18, the values of the D-statistic described by equation (5.13) for assessing the influence of each observation individually, are given in

Table 5.12. These values indicate that the eighth observation is influential. No other observation has such a large value of D. The effect of omitting this observation on the quality of the fit of the model at each of the data points is reflected in the values of $\Delta_i D_8$, for $i = 1, 2, \ldots, 8$, which are also given in Table 5.12. The fit at the eighth

Table 5.12 Values of D_i and $\Delta_i D_8$ for the data from Example 5.18

Unit	D_i	$\Delta_i D_8$
1	0.278	−0.684
2	0.040	−0.178
3	0.034	−0.098
4	0.170	−0.102
5	0.036	−0.205
6	0.129	−0.701
7	0.061	1.246
8	1.840	4.405

observation is clearly the most affected by excluding the eighth observation, but this diagnostic also shows that the fit at the seventh observation is quite severely affected by deleting the eighth observation. These two observations are jointly influential and so the fit of the model can only be improved by deleting both of them. Notice also that the ΔD-statistic is negative for the first six observations, indicating that the omission of the eighth improves the fit at these points, whereas the positive value of the statistic for the other two observations indicates that the fit is now worse at these points. This feature is also apparent from Fig. 5.15.

5.5.8 Treatment of influential values

In Sections 5.5.1–5.5.7, a number of diagnostics have been presented to explore the extent to which observations influence particular aspects of the fit of a linear logistic model. How should one proceed when one or more observations have been found to be influential on the basis of these diagnostics? If at all possible, the raw data should first be checked for errors. If none are found, or if it is not possible to check them, the data should be analysed with and without the influential values to determine their actual effect on the fitted model. This is important, since many of the diagnostics in this section are based on approximations which may not give an accurate picture of the effect of deleting an observation on various aspects of the fit of a model. For example, to examine the effect of the jth observation on the fitted response probabilities, an index plot of the values of $\hat{p}_i - \hat{p}_{i(j)}$ can be informative, where $\hat{p}_{i(j)}$ is the ith estimated response probability when the jth observation has been omitted.

If the influential observations do have a substantial impact on the results, it is a good practice to report the results of both analyses and seek the views of the experimenter on the extent to which the differences are important. It is certainly not good statistical practice to discard influential observations automatically, for although this procedure will perforce lead to a model that is a better fit, the data on which that model is based could be rather different from those which were originally collected.

5.6 CHECKING THE ASSUMPTION OF A BINOMIAL DISTRIBUTION

Suppose that a linear logistic model is fitted to n binomial observations of the form y_i successes out of n_i, $i = 1, 2, \ldots, n$. In order to estimate parameters in the model, and to compare alternative models using an analysis of deviance, the y_i are assumed to have a binomial distribution with mean $n_i p_i$ and variance $n_i p_i(1 - p_i)$, where p_i is the true response probability for the ith observation. The observations y_i have a binomial distribution only when the n_i individual binary observations that lead to y_i are all independent. If they are not, the variance of y_i will be different from $n_i p_i(1 - p_i)$, and the assumption of a binomial distribution for y_i is no longer valid. In particular, if the individual binary observations are positively correlated, the variance will be greater than $n_i p_i(1 - p_i)$, a phenomenon referred to as **overdispersion**. When the individual binary observations are negatively correlated, the variance of a binomial response variable will be less than that for the binomial distribution. However, in practice, this **underdispersion** is much less frequently encountered.

When there is overdispersion, the deviance after fitting an appropriate model will be larger than it would have been under the assumption that the data have a binomial distribution. Consequently, a mean deviance greater than one for a model that is otherwise satisfactory, suggests that the binomial assumption may be incorrect. Since a significantly large deviance can occur when important explanatory variables have not been included in the model, when the data include outlying observations, or when some of the binomial denominators, n_i, are small, it is important to eliminate these possibilities before concluding that the data are overdispersed. The consequences of overdispersion are discussed in the next chapter.

Example 5.20 A toxicological study
In the experiment described in Example 1.4, sixteen pregnant rats were allocated to each of two treatment groups. The number of rat pups in the resulting litters that survived beyond a 21 day period, out of those which were alive four days after birth, was recorded. To compare the two treatments, the model

$$\text{logit}(p_{jk}) = \text{treat}_j$$

is fitted, where p_{jk} is the true probability of survival for rats in the kth litter exposed to the jth treatment group, and treat_j is the effect due to the jth treatment, for $j = 1, 2; k = 1, 2, \ldots, 16$. The deviance for this model is 86.19 on 30 d.f. and the estimated survival probabilities for the rats in the treated and control groups, respectively, are 0.899 and 0.772. Using the subscript i to denote the observational unit, the original data, fitted numbers of rats surviving, $\hat{y}_i$, leverage h_i, standardized deviance residuals, r_{Di}, and values of the D-statistic under this model, are given in Table 5.13. The deviance for the fitted model is rather large, significantly so when judged against a χ^2-distribution on 30 d.f. In the absence of information on additional explanatory variables, there is no other linear logistic model that could

Table 5.13 Observed and fitted numbers of rats surviving, leverage, h_i, standardized deviance residuals, r_{Di}, and values of D_i on fitting a linear logistic model to the data from Example 1.4

Unit	Treatment	y_i	n_i	$\hat{y}_i$	h_i	r_{Di}	D_i
1	Treated	13	13	11.68	0.081	1.738	0.0700
2	Treated	12	12	10.78	0.075	1.664	0.0588
3	Treated	9	9	8.09	0.056	1.427	0.0318
4	Treated	9	9	8.09	0.056	1.427	0.0318
5	Treated	8	8	7.19	0.050	1.341	0.0248
6	Treated	8	8	7.19	0.050	1.341	0.0248
7	Treated	12	13	11.68	0.081	0.316	0.0040
8	Treated	11	12	10.78	0.075	0.220	0.0018
9	Treated	9	10	8.99	0.062	0.014	0.0000
10	Treated	9	10	8.99	0.062	0.014	0.0000
11	Treated	8	9	8.09	0.056	−0.099	0.0003
12	Treated	11	13	11.68	0.081	−0.614	0.0188
13	Treated	4	5	4.49	0.031	−0.666	0.0088
14	Treated	5	7	6.29	0.043	−1.391	0.0621
15	Treated	7	10	8.99	0.062	−1.793	0.1531
16	Treated	7	10	8.99	0.062	−1.793	0.1531
17	Control	12	12	9.27	0.083	2.599	0.1737
18	Control	11	11	8.50	0.076	2.479	0.1438
19	Control	10	10	7.72	0.069	2.355	0.1171
20	Control	9	9	6.95	0.062	2.226	0.0934
21	Control	10	11	8.50	0.076	1.241	0.0519
22	Control	9	10	7.72	0.069	1.090	0.0368
23	Control	9	10	7.72	0.069	1.090	0.0368
24	Control	8	9	6.95	0.062	0.931	0.0245
25	Control	8	9	6.95	0.062	0.931	0.0245
26	Control	4	5	3.86	0.034	0.152	0.0004
27	Control	7	9	6.95	0.062	0.040	0.0000
28	Control	4	7	5.41	0.048	−1.207	0.0428
29	Control	5	10	7.72	0.069	−1.945	0.1677
30	Control	3	6	4.63	0.041	−1.485	0.0570
31	Control	3	10	7.72	0.069	−3.286	0.5045
32	Control	0	7	5.41	0.048	−3.299	0.6324

be fitted, except for the full model which would allow the rats from each of the thirty-two different litters to have different survival probabilities. The fitted probabilities for this model would be equal to those observed and so it does not summarize the original data, it merely reproduces them. The only other way in which the fit of the model might be improved is to omit outlying observations. From Table 5.13, six observations have residuals greater than two, and observations 31 and 32 also have relatively large values of the D-statistic, suggesting that they are influential. When these two observations are omitted from the fit, the estimated survival probability for rats in the control group increases from 0.772 to 0.852, and the deviance for the fitted model becomes 50.38 on 28 d.f., which is still highly significant. Now, five observations have residuals greater than two, all of which are in the control group. However, it is futile to continue excluding observations in an attempt to minimize the deviance. Instead, we recognise that the observations, particularly those in the control group, are more variable than might have been anticipated on the basis of a binomial distribution. There is overdispersion, which in this example is likely to be due to the inter-correlation between the individual binary responses of rat pups in the same litter, as discussed in Example 1.4. The effect of this on inference about the size of the treatment effect, and models for data that exhibit overdispersion are discussed in the next chapter.

5.7 MODEL CHECKING FOR BINARY DATA

In Sections 5.1–5.6, a number of diagnostics have been described for use in situations where the response is binomial, that is, where the binomial denominators all exceed one. For binary data, where the binomial denominators are all equal to one, most of the diagnostics that have already been presented can be applied without modification, but some of them become difficult to interpret. As for binomial data, techniques for examining the adequacy of the linear systematic part of the model and the validity of the link function, and methods for identifying outliers and influential values, are considered in turn in Sections 5.7.1–5.7.4. Additionally, a diagnostic that is designed to assess the effect that each observation in the data set has on the classification of future individuals is described in Section 5.7.5. Overdispersion, as defined in Section 5.6, cannot occur when binary data are being modelled, and so this phenomenon need not be considered further in this section.

5.7.1 Checking the form of the linear predictor

When the data are binary, the deviance residuals become

$$d_i = \text{sgn}\,(y_i - \hat{p}_i)\sqrt{\{-2[y_i\log \hat{p}_i + (1 - y_i)\log (1 - \hat{p}_i)]\}}$$

and the Pearson residuals are

$$X_i = \frac{y_i - \hat{p}_i}{\sqrt{\{\hat{p}_i(1 - \hat{p}_i)\}}}$$

Both these types of residual, and the corresponding standardised versions, r_{Di} and r_{Pi}, take one of two values depending on the observed value of y_i, for each value of i, and will be positive if $y_i = 1$ and negative if $y_i = 0$. For example, $d_i = -\sqrt{\{-2\log(1 - \hat{p}_i)\}}$ if $y_i = 0$, and $d_i = \sqrt{\{-2\log \hat{p}_i\}}$ if $y_i = 1$. This means that the distribution of residuals obtained from modelling binary data will not be approximated by a normal distribution, even when the fitted model is correct. In addition, some plots based on the residuals will contain features that are a direct result of the binary nature of the data. For example, plots of the residuals against the linear predictor, or explanatory variables in the model, will be uninformative, since there will generally be a pattern in these plots, whether the model is correct or not. However, index plots and half-normal plots of the residuals can provide useful information about the adequacy of the linear part of the model. Since residuals from ungrouped binary data are not normally distributed, simulated envelopes for the half-normal plot will provide an essential aid to their proper interpretation.

Example 5.21 Determination of the ESR
The data given in Example 1.7 came from an investigation into the relationship between the disease state of an individual, judged from the value of the ESR, and the level of plasma fibrinogen and γ-globulin. In Example 3.6, the logistic regression model with logit $(\hat{p}_i) = \hat{\eta}_i$ was obtained, where

$$\hat{\eta}_i = -6.845 + 1.827 f_i$$

$\hat{p}_i$ is the fitted probability that the ith individual is diseased and f_i is the fibrinogen level for that individual. The fitted probabilities, $\hat{p}_i$, values of the leverage, h_i, and the standardized deviance residuals, r_{Di}, under this model are given in Table 5.14. An index plot of the standardized deviance residuals, a plot of these residuals against the values of the linear predictor, and a half-normal plot of the residuals with a simulated envelope are given in Figs. 5.16–5.18. From the index plot in Fig. 5.16, two observations stand out as having relatively large residuals, namely, those corresponding to individuals 15 and 23. The responses from these individuals are not well fitted by the model, in that they both have an observed response of unity when the fitted probabilities are 0.076 and 0.046, respectively. The question of whether or not they are influential observations will be taken up later in Example 5.25. The pattern in the plot of r_{Di} against the linear predictor is entirely due to the binary nature of the data, and this makes it hard to interpret. Similarly, a plot of r_{Di} against the omitted variable g_i, the level of γ-globulin for the ith individual will not help in assessing whether there is a relationship between the response probability and g_i, after fitting f_i. In the half-normal plot in Fig. 5.18, the observations corresponding to individuals 15 and 23 occur on the edge of the simulated envelope, again suggesting that these observations are outliers. Apart from this, the plot does not display any unusual features.

Table 5.14 Fitted values, $\hat{p}_i$, leverages, h_i, and standardized deviance residuals, r_{Di}, for the model $\text{logit}(p_i) = \beta_0 + \beta_1 f_i$ fitted to the data from Example 1.7

Unit	$\hat{p}_i$	h_i	r_{Di}	Unit	$\hat{p}_i$	h_i	r_{Di}
1	0.096	0.0404	−0.459	17	0.402	0.1133	1.433
2	0.103	0.0400	−0.475	18	0.125	0.0389	−0.526
3	0.055	0.0426	−0.344	19	0.110	0.0396	−0.492
4	0.054	0.0426	−0.341	20	0.059	0.0426	−0.356
5	0.351	0.0875	−0.973	21	0.170	0.0393	−0.623
6	0.087	0.0410	−0.436	22	0.119	0.0391	−0.513
7	0.276	0.0588	−0.829	23	0.046	0.0425	2.534
8	0.057	0.0426	−0.350	24	0.064	0.0424	−0.372
9	0.252	0.0518	−0.782	25	0.123	0.0389	−0.522
10	0.110	0.0396	−0.492	26	0.065	0.0424	−0.376
11	0.065	0.0424	−0.376	27	0.051	0.0426	−0.332
12	0.072	0.0420	−0.396	28	0.099	0.0402	−0.467
13	0.917	0.2762	0.490	29	0.583	0.2260	1.181
14	0.322	0.0752	1.564	30	0.322	0.0752	−0.918
15	0.076	0.0418	2.319	31	0.201	0.0421	−0.684
16	0.252	0.0518	−0.782	32	0.314	0.0720	−0.902

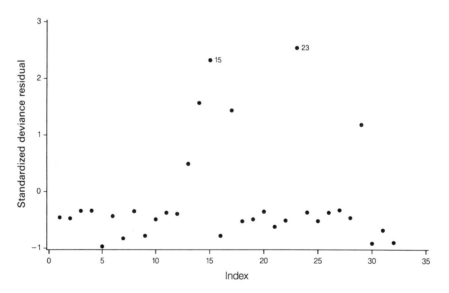

Fig. 5.16 Index plot of the standardized deviance residuals on fitting a logistic regression line to the data from Example 1.7.

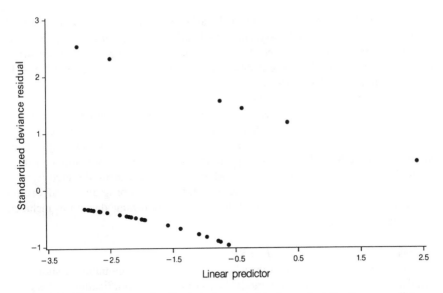

Fig. 5.17 Plot of the standardized deviance residuals against the linear predictor.

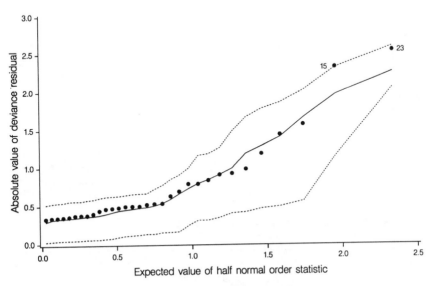

Fig. 5.18 Half-normal plot of the standardized deviance residuals, together with simulated envelopes (⋯⋯) and the means of the simulated values (——).

Added variable plots and constructed variable plots will also be difficult to interpret when the data are binary, since the Pearson residuals used for the vertical axis are negative or positive according to whether the corresponding observation is zero or unity. Typically, these displays show two clouds of points on either side of a horizontal line that corresponds to a residual of zero. A similar pattern is generally apparent in partial residual plots.

It is much easier to interpret each of these plots when a **smoothing algorithm** is used to uncover any structure in them. The method of locally weighted regression described by Cleveland (1979) is particularly suited for this, although other smoothing methods, such as spline smoothing could be used. Cleveland's algorithm is described in section 9.3.6 and a GLIM macro to implement the algorithm is given in Appendix C.3. It would be quite straightforward to program this algorithm for use in other software packages that do not have smoothing algorithms included as a standard facility.

Example 5.22 Determination of the ESR
An added variable plot to identify whether the level of γ-globulin, g, should be included in the logistic regression model fitted to the data from Example 1.7 is given in Fig. 5.19. In this plot, it is difficult to see any relationship between the Pearson residuals on the vertical axis and the g-residuals on the horizontal axis. The smoothed line shown as the solid line in Fig. 5.19 indicates that there is no clear relationship, and it is concluded that the explanatory variable g need not be added to the model. Observations 15 and 23 are separated from the others in this plot, but they are unlikely to have a severe impact on the position of the smoothed line.

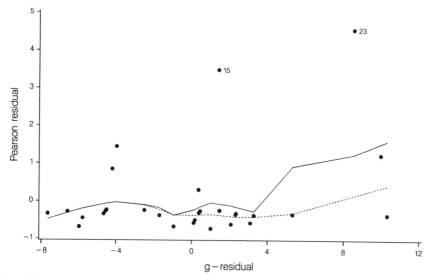

Fig. 5.19 Added variable plot for the inclusion of g (γ-globulin) in a logistic regression model for the data from Example 1.7, with a smoothed line fitted through (i) all the points (——), and (ii) all points excluding 15 and 23 (······).

This impression is confirmed by smoothing the plot after omitting these two observations. The resulting line, shown as a dotted line in Fig. 5.19, is similar to the line based on all the points, particularly for values of the g-residual less than 4. This diagnostic analysis confirms the results of the analysis of deviance given in Example 3.6, which showed that g was not needed in the model.

A constructed variable plot to determine if a non-linear transformation of the term in the model corresponding to the level of fibrinogen, f, is required, is given in Fig. 5.20. Although the plotted points have no obvious pattern, the smoothed

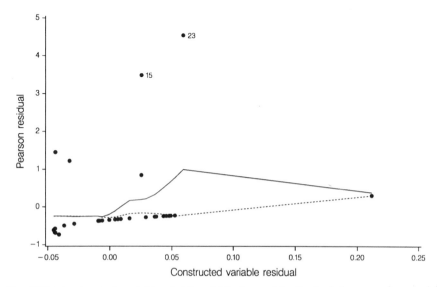

Fig. 5.20 Constructed variable plot for f (fibrinogen) in the logistic regression model fitted to the data from Example 1.7, with a smoothed line fitted through (i) all the points (——), and (ii) all points excluding 15 and 23 (······).

line, shown as a solid line in the figure, indicates that there is an upward trend across values of the constructed variable residuals between -0.02 and 0.06. The observation with the largest constructed variable residual is widely separated from the others, and has a great influence on the position of the smoothed curve. Because the slope of the smoothed line is not constant, it is difficult to use this plot to indicate how f should be transformed.

It does appear that the upward trend in the solid line in Fig. 5.20 may be due to the two unusual observations, 15 and 23. When these two points are left out of the plot, the smoothed line shown as the dotted line in Fig. 5.20 is obtained. This is generally flat, indicating that no transformation of f is now required, and suggests that the evidence for a non-linear term comes from observations 15 and 23.

Since the constructed variable plot suggests that a non-linear term in f is needed when observations 15 and 23 are included in the data set, we go on to examine the effect of including a quadratic term in the model. An analysis of deviance is used to determine if the term f^2 is needed in the model fitted to the complete data set, and that obtained on omitting observations 15 and 23. In summary, for the complete data set, inclusion of the quadratic term reduces the deviance by 7.873 on 1 d.f., a reduction that is significant at the 1% level. However, even after adding the quadratic term to the model, observation 15 still stands out as an outlier, as can be seen from an index plot of the standardized deviance residuals for this model, given in Fig. 5.21.

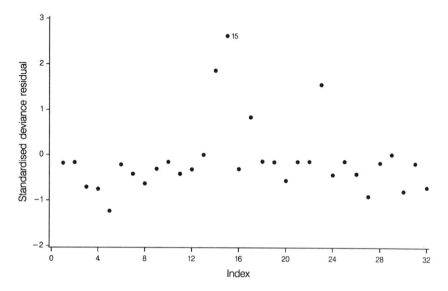

Fig. 5.21 Index plot of the standardised deviance residuals on fitting a quadratic logistic regression model to the data from Example 1.7.

When observations 15 and 23 are omitted from the data set, the fitted logistic model for the relationship between the probability of being diseased and f is

$$\text{logit}(\hat{p}_i) = -58.62 + 17.46 f_i$$

for which the deviance is 5.582 on 28 d.f. When f^2 is added to this model, the deviance is only reduced by 0.003, and so there is no longer any reason to include a quadratic term in the model. From this analysis, we can conclude that the fit of the original model can be improved either by omitting observations 15 and 23, or by including a quadratic term in f.

5.7.2 Checking the adequacy of the link function

The procedure for checking if a particular choice of link function is satisfactory, and for identifying a suitable one if it is not, described in Section 5.3, can be applied without modification when the data are binary. However, an example is included to show that difficulties may be encountered when using this procedure.

Example 5.23 Determination of the ESR
The suitability of the model

$$\text{logit}(\hat{p}_i) = -6.845 + 1.827 f_i$$

for describing the relationship between the probability of disease and plasma fibrinogen has been studied in the two previous examples. Here, the suitability of the link function is examined. The hypothesized link function is the logistic transformation and so the values of the constructed variable z are obtained using equation (5.10). When this variable is added to the logistic regression model, the deviance is reduced from 24.84 on 30 d.f. to 16.79 on 29 d.f. This significantly large reduction suggests that the original model is not satisfactory. However, the coefficient of z in the augmented model is 84.05, which does not indicate any sensible value of the α-parameter to use in a revised link function.

The reason why the procedure has failed to give a sensible result is that the initial model is not satisfactory. This in turn means that the Taylor series approximation to the true link function, used in deriving equation (5.9), is not adequate and so the result given by equation (5.9) no longer holds.

Example 5.22 has shown that observations 15 and 23 are not well fitted by the model, and that in their presence, a quadratic term in f is needed. When these two observations are omitted from the fit, the equation of the linear logistic model that includes f_i alone is

$$\text{logit}(\hat{p}_i) = -59.62 + 17.46 f_i$$

with a deviance of 5.58 on 28 d.f. When the constructed variable z, computed using the fitted values $\hat{p}_i$ under this model, is added, the deviance is only reduced to 5.51 on 27 d.f. In view of this very small change in the deviance, there is no evidence that the logistic link function is unsatisfactory. Similarly, when a quadratic model is fitted to the full set of 32 observations, the deviance on introducing the appropriate constructed variable is reduced by 1.037, also a non-significant amount. The conclusion from this analysis is that for either the quadratic model, or the linear model fitted to the 30 observations omitting numbers 15 and 23, a logistic link function is appropriate.

5.7.3 Identification of outliers

For binary data, an outlier occurs when $y = 1$ and the corresponding fitted probability is near zero, or when $y = 0$ and the fitted probability is near unity. Since fitted probabilities near to zero or one will occur when the linear predictor has a large negative or a large positive value, outliers can only occur at observations that have extreme values of the explanatory variables. These are the observations for which the leverage, h_i, is large. This is in contrast to the situation for binomial data, where it is possible to encounter outliers when the values of the explanatory variables are not extreme.

Outliers can be identified as observations that have relatively large standardized deviance or likelihood residuals from either an index plot or from a half-normal plot of the residuals. The formal test of significance described in section 5.4 cannot be used, since the residuals are not even approximately normally distributed, but large residuals that occur outside the simulated envelope in a half-normal plot correspond to observations that would be adjudged discordant.

Another method for detecting outliers in binary data, proposed by Davison (1988), is to calculate the probability of the ith binary observation, given the set of $(n-1)$ remaining observations, for $i = 1, 2, ..., n$. This probability is called the **cross-validation probability** for the ith observation and is written as $P(y_i | \mathbf{y}_{(i)})$, where $\mathbf{y}_{(i)}$ represents the set of data without the ith observation. It turns out that this probability can be approximated by

$$e^{-r_{Li}^2/2} \sqrt{(1 - h_i)}$$

where the r_{Li} are the likelihood residuals and the h_i are the values of the leverage for the fitted model, and so the approximate values of $P(y_i | \mathbf{y}_{(i)})$ can easily be obtained. An index plot of these approximate probabilities will indicate those observations which have low probabilities, given the remaining observations.

For binary data, outliers generally arise as a result of **transcription errors**, that is, y is recorded as unity when it should be zero, or *vice versa*. For example, in connection with Example 1.9 on the presence of nodal involvement in prostatic cancer patients, an error in diagnosing whether or not a patient had nodal involvement would result in a transcription error.

Misclassification of the binary response can be modelled by letting γ (<0.5) be the probability of a transcription error. In other words, γ is the probability that the observation $y = 0$ is recorded when the true response probability is $p = 1$; it is also the probability that the observation $y = 1$ is recorded when $p = 0$. The probability of a success ($y_i = 1$) for the ith individual is then taken to be

$$p_i^* = (1 - \gamma)p_i + \gamma(1 - p_i) \tag{5.17}$$

rather than p_i, although it is the logistic transform of p_i that is modelled as a linear function of explanatory variables, that is $\text{logit}(p_i) = \eta_i = \Sigma \beta_j x_{ji}$. Note that as $\eta_i \to \infty$, $p_i^* \to 1 - \gamma$ and as $\eta_i \to -\infty$, $p_i^* \to \gamma$. Under this model, an observation of zero, when the true success probability is close to unity, is modelled as a transcription error, with probability γ, rather than being adjudged discordant.

This model can be fitted using the method of maximum likelihood. The likelihood function for the n binary observations is

$$\prod_i (p_i^*)^{y_i} (1 - p_i^*)^{1 - y_i}$$

where p_i^* is given by equation (5.17), and this is maximized over γ and the unknown parameters in the linear systematic component of the model. A non-linear optimization routine can be used for this. Alternatively, since the model can be written in the form

$$\log \left\{ \frac{p_i^* - \gamma}{1 - p_i^* - \gamma} \right\} = \eta_i \tag{5.18}$$

the link function is not the logistic link, but one that includes the unknown parameter γ. This model can be fitted directly using computer packages that allow the user to specify their own choice of link function, such as GLIM. The procedure is to fit the model with the link function given by equation (5.18), for a range of values of γ, and to then adopt that model for which the deviance is minimized. This procedure may be embedded in an algorithm that searches for the value of γ that minimizes the deviance. Further details are given in Section 9.3.9. However, it can be difficult to estimate γ unless there is a relatively large number of observations for which the linear predictor, $\hat{\eta}$, has large positive or large negative values. For this reason, generalizations of the model to allow the two types of transcription error ($1 \rightarrow 0$ and $0 \rightarrow 1$) to have unequal probabilities, or to depend on explanatory variables, will not be considered.

Example 5.24 Determination of the ESR
After fitting the model $\text{logit}(p_i) = \beta_0 + \beta_1 f_i$ to the data originally given in Example 1.7, the index plot of the standardized deviance residuals given in Fig. 5.16 shows that there are two outliers, corresponding to observations 15 and 23. An index plot of the approximate cross-validation probabilities is given in Fig. 5.22.

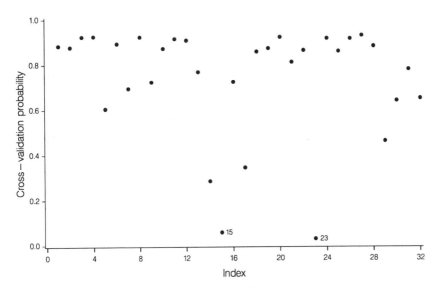

Fig. 5.22 Index plot of the approximate cross-validation probabilities on fitting a linear logistic regression model to the data from Example 1.7.

The two observations that were identified as outliers stand out as having low probabilities, given the remaining data. In fact, observations 15 and 23 have probabilities of 0.057 and 0.029, respectively, conditional on the remaining observations. When the quadratic logistic regression model is fitted, the probability of observation number 15, given the remaining ones is 0.020, while that for observation number 23 increases to 0.264. This shows that including the quadratic term improves the fit of the model to observation number 23, but number 15 remains discordant.

On fitting the model described by equation (5.18) that allows outliers to be modelled as transcription errors, the fitted response probability for the ith individual is such that logit $(\hat{p}_i) = -65.763 + 19.198 f_i$, and the estimated probability of a transcription error is $\hat{\gamma} = 0.0754$. With this probability of a transcription error, about two such errors would be expected in the data set consisting of 32 observations. This agrees with the observation that there are two individuals for which the observed response probability is one when the corresponding fitted values are near zero. The deviance for this model is 19.81 on 30 d.f., which is smaller than that for the original model of 24.84. This shows that allowing for the chance of a transcription error has improved the fit of the model. However, since these two models are not nested, a formal test of the significance of this reduction cannot be made.

5.7.4 Identification of influential values

For binary data, as for binomial data, the leverage measures the extent to which the ith observation is separated from the others in terms of the values of the explanatory variables. An index plot of the values h_i will therefore provide a useful means of identifying points with a high leverage. For the reason given in Section 5.7.1, a plot of standardized residuals against h_i will not be easy to interpret. All of the other statistics described in Sections 5.5.1–5.5.7 can be used directly for binary data. In particular, the square of the ith likelihood residual measures the influence of the ith observation on the overall fit of the model, the D-statistic described by equation (5.13) can be used for assessing the influence of the ith observation on the set of parameter estimates, and the influence of the ith observation on the jth parameter estimate can be detected using the statistic $\Delta_i \hat{\beta}_j$ described by equation (5.14). The statistic L_i described by equation (5.15) can be used to assess whether the change in deviance on including a term in, or excluding a term from, a linear logistic model is unduly influenced by the ith observation, and $\Delta_i D_j$ in equation (5.16) can be used to examine whether the ith and jth observations are jointly influential.

Example 5.25 Determination of the ESR
For the data pertaining to the use of the ESR as a disease indicator, the effect of each observation on the overall set of parameter estimates, for the logistic regression model that includes f alone, is measured by the values of the D-statistic given in Table 5.15. The values of the leverage have already been given in Table

Table 5.15 Values of influence diagnostics for logistic regression models fitted to the data for Example 1.7

Unit	D_i	L_i	$\Delta_i\hat{\beta}_0$	$\Delta_i\hat{\beta}_1$	$\Delta_i\hat{\beta}_2$
1	0.0023	0.174	0.022	−0.023	0.023
2	0.0025	0.194	0.020	−0.020	0.020
3	0.0014	−0.356	0.038	0.031	−0.026
4	0.0013	−0.415	0.058	0.050	−0.045
5	0.0284	−0.473	0.172	0.185	−0.202
6	0.0021	0.141	0.026	−0.027	0.027
7	0.0127	0.502	0.064	−0.063	0.061
8	0.0014	−0.258	0.008	0.003	0.001
9	0.0097	0.506	0.049	−0.048	0.048
10	0.0026	0.214	0.018	−0.019	0.019
11	0.0016	−0.030	0.034	−0.036	0.038
12	0.0018	0.054	0.035	−0.037	0.037
13	0.0239	0.208	0.000	0.000	0.000
14	0.0923	−1.091	0.124	0.109	−0.085
15	0.2762	−2.136	0.849	0.876	−0.890
16	0.0097	0.506	0.049	−0.048	0.048
17	0.1070	1.435	0.367	−0.376	0.387
18	0.0030	0.253	0.016	−0.017	0.017
19	0.0026	0.214	0.018	−0.019	0.019
20	0.0015	−0.181	0.011	−0.016	0.018
21	0.0044	0.360	0.018	−0.019	0.018
22	0.0029	0.238	0.017	−0.017	0.017
23	0.4776	4.861	0.725	−0.696	0.675
24	0.0016	−0.050	0.032	−0.035	0.037
25	0.0029	0.248	0.017	−0.017	0.017
26	0.0016	−0.030	0.034	−0.036	0.038
27	0.0013	−0.652	0.151	0.139	−0.131
28	0.0024	0.184	0.021	−0.021	0.022
29	0.1349	1.288	0.000	−0.000	0.001
30	0.0209	0.227	0.038	−0.033	0.026
31	0.0058	0.426	0.025	−0.025	0.024
32	0.0192	0.320	0.056	−0.053	0.047

5.14. Also given in Table 5.15 are the values of the L-statistic to determine the effect of each observation on the change in deviance on including the quadratic term in the model, and $\Delta_i\hat{\beta}_0$, $\Delta_i\hat{\beta}_1$ and $\Delta_i\hat{\beta}_2$, which measure the impact of each observation on the estimates of the three parameters, β_0, β_1, and β_2, in the quadratic regression model, logit $(p_i) = \beta_0 + \beta_1 f_i + \beta_2 f_i^2$. The values of the leverage, h_i, given in Table 5.14, show that the explanatory variable for individuals 13 and 29 has values that are most separated from those of the other individuals. In particular, both of these individuals have unusually large plasma fibrinogen levels when compared to the fibrinogen levels of the other individuals in the sample. From the tabulated values of the D-statistic, the observations from these two individuals have little effect on the parameter estimates in the fitted logistic regression model, and so they are not in fact influential. On the other hand, large values of the D-statistic are found for individuals 15 and 23, which have already been identified as outliers. Since these two observations do have an undue impact on the parameter estimates in the fitted model, they are influential outliers.

From the values of the L-statistic in Table 5.15, the observation from the twenty-third individual is clearly important in determining whether a quadratic term is needed in the model. The fifteenth observation also has some bearing on this, but its influence is not as great as that of the twenty-third individual. The tabulated values of $\Delta_i\hat{\beta}_0$, $\Delta_i\hat{\beta}_1$, and $\Delta_i\hat{\beta}_2$, which measure the influence of each of the thirty-two observations on the three parameter estimates in the fitted quadratic logistic regression model, also show clearly that observations 15 and 23 have a much greater influence than any of the others on these parameter estimates. Incidentally, the reason why the values of $\Delta_{13}\hat{\beta}_0$, $\Delta_{13}\hat{\beta}_1$ and $\Delta_{13}\hat{\beta}_2$ are zero is that the model is an excellent fit to the thirteenth observation and $y_{13} - \hat{p}_{13}$ used in the numerator of the $\Delta\hat{\beta}$-statistic is very nearly equal to zero. These results support those found from the use of a constructed variable plot in Example 5.22.

To investigate the extent to which observations 15 and 23 affect the fit of the logistic regression model that includes f alone, at the remaining observations, the ΔD-statistic is used. The values of $\Delta_i D_{15}$ and $\Delta_i D_{23}$ are given in Table 5.16. The values of $\Delta_i D_{15}$ and $\Delta_i D_{23}$ show that there is a close connection between observations 15 and 23, in that when either of them is deleted, the fit of the model at the other point deteriorates. These results confirm that both observations 15 and 23 should be considered together, and that as a pair, they have a considerable influence on the adequacy of the fitted model.

The next step in the analysis of these data would be to check the value of the response variable and the fibrinogen levels for the fifteenth and twenty-third subjects, and if either of these values is found to be incorrectly recorded, it should be corrected and the data re-analysed. If they are found to be incorrect, and correction is not possible, the data should be re-analysed without them. On the other hand, if there is no physiological reason why these two observations should be discarded, conclusions would be based on a model that best fits all thirty-two observations. In this example, the primary aim was to examine whether both fibrinogen and γ-globulin should be included in the model. Since observations 15 and 23 have no bearing on this particular issue, it does not really matter whether they are retained or deleted. However, they certainly affect the form of the relationship between the probability of an ESR reading greater than twenty and the plasma fibrinogen level. When these observations are included, a quadratic logistic regression model fits the data better than a linear one, but additional data would need to be collected in order to determine if the quadratic model was really suitable, or whether a more complicated model is needed.

Table 5.16 Influence diagnostics to identify the effect of observations 15 and 23 on the fit of the model at the remaining points

Unit	$\Delta_i D_{15}$	$\Delta_i D_{23}$
1	-0.075	-0.088
2	-0.078	-0.091
3	-0.050	-0.057
4	-0.050	-0.056
5	-0.040	0.022
6	-0.070	-0.082
7	-0.080	-0.060
8	-0.052	-0.058
9	-0.088	-0.078
10	-0.081	-0.094
11	-0.058	-0.066
12	-0.062	-0.072
13	-0.083	-0.088
14	0.129	0.030
15	1.088	1.417
16	-0.088	-0.078
17	-0.002	-0.141
18	-0.086	-0.100
19	-0.081	-0.094
20	-0.053	-0.060
21	-0.096	-0.106
22	-0.084	-0.098
23	1.399	1.880
24	-0.057	-0.065
25	-0.086	-0.100
26	-0.057	-0.066
27	-0.048	-0.053
28	-0.076	-0.089
29	-0.159	-0.309
30	-0.058	-0.014
31	-0.096	-0.102
32	-0.063	-0.023

5.7.5 Influence of the ith observation on the classification of future individuals

When the aim of fitting a linear logistic model to the observed values of a binary response variable is to use the fitted model as a basis for classifying future individuals, as discussed in Section 3.14, it will be important to examine the effect that the observations in the original data set have on this classification. Since a future individual is classified according to the estimated response probability for that individual, it is the influence of each observation in the original data set on this estimated probability that needs to be studied.

Let $\hat{p}_0$ be the success probability for a future individual, estimated from the model that was fitted to the original data set, and let $\hat{p}_{0(i)}$ be the corresponding estimated success probability when the ith observation is omitted. An estimate of the log odds that the true response for the new individual is a success is $\log\{\hat{p}_0/(1 - \hat{p}_0)\}$, that is $\text{logit}(\hat{p}_0)$, and so the difference $\text{logit}(\hat{p}_0) - \text{logit}(\hat{p}_{0(i)})$ measures the effect of the ith observation on the odds of a success for that future individual. For computational convenience, the difference $\text{logit}(\hat{p}_0) - \text{logit}(\hat{p}_{0(i)})$ can be approximated by the statistic C_i, given by

$$C_i = \frac{\mathbf{x}_0'(X'WX)^{-1}\mathbf{x}_i(y_i - \hat{p}_i)}{(1 - h_i)} \tag{5.19}$$

where $\mathbf{x}_0$ is the vector of explanatory variables for the new individual, $(X'WX)^{-1}$ is the variance–covariance matrix of the parameter estimates in the original data set, $\mathbf{x}_i$ is the vector of explanatory variables and h_i is the leverage for the ith observation in that data set. An index plot of the values of C_i will show which observations are likely to have the greatest impact on the classification of the new individual. Values near zero will be obtained for those observations in the original data set that have little or no effect on $\text{logit}(\hat{p}_0)$. On the other hand, large positive values indicate that the future individual is less likely to be classified as a success when the ith observation is excluded from the data set, and large negative values indicate that the future individual is more likely to be so classified. The vector $(X'WX)^{-1}\mathbf{x}_i$ in the numerator of this statistic can be computed in the same way as the corresponding term in the numerator of the $\Delta\hat{\beta}$-statistic referred to in section 5.5.3. A GLIM macro that can be used to obtain the values of the C-statistic is given in Appendix C.6.

Example 5.26 Prostatic cancer
In Example 3.9, the data on the extent of nodal involvement in prostatic cancer patients, given in Example 1.9, were used to illustrate how the probability of nodal involvement could be predicted for a future individual. We can now examine the effect of each observation on this prediction by computing the values of C_i given by equation (5.19) for a new individual with X-ray = 1, size = 1, grade = 1, and acid = 0.7. The fitted probability of nodal involvement for such an individual was given as 0.95 in Example 3.9. An index plot of the values of C_i for this individual is given in Fig. 5.23. From this figure, we see that individual number 37 in the original data set has the greatest effect on the logistic transform of the fitted probability of nodal involvement for the new individual. The value of C_{37} is -1.528, and since this is negative we can deduce that the exclusion of individual 37 from the original data set increases the estimated probability of nodal involvement. On omitting the data from individual 37, and fitting the same model to the

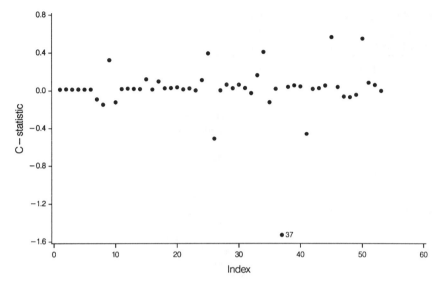

Fig. 5.23 Index plot of the values of the C-statistic to assess the influence of each observation on the classification of a future individual.

remaining 52 observations, the estimated probability of nodal involvement for the new individual is now 0.99. The increase of 0.04 in this estimated probability is unlikely to be of practical importance. The exact value of $\text{logit}(0.95) - \text{logit}(0.99)$ is in fact -1.982, and so the C-statistic which approximates this quantity underestimates the true difference in the two logits. In general, although the C-statistic may be used to highlight those observations in the initial data set which have greatest influence on the estimated response probability for a future individual, it is advisable to estimate this probability from the reduced data set to get an accurate reflection of the extent to which the estimated probability is changed.

5.8 SUMMARY AND RECOMMENDATIONS

This chapter has introduced a considerable number of numerical and graphical procedures that are designed to help in the assessment of whether a fitted model is adequate. For ease of reference, these diagnostics are summarized in Table 5.17 according to their use in detecting different types of inadequacy in the fitted model.

Not all of the diagnostics given in Table 5.17 can be used for both grouped and ungrouped binary data. Diagnostics which are not informative when applied to binary data are labelled with an obelisk (†), while those which are only appropriate for ungrouped binary data are labelled with an asterisk (∗). Those plots and displays that involve residuals of an unspecified type can be based on either standardized deviance residuals defined by equation (5.3), or likelihood residuals defined by equation (5.4). When a number of diagnostics can all be used for the

Table 5.17 Summary of model checking diagnostics for grouped and ungrouped binary data. Diagnostics marked † are uninformative when used with binary data, while those marked * are appropriate only for ungrouped binary data

Type of inadequacy	Diagnostic	Section
Incorrect linear predictor ($\hat{\eta}$)	Display of residuals	5.2
	Index plot of residuals	5.2.1
	Plot of residuals against $\hat{\eta}$	5.2.1†
	Half-normal plot of residuals	5.2.2
Inclusion of x-variable	Plot of residuals against x	5.2.1†
	Added variable plot	5.2.3
Transformation of x	Plot of residuals against x	5.2.1†
	Partial residual plot	5.2.4
	Constructed variable plot	5.2.5
Adequacy of link function	Reduction in deviance on adding a constructed variable	5.3
Overdispersion	Residual mean deviance	5.6†
Outliers	Display of residuals	5.2
	Index plot of residuals	5.2.1
	Plot of residuals against $\hat{\eta}$	5.2.1†
	Half-normal plot of residuals	5.2.2
	Discordancy test	5.4†
	Index plot of cross-validation probabilities	5.7.3*
Transcription error	Modelling transcription error	5.7.3*
Extreme values of x-variables	**Index plot of leverage**	5.5
	Plot of leverage against residuals	5.5†
Influence of an observation on:		
overall fit of model	Index plot of likelihood residuals	5.5.1
set of parameter estimates	Index plot of D-statistic	5.5.2
single parameter estimate	Index plot of $\Delta\hat{\beta}$-statistic	5.5.3
comparison of deviances	**Index plot of L-statistic**	5.5.4
	Influence on slope of added variable plot	5.5.4
goodness of link test	L-statistic for constructed variable	5.5.5
transformation of x	Influence on slope of constructed variable plot	5.5.6
classification of new unit	Index plot of C-statistic	5.7.5*
Joint influence of two observations	Index plot of ΔD-statistic	5.5.7

same purpose, those which are likely to prove the most useful are given in boldface type. The final column of Table 5.17 gives a reference to the section of this chapter in which the diagnostic procedure is first described.

While each of the methods catalogued in Table 5.17 is likely to be useful in particular situations, some are more frequently useful than others. It is therefore convenient to distinguish those diagnostics which are likely to be useful in the vast majority of cases from the more specialized techniques which are useful only in specific circumstances. In the analysis of both grouped and ungrouped binary data, plots that should be used routinely in examining model adequacy are index plots of the residuals, leverage and the D-statistic. These could be supplemented by a half-normal plot of the residuals, with accompanying simulated envelopes. The simulated envelopes are particularly important in the interpretation of half-normal plots for ungrouped binary data. When interest centres on the values of particular parameter estimates, as in the analysis of data from epidemiological studies to be discussed in Chapter 7, an index plot of the appropriate $\Delta\hat{\beta}$-statistics will be useful. In addition, for grouped binary data, the residual mean deviance provides information about the adequacy of the fitted model, while possible transcription errors in binary data can be detected from an index plot of the cross-validation probabilities.

In some situations, it may be more important to be able to detect some types of inadequacy than others, and so these general recommendations should not be adopted uncritically. Moreover, the results of some diagnostic procedures might suggest further analyses which will throw light on the extent to which a given model provides an appropriate description of the observed data.

5.9 A FURTHER EXAMPLE ON THE USE OF DIAGNOSTICS

As a further illustration of the use of diagnostics to assess model adequacy, consider once more the data relating to the occurrence of nodal involvement in patients presenting with prostatic cancer, first given as Example 1.9. In section 3.13, a linear logistic model with explanatory variables X-ray, size, grade, log (acid), size × grade and grade × log (acid) seemed to provide a satisfactory description of the observed binary data. In this section, we examine whether this model is indeed appropriate.

An index plot of the standardized deviance residuals, and a half-normal plot of these residuals with an accompanying simulated envelope, are given in Figs. 5.24

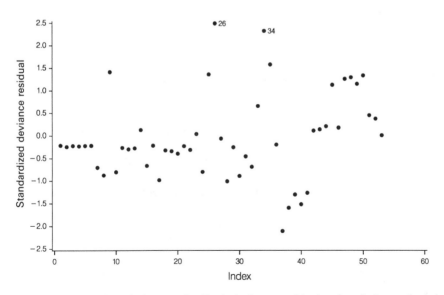

Fig. 5.24 Index plot of the standardized deviance residuals after fitting a logistic regression model to the data from Example 1.9.

and 5.25. Observations 26 and 34 have the largest residuals, and correspond to individuals who have nodal involvement, although the corresponding model-based estimates of the probability of nodal involvement are 0.05 and 0.28, respectively. These two observations do not lie outside the simulated envelope for the half-normal plot, which means that although they are unusual observations, they are not sufficiently extreme to be adjudged discordant. The other feature shown in the half-normal plot is that some of the plotted points are quite close to the upper envelope, and some actually fall outside it. This suggests that the fitted model is not satisfactory.

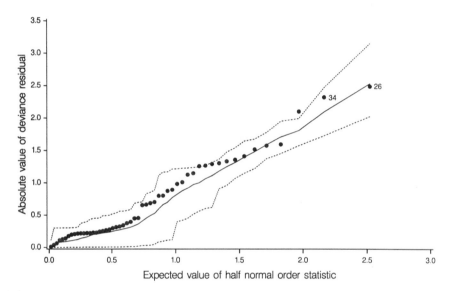

Fig. 5.25 Half-normal plot of the standardized deviance residuals, together with simulated envelopes (······) and means of the simulated values (———).

An index plot of the values of the leverage given in Fig. 5.26 shows that the set of explanatory variables for individual number 26 are not at all extreme, relative to those of the other individuals in the sample, but that they are for individual number 34 and also for individual number 9. To help identify whether the observations from individuals 9, 26 and 34 are influential, an index plot of the values of the D-statistic is given in Fig. 5.27. Observation number 34 clearly has a great influence on the complete set of parameter estimates, and has much more of an impact on them than either of observations 9 and 26. No other observations appear to be influential in this respect.

To examine the influence of each observation on the seven individual parameter estimates, the values of the $\Delta \hat{\beta}$-statistic, given by equation (5.14), are computed and an index plot is constructed from them. These plots (which are not shown here) indicate that the thirty-fourth observation has a particularly strong influence on

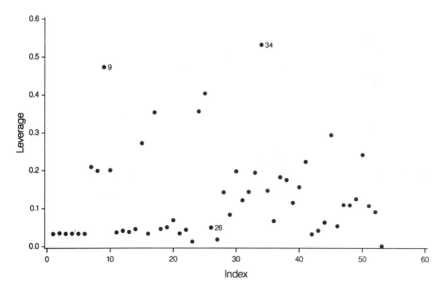

Fig. 5.26 Index plot of the values of the leverage.

the parameter estimate for the product grade × log (acid), and a lesser impact on the estimates for size × grade and grade. In addition, they show that individual number 9 has a relatively large influence on the estimate corresponding to the size × grade interaction and individual number 26 influences size.

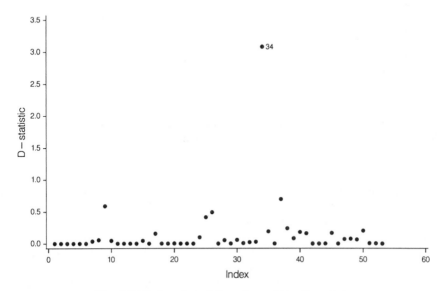

Fig. 5.27 Index plot of the values of the D-statistic.

The next step is to see whether any of these influential observations are affecting the results of the analysis of deviance, described in section 3.13, to such an extent that terms are included in the model simply because they improve the fit of the model to one particular observation. From Section 3.13, when grade × log (acid) is added to the model that already includes X-ray, size, grade, log (acid) and size × grade, the deviance is reduced by 4.17 on 1 d.f., and since this reduction is significant at the 5% level, at that stage it was decided to include it in the model. Using the L-statistic described by equation (5.15), we can obtain an approximation to the amount by which this reduction in deviance would be changed when the ith observation is omitted from the data set, for $i = 1, 2, ..., 53$. An index plot of the L-statistic for assessing the influence of each observation on the change in deviance due to including grade × log (acid) in the model is presented in Fig. 5.28. This

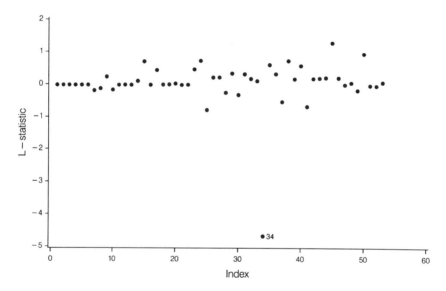

Fig. 5.28 Index plot of the values of the L-statistic for assessing the impact of each observation on the reduction in deviance on including grade × log (acid) in the model.

figure suggests that the thirty-fourth observation is solely responsible for the significance of the term grade × log (acid). The value of L_{34} is -4.68 and so the change in deviance on including grade × log (acid) in the model fitted to the 52 observations, excluding number 34, is estimated to be near zero. To check this, observation number 34 is omitted from the data set and the model with X-ray, size, grade, log (acid) and size × grade is fitted, giving a deviance of 39.73 on 46 d.f. The parameter estimates and their standard errors under this model, obtained using GLIM, are given below.

	estimate	s.e.	parameter
1	−2.315	1.008	1
2	2.311	0.9789	XRAY
3	3.392	1.215	SIZE
4	3.933	1.533	GRAD
5	2.938	1.290	LACD
6	−4.668	1.855	SXG

Notice that none of these parameter estimates are small relative to their standard error, and so all five terms are needed in the model. When grade × log(acid) is added to the model, the deviance converges to 23.85 on 45 d.f., after 16 cycles of the iterative process, a rather unexpected result. The parameter estimates under this model, reproduced from GLIM output are as follows.

	estimate	s.e.	parameter
1	−2.606	1.121	1
2	2.448	1.326	XRAY
3	3.185	1.235	SIZE
4	804.6	9010.	GRAD
5	1.725	1.435	LACD
6	−321.5	8455.	SXG
7	1184.	4928.	GXL

Three of these estimates have extremely large standard errors. In particular, the standard error for grade × log(acid) is more than four times the value of the parameter estimate, indicating that this term is not needed in the model. The large standard errors in the GLIM output are due to the fact that the fitted model contains too many parameters. As discussed in Section 3.12, this overfitting means that the difference between two deviances will not be well approximated by a χ^2-distribution, and cannot be used to guide model selection. So in spite of the results of the analysis of deviance, the term grade × log(acid) is omitted from the model fitted to all 53 observations on the grounds that the need for this term is solely determined by just one observation.

Having amended the model by excluding the term grade × log(acid), we now consider the impact of each of the 53 observations on the other terms in the model. The L-statistic is first used to study the influence of each observation on the size × grade interaction. An index plot of the values of L_i for the change in deviance on including this interaction term in the model after fitting X-ray, size, grade and log(acid) is given in Fig. 5.29. Two observations stand out as influencing the reduction in deviance on including size × grade in the model. Omission of the ninth observation, for which $L_9 = 2.85$, would cause the deviance due to this term to be even greater, while omission of the twenty-sixth observation would result in the change in deviance being reduced by approximately 2.77. With this reduction, the change in deviance would remain significant, and so the size × grade interaction is retained. Since an interaction term is generally only included in a model

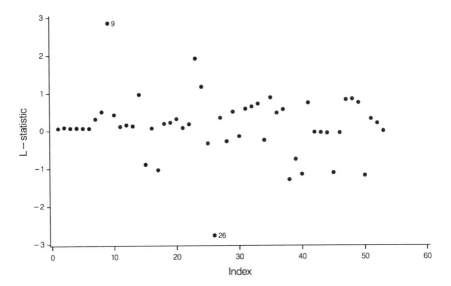

Fig. 5.29 Index plot of the values of the L-statistic for assessing the impact of each observation on the reduction in deviance on including size × grade in the model.

when the corresponding main effects are present, the only other terms that might be excluded are X-ray and log (acid). The increase in deviance on omitting X-ray from the model is 8.26 on 1 d.f., and that obtained when log (acid) is excluded is 5.33 on 1 d.f. Index plots of the L-statistics show that there is no single observation that is particularly responsible for these significant changes in deviance.

After this analysis, we conclude that the model need only include the variables X-ray, size, grade, log (acid) and size × grade. A goodness of link test indicates that there is no reason to doubt that a logistic link function is appropriate, and, on fitting this model, the estimated response probability for the ith observation, $i = 1, 2, \ldots, 53$, is such that

$$\text{logit} (\hat{p}_i) = -2.417 + 2.501 \text{ X-ray} + 3.432 \text{ size} + 3.970 \text{ grade} + 2.842 \log (\text{acid})$$
$$- 4.600 \text{ (size × grade)}$$

Under this model, observation number 34 does not stand out as an outlier, showing that this observation is well-fitted by this model, even though the grade × log (acid) term has been excluded. However, observation number 26 has both a large standardized deviance residual and an unusually large value of the D-statistic. The twenty-sixth observation is therefore declared to be an outlier that has a substantial impact on the parameter estimates. This observation is from an individual who has a negative X-ray, a small tumour of a less serious grade and an acid level that is not very high, but who nevertheless has been diagnosed as having nodal involvement. Although it has already been found that this observation does not influence the structure of the model, its effect on the fitted probabilities can be assessed from an index plot of the values of $\hat{p}_i - \hat{p}_{i(26)}$, where $\hat{p}_i$ is the fitted probability of nodal involvement for the ith individual, and $\hat{p}_{i(26)}$ is the fitted probability when the twenty-sixth observation is omitted from the data set. Such a plot is shown in Fig. 5.30. The fitted probabilities are most different for observation

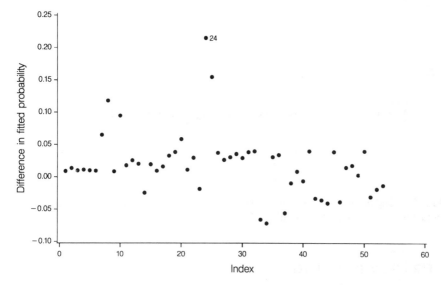

Fig. 5.30 Index plot of the values of $\hat{p}_i - \hat{p}_{i(26)}$ for assessing the effect of deleting observation number 26 on the fitted probabilities.

number 24, for whom the estimated probability of nodal involvement is reduced from 0.35 to 0.13. About a quarter of the 53 fitted probabilities are increased as a result of omitting observation number 26, while the others are all decreased.

Finally, the ΔD-statistic given by equation (5.16) can be used to identify whether the fit of the model at any other point is affected by the twenty-sixth individual. An index plot of $\Delta_i D_{26}$, given in Fig 5.31, shows that there are no observations that are jointly influential with number 26.

If there is no evidence that any of the variables measured for the twenty-sixth individual have been incorrectly recorded, and if there is no medical reason to exclude this particular individual from the data set, it should be retained. The probability of nodal involvement for future patients is then estimated from the model that contains the explanatory variables X-ray, size, grade, log (acid) and size × grade fitted to the complete set of 53 observations. On the other hand, if the data recorded for this individual is judged to be so unusual that it should be omitted, the same model fitted to all 52 observations is used as a basis for prediction. Of course, if the model fitted to all 53 observations is used, the effect of each of them on the estimated response probability for a future individual can be examined using the C-statistic described in Section 5.7.5.

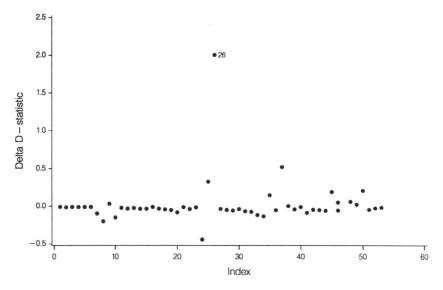

Fig. 5.31 Index plot of the ΔD-statistic for detecting observations that are jointly influential with observation number 26.

FURTHER READING

Methods for examining the adequacy of linear regression models fitted to continuous response data are included in a number of standard textbooks on this subject, including Draper and Smith (1981) and Montgomery and Peck (1982). Atkinson (1985) describes diagnostic methods for linear regression models from a practical viewpoint, while Cook and Weisberg (1982) provide a more advanced treatment of the analysis of residuals and measures of influence. In a review article, Chatterjee and Hadi (1986) summarize methods for identifying outliers and influential observations in linear regression analysis, and discuss the problem of detecting more than one outlier or influential value in some detail.

A number of diagnostics for linear logistic modelling, based on statistics that are analogous to those used for the general linear model, were given by Pregibon (1981), who also introduced deviance residuals. Likelihood residuals were defined by Williams (1987). Anscombe residuals were proposed by Anscombe (1953), and the form of this type of residual for binomial data was given by Cox and Snell (1968). The properties of residuals in generalised linear models, and the linear logistic model for binary data in particular, are discussed by Pierce and Schafer (1986), Jennings (1986) and Williams (1984, 1987). Barnett (1975) discusses the choice of plotting positions in normal probability plots and the use of simulation envelopes in half-normal probability plots of residuals was proposed by Atkinson (1981, 1982).

Wang (1985), Landwehr, Pregibon and Shoemaker (1984) and Wang (1987), respectively, show how added variable, partial residual and constructed variable plots can be used in logistic regression modelling to assess the goodness of fit of the linear part of the model. A method that is designed to show how transforming the explanatory variables in a model for binary data can improve its fit is given by Kay and Little (1987). Their approach is to use information about the distributions of the values of an explanatory variable, at a response of zero and unity, to determine the required transformation.

The goodness of link test described in Section 5.3 was first given by Pregibon (1980). Alternative general families of link functions to that given by equation (5.8) have been suggested by Prentice (1976), Aranda-Ordaz (1981), Guerrero and Johnson (1982) and Stukel (1988).

A comprehensive review of methods for identifying and handling outliers in a variety of situations is the book by Barnett and Lewis (1985). Copas (1988) includes an account of the nature of outliers in binary data, and describes models for misclassification.

Although techniques for identifying influential values in linear logistic modelling were first given by Pregibon (1981), other important papers are those of Landwehr, Pregibon and Shoemaker (1984), Johnson (1985) and Williams (1987). Much of the material of Sections 5.1, 5.5 and 5.7 is based on these articles. All of these papers include illustrative examples, different from those used in this chapter. In addition, the case study presented by Kay and Little (1986) illustrates the application of a number of diagnostic methods to binary data arising from a study of the effectiveness of treatment for haemolytic uraemic syndrome in children. The plot of residuals against leverage introduced in Section 5.5 was suggested as a diagnostic aid by Williams (1982b). Lee (1988) gives an illustration of the use of the L-statistic defined by equation (5.15) for examining the influence of each observation on explanatory variables and constructed variables in a logistic regression model. Johnson (1985) introduces diagnostics for measuring the influence that each observation has on the classification of a set of future observations.

Methods for fitting logistic models that are relatively insensitive to outlying observations have been proposed and illustrated by Pregibon (1982) and Copas (1983). Known collectively as **resistant methods**, these techniques can also be used to highlight individual outlying observations. Their role in model checking is discussed by Copas (1988), Fowlkes (1987) and Azzalini, Bowman and Härdle (1989).

6

Overdispersion

When a linear logistic model is used in the analysis of binary data, or data in the form of proportions, the logistic transformation of the response probabilities is assumed to be linearly dependent on measured explanatory variables, and the observed numbers of successes are generally assumed to have a binomial distribution. If the fitted linear logistic model is to be satisfactory, the model must adequately fit the observed response probabilities, and, in addition, the random variation in the data must be modelled appropriately. When a linear logistic model fitted to n binomial proportions is satisfactory, the residual deviance has an approximate χ^2-distribution on $(n - p)$ degrees of freedom, where p is the number of unknown parameters in the fitted model. Since the expected value of a χ^2 random variable on $(n - p)$ degrees of freedom is $(n - p)$, it follows that the residual deviance for a well-fitting model should be approximately equal to its number of degrees of freedom, or, equivalently, the mean deviance should be close to one. If the fitted model does not adequately describe the observed proportions, the residual mean deviance is likely to be greater than one. Similarly, if the variation in the data is greater than that under binomial sampling, the residual mean deviance is also likely to be greater than one.

When the linear logistic model is thought to be correct, but the residual mean deviance nonetheless exceeds unity, the assumption of binomial variability may not be valid. The data are then said to exhibit **overdispersion**, a phenomenon that is also known as **extra binomial variation**, or simply **heterogeneity**. A much less common occurrence is to find that the deviance on fitting a linear logistic model to proportions is considerably less than its number of degrees of freedom. This is referred to as **underdispersion**. Because this phenomenon rarely arises in practice, the emphasis of this chapter is on overdispersion. Moreover, this chapter will only consider overdispersion in the context of linear logistic modelling, although most of the points that are made apply equally to models based on the probit and complementary log–log transformations of the response probability.

6.1 POTENTIAL CAUSES OF OVERDISPERSION

In modelling grouped binary data expressed as proportions, there are a number of different circumstances that can lead to a residual mean deviance that is substantially greater than one. The first explanation that must be considered is that the systematic component of the model is inadequate in some way. For example, in modelling proportions from a factorial experiment, a large residual deviance, relative to its number of degrees of freedom, may be the result of not including a sufficient number of interaction terms in the model. Similarly, in modelling dose–response data, a linear model for the relationship between the logistic transform of the success probability and dose might have been assumed when the actual relationship is quadratic, or when the logarithm of the dose should have been taken as the explanatory variable.

Example 6.1 Anti-pneumococcus serum
Consider the data on the proportion of mice which dies from pneumonia after being inoculated with different doses of a protective serum, first given as Example 1.6. From Example 4.1, the deviance on fitting a linear logistic model with dose as the explanatory variable is 15.90 on 3 d.f. However, a plot of the original data and the fitted model in Fig. 4.1 shows that this model does not fit the data. If the explanatory variable log (dose) is used in the model in place of dose, the residual deviance is 2.81 on 3 d.f., which suggests that this model is satisfactory. The apparent overdispersion in the linear logistic model with dose as the explanatory variable was due to a misspecification of the systematic component of the model.

Apparent overdispersion may also be caused by the presence of one of more outliers in the observed data set. Modification, or omission, of such observations may lead to the deviance being reduced to such an extent that the apparent overdispersion disappears. Then again, it may be that the logistic link function is inappropriate and that a model based on, say, the complementary log–log transformation would provide a better fit. Defects in the specification of the systematic component of the model, the occurrence of outliers and the adequacy of the chosen link function can be detected from an index plot of the standardized deviance or likelihood residuals, or from other diagnostics described in Chapter 5.

In Section 3.8.2, it was pointed out that when the data are sparse, that is, when there are proportions based on small numbers of individual experimental units, the chi-squared approximation to the distribution of the deviance breaks down. In this event, the size of the residual deviance cannot be compared with percentage points of the χ^2-distribution. A large residual mean deviance may not then require special attention. In the extreme case where the observations are ungrouped binary data, the deviance does not have a χ^2-distribution, and its magnitude depends solely on the values of the fitted probabilities, as shown in Section 3.8.1. This means that large values of the residual mean deviance for binary data cannot be taken to be an indication of overdispersion. Overdispersion may still occur in binary data, but it will not be possible to detect it from the value of the residual mean deviance.

It is vital that all of these possible explanations of apparent overdispersion be eliminated before concluding that the data are overdispersed. Once the possibilities that an uncomfortably large residual mean deviance is due to an incorrectly specified linear systematic component, an incorrect link function, the occurrence of outliers or sparseness of the data have been eliminated, the overdispersion can be explained by the occurrence of **variation between the response probabilities** or **correlation between the binary responses**. These two phenomena amount to the same thing, in that correlation between the binary responses leads to variation between the response probabilities, and *vice versa*. However, it is sometimes easier to be able to interpret overdispersion in terms of one or other of them, and for this reason, they will be discussed under separate headings.

6.1.1 Variation between the response probabilities

When a number of batches of experimental units have been observed under the same conditions, the response probabilities may nevertheless differ from batch to batch. To fix ideas, consider an investigation that is designed to assess the effect of exposure to a particular chemical on the formation of developmental abnormalities in a foetus. This type of study is known as a **teratogenic study**. Suppose that in such a study, female rabbits are exposed to a particular chemical during pregnancy, and are sacrificed just before the birth of the offspring. Each foetus is then examined to determine whether a particular malformation is present or absent. This is the binary response variable. Because of genetic influences, the probability that a foetus has a particular abnormality is unlikely to be the same for the litters of different rabbits, even when the rabbits have been exposed to the same experimental conditions. This variation in the response probabilities leads to the observed proportions of abnormal foetuses in a number of rabbits having a variance that is greater than would be the case if the response probabilities were constant.

Generally speaking, observed variation amongst the response probabilities is due to the absence of relevant explanatory variables in the model, or to the inclusion of explanatory variables that have not been adequately measured or controlled. Had the investigator been aware of additional explanatory variables before the data were collected, and had he or she been able to measure them, there would have been the potential for reducing the residual deviance to a value that is not indicative of unexplained overdispersion. Similarly, when the recorded explanatory variables have not been controlled sufficiently precisely or measured sufficiently accurately, the residual deviance will be larger than it would have been if the actual values, rather than the target values, of the explanatory variables had been used in the modelling process.

Example 6.2 Germination of Orobanche
Consider the data described in Example 1.3 on the proportion of seeds of two species of *Orobanche* that germinate in bean and cucumber root extracts. The deviances obtained on fitting a number of linear logistic models to the data are summarized in Table 6.1, where species$_j$ and extract$_k$ refer to the effects of the jth species, $j = 1, 2$, and the kth root extract, $k = 1, 2$.

The change in deviance on adding the interaction term into the model is 6.41 on 1 d.f., which, relative to percentage points of the χ^2-distribution on 1 d.f., is almost significant at the 1% level ($P = 0.011$). This indicates that the interaction term

Table 6.1 Deviances on fitting linear logistic models to the data on the germination of *Orobanche*

Terms fitted in model	Deviance	d.f.
β_0	98.72	20
β_0 + species$_j$	96.18	19
β_0 + extract$_k$	42.75	19
β_0 + species$_j$ + extract$_k$	39.69	18
β_0 + species$_j$ + extract$_k$ + (species × extract)$_{jk}$	33.28	17

should be retained in the model. Because of this significant interaction, we need not go on to look at the significance of the main effects of species and extract.

The residual deviance for the model that includes the main effects of species and extract, as well as their interaction, is 33.28 on 17 d.f., which is significantly large. It is not clear how the systematic component of the model can be amended, and an examination of the residuals reveals that no single observation can be regarded as an outlier. We may therefore conclude that the data are overdispersed. One explanation for this overdispersion is that the batches of seed of a particular species, germinated in a particular root extract, are not homogeneous, so that the germination probabilities differ between the batches. An alternative explanation for the apparent variation in the germination probabilities between the different batches of seed is that the different batches of seed were not germinated under similar experimental conditions. For example, the temperature, or relative humidity, of the chamber in which the different batches of seeds were germinated may not have been kept constant. In practice, the knowledge of the experimenter would be expected to throw light on which of these explanations was the more plausible.

6.1.2 Correlation between binary responses

Suppose that in modelling binary response data, the individual binary observations that make up the observed proportions are not independent. Since the observed number of successes can only be assumed to have a binomial distribution when the component binary observations are independent (see Section 2.1), non-independence will lead to a variance that is greater than, or less than, that on the assumption of binomial variability. A variance greater than the binomial variance would be indicative of overdispersion.

For illustrative purposes, consider again a teratogenic experiment. If one of the foetuses from a particular rabbit has a particular abnormality, that same abnormality is more likely to be found in other foetuses from that same rabbit. Similarly, the absence of the abnormality in the foetus of another rabbit makes it less likely that the remaining foetuses are malformed. As a result of this **positive correlation** between the responses of each foetus within a litter, the proportions of abnormal foetuses from each pregnant rabbit will again tend to be more variable than would have been the case if these binary responses had been independent. Because independent binary responses lead to binomial variability, the proportions of abnormal foetuses will be more variable than they would have been under the assumption that the numbers of malformed foetuses are binomially distributed. Consequently, the mean deviance after fitting a suitable model to the proportions can be expected to be greater than unity.

Example 6.3 Germination of Orobanche
In Example 6.2, we saw that there is overdispersion in the data on the germiation of *Orobanche*. Another possible explanation for the overdispersion is that when a seed in a particular batch germinates, a chemical is released which promotes germination in the remaining seeds of that batch. This means that the seeds in a given batch do not germinate independently of one another, and so there is a correlation between the binary responses associated with whether or not a particular seed germinates.

Although different explanations of the overdispersion have been given in Examples 6.2 and 6.3, there is really no difference between them. For example, if different amounts of a chemical that promotes germination are released by seeds in different batches, there would be variation in the proportions of seed germinating in the different batches. Similarly, variation in the response probabilities between batches that results from different experimental conditions in the germination chambers can be explained in terms of there being a positive correlation between the binary responses of the seeds within a batch. If it is possible to identify physiological, biochemical, or other reasons for one binary response being influenced by another, overdispersion might more easily be explained in terms of intercorrelation between the individual binary responses. On the other hand, if there is the suspicion that different sets of binary data have not been treated in a similar manner, overdispersion could be explained in terms of there being variability between the batches. Since there is no logical distinction between these two explanations of overdispersion, it is not surprising that they lead to the same statistical model. This is shown in the following two sections.

6.2 MODELLING VARIABILITY IN RESPONSE PROBABILITIES

When the response probabilities vary amongst groups of experimental units exposed to similar experimental conditions, some assumption has to be made about the form of this variation. In this section, we consider a general model, described by Williams (1982a).

Suppose that the data consist of n observed proportions, y_i/n_i, $i = 1, 2, \ldots, n$, and suppose that the corresponding response probability for the ith unit depends on k explanatory variables, $x_1, x_2, \ldots, x_k$, through a linear logistic model. To introduce variability in the response probabilities, the actual response probability for the ith observation, ϑ_i, say, will be assumed to vary about a mean of p_i. The actual response probability, ϑ_i, is therefore a random variable where $E(\vartheta_i) = p_i$. The variance of ϑ_i must be zero when p_i is either zero or unity, and the simplest function for which this is true is such that

$$\text{Var}(\vartheta_i) = \phi p_i(1 - p_i) \tag{6.1}$$

where $\phi(\geqslant 0)$ is an unknown scale parameter. The quantity ϑ_i is an unobservable random variable, or **latent variable.** However, given a particular value of ϑ_i, the observed number of successes for the ith unit, y_i, will have a binomial distribution with mean $n_i\vartheta_i$ and variance $n_i\vartheta_i(1 - \vartheta_i)$. The mean of y_i, **conditional** on ϑ_i, is therefore given by

$$E(y_i|\vartheta_i) = n_i\vartheta_i$$

and the **conditional variance of** y_i is

$$\text{Var}(y_i|\vartheta_i) = n_i\vartheta_i(1 - \vartheta_i)$$

where the notation '|' is read as 'conditional on'. Since ϑ_i cannot be estimated, the observed proportion y_i/n_i has to be taken to be an estimate of p_i, the expected value of ϑ_i, and so the unconditional mean and variance of y_i will be required. This leads us to investigate the effect of the assumption about the random variability in the response probabilities in equation (6.1) on $E(y_i)$ and $\text{Var}(y_i)$.

According to a standard result from conditional probability theory, the unconditional expected value of a random variable Y can be obtained from the conditional expectation of Y given X using

$$E(Y) = E\{E(Y|X)\}$$

and the unconditional variance of Y is given by

$$\text{Var}(Y) = E\{\text{Var}(Y|X)\} + \text{Var}\{E(Y|X)\}$$

Application of these two results gives

$$E(y_i) = E\{E(y_i|\vartheta_i)\} = E(n_i\vartheta_i) = n_iE(\vartheta_i) = n_ip_i$$

and

$$\text{Var}(y_i) = E(\text{Var}(y_i|\vartheta_i)) + \text{Var}\{E(y_i|\vartheta_i)\}$$

Now,

$$
\begin{aligned}
E\{\text{Var}(y_i|\vartheta_i)\} &= E\{n_i\vartheta_i(1 - \vartheta_i)\} \\
&= n_i\{E(\vartheta_i) - E(\vartheta_i^2)\} \\
&= n_i\{E(\vartheta_i) - \text{Var}(\vartheta_i) - [E(\vartheta_i)]^2\} \\
&= n_i\{p_i - \phi p_i(1 - p_i) - p_i^2\} \\
&= n_i p_i(1 - p_i)[1 - \phi]
\end{aligned}
$$

Also,

$$\text{Var}\{E(y_i|\vartheta_i)\} = \text{Var}(n_i\vartheta_i) = n_i^2 \text{Var}(\vartheta_i) = n_i^2 \phi p_i(1 - p_i)$$

and so

$$\text{Var}(y_i) = n_i p_i(1 - p_i)[1 + (n_i - 1)\phi] \tag{6.2}$$

In the absence of random variation in the response probabilities, y_i would have a binomial distribution, $B(n_i, p_i)$, and in this case, $\text{Var}(y_i) = n_i p_i(1 - p_i)$. This corresponds to the situation where $\phi = 0$ in equation (6.1) and leads to $\text{Var}(y_i) = n_i p_i(1 - p_i)$ in equation (6.2), as required. If on the other hand there is variation amongst the response probabilities, so that ϕ is greater than zero, the variance of y_i will exceed $n_i p_i(1 - p_i)$, the variance under binomial sampling, by a factor of $[1 + (n_i - 1)\phi]$. Thus variation amongst the response probabilities causes the variance of the observed number of successes to be greater than it would have been if the response probabilities did not vary at random, resulting in overdispersion.

In the special case of ungrouped binary data, $n_i = 1$, for all values of i, and the variance in equation (6.2) becomes $p_i(1 - p_i)$, which is exactly the variance of a binary response variable. Consequently, binary data can provide no information about the parameter ϕ, and so it is not then possible to use equation (6.1) to model variation amongst the response probabilities.

6.3 MODELLING CORRELATION BETWEEN BINARY RESPONSES

In Section 6.1.2, an intuitive explanation was given for the occurrence of a mean deviance in excess of unity when there is positive correlation between the binary responses. In this section, we see how a particular model for this correlation leads to this conclusion.

Suppose that the ith of n sets of binary data consists of y_i successes in n_i observations. Let $R_{i1}, R_{i2}, \ldots, R_{in_i}$, be the random variables associated with the n_i observations in this set, where $R_{ij} = 1$, for $j = 1, 2, \ldots, n_i$, corresponds to a success, and $R_{ij} = 0$ to a failure. Now suppose that the probability of a success is p_i, so that $P(R_{ij} = 1) = p_i$. Since R_{ij} is a Bernoulli random variable, $E(R_{ij}) = p_i$ and $\text{Var}(R_{ij}) = p_i(1 - p_i)$ from the results in section 2.1. The number of successes, y_i, is then the observed value of the random variable $\sum_{j=1}^{n_i} R_{ij}$, and so $E(y_i) = \sum_{j=1}^{n_i} E(R_{ij}) = n_i p_i$, and the variance of y_i is given by

$$\text{Var}(y_i) = \sum_{j=1}^{n_i} \text{Var}(R_{ij}) + \sum_{j=1}^{n_i} \sum_{k \neq j} \text{Cov}(R_{ij}, R_{ik})$$

where $\text{Cov}(R_{ij}, R_{ik})$ is the covariance between R_{ij} and R_{ik}, for $j \neq k$, and

$k = 1, 2, ..., n_i$. If the n_i random variables $R_{i1}, R_{i2}, ..., R_{in_i}$ were mutually independent, each of these covariance terms would be zero. However, we will suppose that the correlation between R_{ij} and R_{ik} is δ, so that from the definition of a correlation coefficient,

$$\delta = \frac{\text{Cov}(R_{ij}, R_{ik})}{\sqrt{\{\text{Var}(R_{ij})\text{Var}(R_{ik})\}}}$$

Since $\text{Var}(R_{ij}) = \text{Var}(R_{ik}) = p_i(1 - p_i)$, it follows that $\text{Cov}(R_{ij}, R_{ik}) = \delta p_i(1 - p_i)$, and so

$$
\begin{aligned}
\text{Var}(y_i) &= \sum_{j=1}^{n_i} p_i(1 - p_i) + \sum_{j=1}^{n_i} \sum_{k \neq j} \delta p_i(1 - p_i) \\
&= n_i p_i(1 - p_i) + n_i(n_i - 1)[\delta p_i(1 - p_i)] \\
&= n_i p_i(1 - p_i)[1 + (n_i - 1)\delta]
\end{aligned}
\tag{6.3}
$$

When there is no correlation between pairs of binary observations, δ is equal to zero and $\text{Var}(y_i) = n_i p_i(1 - p_i)$, the variance of y_i under binomial sampling. On the other hand, if the binary responses are positively correlated, $\delta > 0$, and $\text{Var}(y_i) > n_i p_i(1 - p_i)$. We may therefore conclude that positive correlation amongst the binary observations leads to greater variation in the numbers of successes than would be expected if they were independent.

On the face of it, equation (6.3) is the same as equation (6.2). However, there is one important difference. The model used in Section 6.2 can only be used to model overdispersion, since ϕ must be non-negative. In equation (6.3), δ is a correlation, and so it can in principle be negative, allowing underdispersion to be modelled. However, the requirement that $1 + (n_i - 1)\delta$ be positive means that $-(n_i - 1)^{-1} \leqslant \delta \leqslant 1$, and so unless n_i is small, the lower bound on δ will effectively be zero. This means that it will only be possible to use this model for underdispersion when the number of binary observations that make up each proportion is small. If overdispersion is being modelled, $\delta > 0$, and now equations (6.3) and (6.2) are identical. This confirms that the effects of correlation between the binary responses and variation between the response probabilities cannot be distinguished.

6.4 MODELLING OVERDISPERSED DATA

Suppose that evidence of overdispersion is found after fitting a linear logistic model to n observations of the form y_i/n_i, $i = 1, 2, ..., n$. In order to model this overdispersion, the variance of y_i will be taken to be $n_i p_i(1 - p_i)\sigma_i^2$, where, from equation (6.2), $\sigma_i^2 = 1 + (n_i - 1)\phi$. This function includes an unknown parameter, ϕ, which will have to be estimated before we can proceed with the modelling process.

Williams (1982a) shows how an estimate, $\hat{\phi}$, of the parameter ϕ can be found by equating the value of Pearson's X^2-statistic for the model to its approximate expected value. The value of X^2 for a given model depends on the value of $\hat{\phi}$, and so this procedure is iterative. The estimate of the parameter ϕ will also depend on the actual explanatory variables in the fitted model, but ϕ is best estimated for the model that contains the largest number of terms that one is prepared to consider, that is the full model. The actual algorithm used for estimating ϕ is quite complicated and so computational details will be postponed until section 9.3.11. A GLIM macro that implements the algorithm is given in Appendix C.7.

When this method is used to accommodate overdispersion, the value of the X^2-statistic for the fitted model will be equal to the number of degrees of freedom for that model. Because of the similarity of the deviance to the X^2-statistic, this in turn means that the mean deviance for the model will be very close to unity and all the overdispersion will have been explained by the model.

Once ϕ has been estimated from fitting the full model, different linear logistic models can be fitted using the standard algorithm described in Appendix B.1, with the observations weighted by $w_i = 1/\hat{\sigma}_i^2$, where $\hat{\sigma}_i^2 = 1 + (n_i - 1)\hat{\phi}$. This is readily accomplished by the declaration of **weights** in packages such as GLIM. After fitting models with different linear components to the observed proportions, using weights w_i obtained from the estimate of ϕ for the full model, two alternative (nested) models can be compared in the usual way. That is, the difference in the deviance for two models is compared with percentage points of the χ^2-distribution; a non-significant result means that the two models cannot be distinguished.

The effect of introducing the weights, w_i, is to inflate the standard errors of the estimated values of the β-parameters. This in turn means that quantities derived from these estimates, such as fitted probabilities or *ED50* values, will have larger standard errors than they would have had in the absence of overdispersion. The corresponding confidence intervals for these quantities will then be wider than they would have been if no adjustment were made for overdispersion. One further consequence of having to estimate an additional parameter in the variance of the response variable is that confidence intervals for parameter have now to be constructed from percentage points of the t-distribution on ν degrees of freedom, where ν is the number of degrees of freedom of the deviance for the fitted model, instead of the standard normal distribution. This method for modelling overdispersion will be referred to as the **Williams procedure** and is illustrated in the following example.

Example 6.4 Germination of Orobanche
In Example 6.2, the probability, p_{jk}, that a seed of the jth species of *Orobanche*, $j = 1, 2$, germinates when exposed to the kth type of root extract, $k = 1, 2$, was found to be described by a linear logistic model, where

$$\text{logit}\,(p_{jk}) = \beta_0 + \text{species}_j + \text{extract}_k + (\text{species} \times \text{extract})_{jk} \qquad (6.4)$$

The deviance on fitting this model is 33.28 on 17 d.f., which suggests that there is overdispersion. To model this overdispersion, suppose that the actual germination probability for seeds of the jth species when germinated in the kth root extract varies about a mean response probability p_{jk}. The variance of the actual number of seeds in the lth batch which germinate, y_{jkl}, will be taken as $n_{jkl}p_{jk}(1 - p_{jk})\sigma_{jkl}^2$, where

$$\sigma_{jkl}^2 = 1 + (n_{jkl} - 1)\phi$$

and n_{jkl} is the number of seeds in the lth batch of seeds of the jth species germinated in the kth root extract. Using the algorithm of Williams for estimating ϕ, after fitting the full model in equation (6.4), it is found that $\hat{\phi} = 0.0249$, after three cycles of the iterative process.

The next step is to weight the individual binomial observations, y_{jkl}/n_{jkl}, by a factor of $1/\hat{\sigma}_{jkl}^2$, and to fit linear logistic models with various terms along the lines of Example 6.2. The resulting deviances are given in Table 6.2.

Table 6.2 Deviances on fitting weighted linear logistic models to the data on the germination of *Orobanche*

Terms fitted in model	Deviance	d.f.
β_0	47.26	20
$\beta_0 + \text{species}_j$	44.94	19
$\beta_0 + \text{extract}_k$	24.63	19
$\beta_0 + \text{species}_j + \text{extract}_k$	21.99	18
$\beta_0 + \text{species}_j + \text{extract}_k + (\text{species} \times \text{extract})_{jk}$	18.45	17

Notice that the deviance for the model that contains the main effects of species and extract, and the interaction between species and extract, is very close to its number of degrees of freedom, as it must be. Indeed, the only reason why the mean deviance is slightly different from unity is that the iterative scheme for estimating ϕ is continued until Pearson's X^2-statistic, rather than the deviance, is equal to the corresponding number of degrees of freedom. This means that the final residual mean deviance can no longer be used as a measure of the extent to which the model that incorporates overdispersion fits the observed data. Instead, diagnostics for model checking described in Chapter 5 will have to be adopted.

From Table 6.2, the reduction in deviance on including the term corresponding to the interaction between species and extract in the model that contains the main effects of species and extract is 3.54 on 1 d.f., which is nearly significant at the 5% level ($P = 0.060$). Although the P-value here is greater than that found when no allowance is made for variation amongst the germination probabilities, this analysis also leads to the conclusion that there is an interaction between species and root extract.

The parameter estimates for this model, reproduced from GLIM output, are as follows.

	estimate	s.e.	parameter
1	−0.5354	0.1937	1
2	0.0701	0.3114	SPEC(2)
3	1.330	0.2781	EXTR(2)
4	−0.8195	0.4351	SPEC(2).EXTR(2)

These estimates are very similar to those obtained after fitting the model described by equation (6.4) without allowing for overdispersion, but the standard errors are larger. The parameter estimates can be used to obtain the fitted germination probabilities, $\hat{p}_{jk}$, from

$$\text{logit}\,(\hat{p}_{jk}) = \hat{\beta}_0 + \widehat{\text{species}}_j + \widehat{\text{extract}}_k + (\text{species} \,\hat{\times}\, \text{extract})_{jk}$$

and the resulting estimated probabilities are presented in Table 6.3.

Table 6.3 Estimated germination probabilities for different species of *Orobanche* germinated in different root extracts, after allowing for overdispersion

Species	Root extract	Estimated probability
O.a. 75	Bean	0.37
	Cucumber	0.69
O.a. 73	Bean	0.39
	Cucumber	0.51

The difference between the estimated germination probabilities for seeds germinated in bean and cucumber root extract is greater for the species *O. aegyptiaca* 75 than it is for *O. aegyptiaca* 73, due to the presence of the interaction term in the model. If the aim of this study is to identify methods that can be used to prevent germination of the parasite, bean root extract effectively reduces the germination rate of each of the two species of *O. aegyptiaca*. For comparison purposes, the estimated germination probabilities when no allowance is made for overdispersion are 0.36, 0.68, 0.40 and 0.53, which are only slightly different from those given in Table 6.3.

A 95% confidence interval for the true germination probability for seed of the species *O. aegyptiaca* 75, when germinated in extract of bean root will now be obtained. When using GLIM, the parameter estimates species_1, extract_1, and $(\text{species} \,\hat{\times}\, \text{extract})_{11}$ are set equal to zero, and so $\text{logit}\,(\hat{p}_{11}) = \hat{\beta}_0 = -0.535$. The standard error of $\hat{\beta}_0$ is 0.194 and, using the two-sided 5% point of a t-distribution on 17 d.f., a 95% confidence interval for the logistic transform of the true germination probability, logit (p_{11}), for the first species (*O.a.* 75) germinated in the first root extract (bean) is $-0.535 \pm (2.11 \times 0.194)$, that is, the interval from -0.944 to -0.126. The corresponding interval for the germination probability itself is from 0.28 to 0.47. When no allowance is made for overdispersion, $\hat{\beta}_0 = -0.558$ with a standard error of 0.126. The resulting 95% confidence interval for p_{11} ranges from 0.31 to 0.42, which is somewhat narrower than the interval formed after adjusting for overdispersion. Confidence intervals for the remaining fitted probabilities can be found using the procedure described in section 3.14.

6.5 THE SPECIAL CASE OF EQUAL n_i

In the particular case where each observed proportion is based on the same number of binary observations, and is equal to n_0, say, the formula for the variance of the observed number of successes in equation (6.2) becomes

$$\text{Var}(y_i) = n_0 p_i (1 - p_i)[1 + (n_0 - 1)\phi]$$

This variance can be written as $n_0 p_i (1 - p_i)\sigma^2$, where σ^2, where $\sigma^2 = 1 + (n_0 - 1)\phi$ is an unknown constant. In this special case, the expected value of Pearson's X^2-statistic for the full model can be approximated by

$$(n - p)[1 + (n_0 - 1)\phi]$$

which is $(n - p)\sigma^2$, where p is the number of unknown parameters in the full model fitted to the n proportions. The parameter σ^2 may therefore be estimated by $X^2/(n - p)$, where X^2 is the value of Pearson's X^2-statistic for the full model, and the iterative estimation procedure is no longer required. The estimate of σ^2 is sometimes known as a **heterogeneity factor**. Since the deviance will tend to be very similar to the X^2-statistic, σ^2 may also be estimated by the mean deviance for the full model. This will generally be more convenient.

The result that $E(X^2) \approx (n - p)\sigma^2$ suggests that now X^2 has a $\sigma^2 \chi^2_{n-p}$-distribution, where χ^2_{n-p} denotes a chi-squared random variable with $(n - p)$ degrees of freedom. Again using the result that the deviance will usually be approximately equal to X^2, we might expect that in this situation the deviance will also have an approximate $\sigma^2 \chi^2_{n-p}$-distribution. This result is confirmed by a more technical argument, given in McCullagh and Nelder (1989).

After fitting linear logistic models in the usual way, that is without using weights, two nested models can be compared by examining the ratio of the difference in deviances to the deviance for the full model. This ratio will have a distribution that is independent of σ^2. In fact, the ratio of the change in deviance, divided by the change in degrees of freedom, to the mean deviance for the full model will have an F-distribution, and so models can be compared using F-tests as in the analysis of variance for continuous response variables. Specifically, if the deviance for a particular model, model (1), is D_1 on v_1 degrees of freedom, and that for model (2), which contains a subset of the terms in model (1), is D_2 on v_2 degrees of freedom, the ratio

$$\frac{(D_1 - D_2)/(v_1 - v_2)}{D_0/v_0}$$

has an F-distribution on $(v_1 - v_2)$, v_0 degrees of freedom, where D_0 is the deviance for the full model on v_0 degrees of freedom. A non-significant value of the F-statistic indicates that model (1) and model (2) cannot be distinguished.

A further consequence of the variance of y_i being inflated by σ^2 is that the standard errors of parameter estimates after fitting a linear logistic model will have to be multiplied by a factor of $\sqrt{(\hat{\sigma}^2)}$. In addition, confidence intervals based on parameter estimates will need to be constructed from percentage points of the t-distribution. For example, if Fieller's theorem, described in section 4.2.2, is to be used to compute a confidence interval for the true *ED50* value, after fitting a linear logistic model to dose–response data, the standard errors used in equation (4.7) have to be multiplied by a factor of $\sqrt{(\hat{\sigma}^2)}$, and $z_{\alpha/2}$ is replaced by the upper $(100\alpha/2)\%$ point of the t-distribution on $(n - p)$ d.f.

In the terminology of the statistical package GLIM, σ^2 is the **scale parameter**. When there is no overdispersion, and the number of binary observations in the ith data set is n_i, $\text{Var}(y_i) = n_i p_i (1 - p_i)$, and $\sigma_i^2 = 1$. This accounts for the message 'scale parameter taken as 1.000' which appears in standard GLIM output. When the n_i are equal, so that σ_i^2 is constant, the value of the scale parameter can be set equal to the mean deviance for the full model, which leads to the correct standard errors being output. No difference is made to the values of the deviances, but as pointed out above, they now have to be compared using F-tests.

Circumstances in which each proportion is based on the same number of binary observations are comparatively rare. However, the procedure described in this section is not too sensitive to differences in the values of the n_i, and so this method of allowing for overdispersion can be used as a first approximation even when the n_i are not all equal. This general method for adjusting the analysis to take account of overdispersion was first proposed by Finney (1947), and is illustrated in the following example.

Example 6.5 Grain beetles
The results of investigations into the effectiveness of ethylene oxide as a fumigant against common insect pests were presented by Busvine (1938). In an experiment to determine the toxicity of ethylene oxide to the grain beetle, *Calandra granaria*, batches of about thirty insects were exposed to various concentrations of the fumigant, measured in mg/l, for a period of one hour. Because the results of the fumigation are not apparent until several days after exposure, the insects were subsequently transferred to clean glass tubes containing suitable food, and incubated at 25 °C. The numbers of insects affected by the poison were included in Bliss (1940), and are given in Table 6.4.

Table 6.4 Numbers of grain beetles affected by exposure to various concentrations of ethylene oxide

Concentration	Number affected	Number exposed
24.8	23	30
24.6	30	30
23.0	29	31
21.0	22	30
20.6	23	26
18.2	7	27
16.8	12	31
15.8	17	30
14.7	10	31
10.8	0	24

Let conc be the explanatory variable corresponding to the concentration of ethylene oxide used in the experiment. The values of the deviance on fitting linear logistic models to these data, using log (conc) as the explanatory variable, are given in Table 6.5.

Table 6.5 Deviances on fitting linear logistic models to the data on the toxicity of ethylene oxide to the grain beetle

Terms fitted in model	Deviance	d.f.
β_0	138.00	9
$\beta_0 + \log(\text{conc})$	36.44	8

The deviance on fitting a logistic regression on log(conc) is 36.44 on 8 d.f., indicating that there is overdispersion. Figure 6.1 shows the empirical logits of the observed proportions, defined in Example 3.8 of section 3.11, plotted against log(conc). The fitted logistic regression line has been superimposed. This figure shows that there are no outliers and that the observations do not differ systematically from a straight line relationship between the logistic transform of the probability of a beetle being affected by the fumigant and log(conc). We may therefore conclude that the overdispersion is the result of there being a greater degree of variability about the fitted line than would have been expected on the basis of binomial variation. One possible explanation for this overdispersion is that there were differences in the way that the ten batches of insects were managed during the course of the experiment.

Although the number of insects in each of the ten batches does vary slightly, the batch sizes are so similar that the method described in Section 6.5 can safely be used to allow for the overdispersion. In this approach, an F-test is used to assess the significance of the regression of the logistic transform of the response probability on log(conc). The relevant analysis of deviance table is given as Table 6.6.

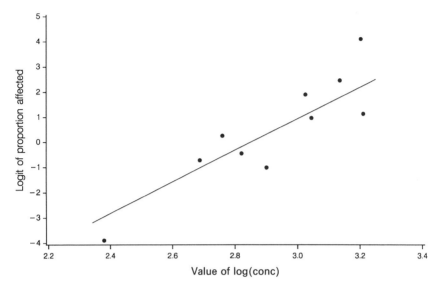

Fig. 6.1 Plot of the data from Example 6.5, with the fitted logistic regression line superimposed.

Table 6.6 Analysis of deviance table to determine the significance of the logistic regression on log(conc)

Source of variation	d.f.	Deviance	Mean deviance	F-ratio
Regression on log(conc)	1	101.56	101.56	22.3
Residual	8	36.44	4.56	
Total	9	138.00		

When compared with percentage points of the F-distribution on 1, 8 d.f., the observed value of the F-statistic is significant at the 1% level ($P = 0.002$). There is therefore very strong evidence of a linear dose–response relationship in these data.

The equation of the fitted model, with the standard errors of the parameter estimates, adjusted to take account of the overdispersion by multiplication by $\sqrt{(36.44/8)} = 2.13$, given in parentheses, is

$$\text{logit}(\hat{p}) = -17.87 + 6.27 \log(\text{conc})$$
$$(4.84) \quad (1.66)$$

The parameter estimates and their standard errors can now be used to estimate the *ED50* value and to obtain a 95% confidence interval for the true *ED50* using Fieller's theorem, given in Section 4.2.2. The values of $\hat{\beta}_0$ and $\hat{\beta}_1$ are -17.87 and 6.27, respectively, and so the estimated *ED50* value is given by

$E\hat{D}50 = \exp(-\hat{\beta}_0/\hat{\beta}_1) = 17.29$ mg/l. To obtain a 95% confidence interval for the true $ED50$ value using the result in equation (4.7), $z_{\alpha/2}$ has to be replaced by the upper $\alpha/2$ point of the t-distribution on v d.f., where v is the number of degrees of freedom for the full model. Here, this is the model that contains a constant term and log(conc), for which $v = 8$, and so the appropriate percentage point of the t-distribution is 2.31. The variance–covariance matrix of the parameter estimates, adjusted to take account of the overdispersion, is

$$\begin{array}{c} \text{Constant} \\ \log(\text{conc}) \end{array} \begin{bmatrix} 23.423 & \\ -8.016 & 2.754 \end{bmatrix}$$

$$\qquad\qquad \text{Constant} \quad \log(\text{conc})$$

from which $\hat{v}_{00} = 23.42$, $\hat{v}_{11} = 2.75$, and $\hat{v}_{01} = -8.02$. Using equation (4.7), 95% confidence limits for the true log($ED50$) value are 2.67 and 2.97, from which a 95% confidence interval for the $ED50$ value itself ranges from 14.37 to 19.46.

Example 6.6 Germination of Orobanche
In Example 6.4, variation in the germination probabilities of *Orobanche* was modelled by taking the variance of the number of seeds germinating out of n_{jkl} to be $n_{jkl}p_{jk}(1 - p_{jk})\sigma_{jkl}^2$. The estimated values of the scale parameter σ_{jkl}^2, given by $\hat{\sigma}_{jkl}^2 = 1 + (n_{jkl} - 1)\hat{\phi}$, range from 1.075 when $n_{jkl} = 4$, to 2.992 when $n_{jkl} = 81$, which suggests that there is a need for a model under which $\hat{\sigma}_{jkl}^2$ is not assumed to be constant. However, suppose that the method in section 6.5 is used to adjust for overdispersion. In this approach, σ_{jkl}^2 is assumed to be constant, and can be estimated by the residual mean deviance for the full model, which is 1.958. The deviances on fitting various models to these data were given in Table 6.1, from which the analysis of deviance in Table 6.7 can be formed.

Table 6.7 Analysis of deviance for the data on the germination of *Orobanche*

Source of variation	d.f.	Deviance	Mean deviance	F-ratio
Species adjusted for extract	1	3.06	3.06	1.56
Extract adjusted for species	1	56.49	56.49	28.86
Species × extract interaction	1	6.41	6.41	3.27
Residual	17	33.28	1.96	

The deviances in this table are not additive and so the 'Total' line has been omitted. The observed value of the F-statistic for the interaction between species and extract is significant at the 10% level ($P = 0.088$). The P-value for this interaction is a little larger than the value of 0.060 obtained in Example 6.4, when the Williams procedure was used to adjust for the overdispersion. However, we still draw the conclusion that there is evidence of an interaction between species and root extract.

Example 6.6 shows that even when the numbers of binary observations that make up the observed proportions vary considerably, the simple method of adjusting for overdispersion described in section 6.5 can lead to results that are not much different from those obtained using the more complicated procedure described in Section 6.4. However, the Williams procedure has the advantage that correct standard errors can be obtained directly, and since it gives the same results as the method of section 6.5 when the n_i are all equal, this procedure is recommended for general use.

6.6 THE BETA-BINOMIAL MODEL

So far, a rather general assumption has been made about the form of variation about the response probabilities, but there are alternative models based on more specific assumptions. For example, we could assume that the variability about the mean response probability can be modelled by a **beta distribution.** A random variable X is said to have a beta distribution with parameters a and b if the probability density function of X is

$$\frac{1}{B(a, b)} x^{a-1}(1-x)^{b-1}, \quad 0 \leqslant x \leqslant 1$$

where $B(a, b)$, a function known as a beta function, is such that the integral of this density function is unity. The beta distribution is defined over the range $(0, 1)$, and can take a variety of forms, depending on the values of the two parameters, a and b. The mean and variance of X are $a/(a + b)$ and $ab/\{(a + b)^2(a + b + 1)\}$ respectively, and the distribution is unimodal if a and b are both greater than unity.

We now assume that the variance of ϑ_i about the mean response probability p_i, for the ith observation, can be modelled by a beta distribution with parameters a_i and b_i. Since $E(\vartheta_i) = p_i$, we have that $p_i = a_i/(a_i + b_i)$ and the variance of ϑ_i can be written as

$$\text{Var}(\vartheta_i) = p_i(1 - p_i)/[a_i + b_i + 1] = p_i(1 - p_i)\tau_i$$

where $\tau_i = 1/[a_i + b_i + 1]$. This expression has the same form as that in equation (6.1), except that the constant term ϕ has been replaced by the term τ_i that takes different values for each observation.

If there is reason to believe that values of ϑ_i near zero or unity are unlikely, the density function of the beta distribution must be unimodal, and zero at both zero and unity. Then, $a_i > 1$ and $b_i > 1$, and so the variance of ϑ_i cannot exceed $p_i(1 - p_i)/3$. This could be rather restrictive. Moreover, in fitting this model it is often assumed that a_i and b_i, and therefore τ_i, are constant for all i. This model then reduces to that described by equation (6.1), but with the scale parameter restricted to be less than one third.

This model is called the **beta-binomial model** for the analysis of proportions and can be fitted using the method of maximum likelihood, described by Williams (1975) and Crowder (1978). Standard statistical software cannot be used to fit this model, although the package EGRET, distributed by the Statistics and Epidemiology Research Corporation in Seattle, USA, does have the appropriate facilities. Additional information about this package can be found in Chapter 9.

In beta-binomial modelling, differences in deviance will have an approximate χ^2-distribution. Accordingly, the assumption of a common value of τ can be tested in any particular application by comparing the deviance for the model with individual values of τ_i for each observation with that for the model where τ is constant. However, since the model is less flexible than that in equation (6.1) when τ is assumed to be constant, and because it can be difficult to fit the model when the τs are not constant, this model for variation in the response probabilities will not be considered in greater detail.

Example 6.7 Germination of Orobanche
In this example, the overdispersion in the data on seed germination is modelled by assuming a beta-binomial distribution for the variation in the response probabilities. Denote by y_{jkl} the number of seeds that germinate out of n_{jkl} in the lth batch of seeds of the jth species exposed to the kth root extract. The full model for the probability that seeds of the jth species germinate in the kth root extract is then

$$\text{logit}\,(p_{jk}) = \beta_0 + \text{species}_j + \text{extract}_k + (\text{species} \times \text{extract})_{jk}$$

with $\text{Var}\,(y_{jkl}) = n_{jkl}p_{jk}(1 - p_{jk})\tau_{jk}$, where $\tau_{jk} = 1/[a_{jk} + b_{jk} + 1]$, and a_{jk}, b_{jk} are the parameters of a beta distribution for the variation in the response probabilities for the jth species and the kth root extract. Using EGRET, the deviance for this model is 30.74 on 13 d.f. When the τs are constrained to be equal, the deviance is increased by 0.20 to 30.94 on 16 d.f. This increase is certainly not significant when judged against percentage points of the χ^2-distribution on 3 d.f., and so the τs may be assumed to be equal. Under this assumption, the deviance on fitting the linear logistic model that contains the main effects of species and extract alone is 35.07 on 17 d.f., and so the change in deviance on including the interaction term in the model is 4.13 on 1 d.f. This is significant at the 5% level ($P = 0.042$), and the P-value for this test is very similar to that in Example 6.4. The fitted germination probabilities under the model that includes the interaction term are 0.37, 0.69, 0.39 and 0.52, which are again very close to those reported in Table 6.3.

6.7 RANDOM EFFECTS IN A LINEAR LOGISTIC MODEL

We have seen that overdispersion can be explained by variation in the response probabilities for different batches of identically treated experimental units. As explained in Section 6.1, this variation could be attributed to relevant explanatory variables not having been recorded, or to the inclusion in the model of certain explanatory variables that have not been adequately measured or controlled. The manner in which each of these possible explanations for overdispersion can be incorporated in a linear logistic model will now be considered in this section, although the resulting model can be regarded as a general model for overdispersion.

First, suppose that a linear logistic model is adopted for the relationship between the true response probability for the ith observation, p_i, $i = 1, 2, \ldots, n$, and a single explanatory variable x, so that

$$\text{logit}(p_i) = \beta_0 + \beta_1 x_i \tag{6.5}$$

However, if the true response probability also depends linearly on r unknown explanatory variables, $u_1, u_2, \ldots, u_r$, whose values have not actually been recorded, the correct model for $\text{logit}(p_i)$ is

$$\text{logit}(p_i) = \beta_0 + \beta_1 x_i + \lambda_1 u_{1i} + \lambda_2 u_{2i} + \cdots + \lambda_r u_{ri}$$

In the absence of information about the values of $u_1, u_2, \ldots, u_r$, variation in the response probabilities, p_i, that would have been attributed to the effect of the u-variables, will be interpreted as overdispersion. The effect of these unknown explanatory variables can be assumed to be such that their joint effect can be modelled by the addition of a single variable to the linear logistic model. The value of this variable will be different for each observation, and will be denoted by δ_i. Numerical values cannot be assigned to the quantities $\delta_1, \delta_2, \ldots, \delta_n$, and so they will be taken to be random variables with common mean and common variance σ_δ^2. Since a constant term has been included in the model, we can without loss of generality assume that the δ_i have zero mean. As a consequence of this assumption, the true response probability is a random variable, ϑ_i, say, whose expected value is p_i. The resulting model for $\text{logit}(\vartheta_i)$ is then

$$\text{logit}(\vartheta_i) = \beta_0 + \beta_1 x_i + \delta_i \tag{6.6}$$

and the term δ_i is known as a **random effect**. An alternative way of expressing the model in equation (6.6) is to suppose that z_i is a random variable with zero mean and unit variance, and to write the model as

$$\text{logit}(\vartheta_i) = \beta_0 + \beta_1 x_i + \gamma z_i \tag{6.7}$$

where γ, the coefficient of the random term z_i, is the standard deviation of δ_i, that is $\gamma \equiv \sigma_\delta$. Note that when $\gamma = 0$ in equation (6.7), $\sigma_\delta = 0$ and there is no random variation about the logistic transform of the response probability, leading back to the model in equation (6.5). The assumption that $E(z_i) = 0$, for $i = 1, 2, \ldots, n$, means that the logistic transform of ϑ_i varies randomly about a mean of $\beta_0 + \beta_1 x_i$. This is in contrast to the model in Section 6.2, in which the response probabilities themselves were assumed to vary about a mean of p_i.

Now consider the situation where the variation in the response probabilities is due to the fact that one or more of the explanatory variables included in a linear logistic model have not been adequately controlled. For example, consider the model

$$\text{logit}(\vartheta_i) = \beta_0 + \beta_1 \xi_i$$

which relates the response probability for the ith observation to the true, or exact, value of an explanatory variable ξ, for the ith observation. Suppose that the value of ξ cannot be accurately measured, and that the recorded value of this explanatory variable for the ith observation is actually x_i. We can then write $x_i = \xi_i + \varepsilon_i$, where ε_i is uncorrelated with the true value ξ_i and represents random deviation of the observed value of the explanatory variable from the true value. The model for $\text{logit}(\vartheta_i)$ then becomes

$$\text{logit}(\vartheta_i) = \beta_0 + \beta_1 x_i - \beta_1 \varepsilon_i \tag{6.8}$$

If the values ε_i are assumed to be random variables with zero mean and common variance σ_ε^2, the model described in equation (6.8) can be expressed as in equation (6.7), where γ in equation (6.7) corresponds to $-\beta_1 \sigma_\varepsilon$ in equation (6.8). Consequently, the effect of including explanatory variables that have not been properly controlled in the model can again be modelled by including a random effect.

More generally, the response probability for the ith observation may depend on the values of k explanatory variables, $x_1, x_2, \ldots, x_k$, and so when a random effect δ_i is included in the model, we have that

$$\text{logit}(\vartheta_i) = \eta_i + \delta_i = \eta_i + \gamma z_i \tag{6.9}$$

where $\eta_i = \beta_0 + \beta_1 x_{1i} + \cdots + \beta_k x_{ki}$, $\gamma \geq 0$, and z_i is a random variable with zero mean and unit variance. This model can be used as a general model for overdispersion, irrespective of the underlying cause of variation in the response probabilities. Moreover, the model described in equation (6.9) can be used in the analysis of ungrouped binary data, as well as for grouped binary data expressed in the form of proportions. The model for overdispersion developed in section 6.2 can only be used in modelling grouped binary data.

6.7.1 Fitting a model that contains a random effect

The method of maximum likelihood can be used to fit a linear logistic model that contains a random effect. The likelihood of the unknown parameters in the model underlying equation (6.9) is given by

$$L(\boldsymbol{\beta}, \gamma, z_i) = \prod_{i=1}^{n} \binom{n_i}{y_i} \vartheta_i^{y_i} (1 - \vartheta_i)^{n_i - y_i}$$

$$= \prod_{i=1}^{n} \binom{n_i}{y_i} \frac{[\exp(\eta_i + \gamma z_i)]^{y_i}}{[1 + \exp(\eta_i + \gamma z_i)]^{n_i}}$$

This likelihood function depends on the $k + 2$ unknown parameters $\beta_0, \beta_1, \ldots, \beta_k$ and γ, and also on the random variables, $z_1, z_2, \ldots, z_n$. In order to be able to estimate the values of the unknown parameters, we will have to be more specific about the distribution of the random variables, z_i. Here, we will assume that the z_i have independent standard normal distributions, although a number of alternative distributions could have been adopted.

The standard method of handling a likelihood function that involves random variables that have a fully specified probability distribution is to integrate the likelihood function with respect to the distribution of these variables. After 'integrating out' the z_i, the resulting function is termed a **marginal likelihood function**, and depends only on $\beta_0, \beta_1, \ldots, \beta_k$ and γ. The maximum likelihood estimates of these parameters are then those values which maximise the marginal likelihood function. Specialist software is needed for this, such as the statistical package EGRET, mentioned in section 6.6.

In the absence of a package such as EGRET, a computer program for function maximization is needed to obtain the values of $\beta_0, \beta_1, \ldots, \beta_k$ and γ which maximize the marginal likelihood function. This function is actually

$$\prod_{i=1}^{n} \int_{-\infty}^{\infty} \binom{n_i}{y_i} \frac{[\exp(\eta_i + \gamma z_i)]^{y_i}}{[1 + \exp(\eta_i + \gamma z_i)]^{n_i}} \frac{\exp\{-z_i^2/2\}}{\sqrt{(2\pi)}} \, dz_i \tag{6.10}$$

but the integral can only be evaluated numerically. One way of carrying out this numerical integration is to use the **Gauss–Hermite** formula for numerical integration (or quadrature), according to which

$$\int_{-\infty}^{\infty} f(u) e^{-u^2} \, du \approx \sum_{j=1}^{m} c_j f(s_j)$$

where the values of c_j and s_j are given in standard tables, such as those of Abramowitz and Stegun (1972). For most practical purposes, m need not be greater than 20, although some authors suggest using even smaller values. The integral in equation (6.10) can then be expressed as a summation, and so the marginal likelihood becomes

$$\pi^{-n/2} \prod_{i=1}^{n} \binom{n_i}{y_i} \sum_{j=1}^{m} c_j \left\{ \frac{[\exp\{\eta_i + \gamma s_j \sqrt{(2)}\}]^{y_i}}{[1 + \exp\{\eta_i + \gamma s_j \sqrt{(2)}\}]^{n_i}} \right\}$$

The values of $\hat{\beta}$ and $\hat{\gamma}$ which maximize this expression can then be determined numerically.

After a linear logistic model with a random effect has been fitted, the deviance for the model can be computed. Alternative linear logistic models that contain a random effect can be compared by examining differences in the deviances for models with different linear systematic components. These differences can be compared with percentage points of the χ^2-distribution in the usual way. The same procedure cannot be used to test for the significance of the overdispersion by comparing the deviance for a model with a given linear component η_i, and a random effect, δ_i, with that for the model that contains η_i alone, since the difference between these two deviances cannot be assumed to be even approximately distributed as χ^2. The reason for this is that the hypothesis that is being tested here is that $\gamma = 0$, and since the value of γ under the null hypothesis is in fact the smallest permissible value of γ, the distribution theory associated with likelihood ratio testing no longer holds.

Even after including a random effect in the model, the residual deviance may remain high relative to the number of degrees of freedom. The distribution of the deviance for a model that includes a random effect is unknown, and since this deviance will not necessarily have a χ^2-distribution, it follows that the deviance for a satisfactory model need not be approximately equal to the number of degrees of freedom. This means that after a random effect has been fitted, it is not possible to judge whether there is an additional source of overdispersion in the data.

When a model that includes a random effect is fitted to grouped binary data, the resulting parameter estimates can be used to obtain fitted values of the response probabilities, or estimates of functions of the parameter estimates, such as *ED50* values. Standard errors of the parameter estimates are given by packages such as EGRET, and can be used in the construction of confidence intervals for functions of these parameters. Although little is known about the properties of parameter estimates in linear logistic models that include a random effect, it would be prudent to use percentage points of the t-distribution, rather than the standard normal distribution, in deriving such confidence intervals.

The estimate of the parameter γ in the model underlying equation (6.9), can be used to provide information about the range of likely success probabilities for a particular individual. Since z_i has a standard normal distribution, its likely range is from -1.96 to $+1.96$. This means that the estimated value of the logistic transform of the response probability for a particular individual for whom the linear predictor is $\hat{\eta}$, could range from $\hat{\eta} - 1.96\hat{\gamma}$ to $\hat{\eta} + 1.96\hat{\gamma}$, from which a corresponding interval estimate for the response probability can be found. Note that this is not a confidence interval, since the variance of $\hat{\eta}$ has been ignored. Moreover, these intervals tend to be rather wide in practice, and so they are not particularly informative.

Example 6.8 A toxicological study
Consider the data from Example 1.4 on the proportion of surviving rat pups in the litters of mothers which have been either exposed or not exposed to a particular

substance. In Example 5.20, the deviance on fitting a linear logistic model that contains a term representing the treatment effect was found to be 86.19 on 30 d.f. Although this deviance is very large, there was no evidence that this was due to an inappropriate choice of model, or to the presence of outliers. The conclusion is that the data exhibit overdispersion. The most likely explanation for this overdispersion is that because of genetic, social or environmental influences, the binary responses of the rat pups in a given litter are correlated.

Now suppose that the overdispersion in these data is modelled using a linear logistic model that contains a random effect. The model will then include a term that represents the treatment effect, treat_j, say, $j = 1, 2$, and a term δ_{jk}, which will correspond to the effect of the kth litter in the jth treatment group, $k = 1, 2, \ldots, 16$. If the term δ_{jk} were included in the linear systematic component of the model, the resulting model would contain as many unknown parameters as there are observations, and the model would fit the data perfectly. Because this model is uninformative, δ_{jk}, will be taken as a random effect. This term will then represent within-treatment variation in ϑ_{jk}, the response probability for rats in the kth litter exposed to the jth treatment. If we assume that $\delta_{jk} \sim N(0, \gamma^2)$, the resulting model can be written

$$\text{logit}(\vartheta_{jk}) = \text{treat}_j + \delta_{jk} = \text{treat}_j + \gamma z_{jk} \tag{6.11}$$

where the z_{jk} are independently distributed standard normal random variables. The estimated expected value of the logistic transform of the survival probabilities of the rats in each treatment group is simply $\widehat{\text{treat}}_j$, since the z_{jk} have zero expectation. Consequently,

$$\frac{\exp(\widehat{\text{treat}}_j)}{1 + \exp(\widehat{\text{treat}}_j)}$$

is an estimate of p_j, the survival probability under the jth treatment. On fitting the model depicted in equation (6.11), using the facilities of the statistical package EGRET, we find that $\widehat{\text{treat}}_1 = 2.624$ and $\widehat{\text{treat}}_2 = 1.544$, so that the estimated survival probabilities are 0.93 and 0.82, respectively. The estimated value of γ in equation (6.11) is 1.345, and the deviance for this model is 63.44 on 29 d.f.

Using the value of $\hat{\gamma}$, the likely range of $\text{logit}(\hat{p}_1)$, where $\hat{p}_1$ is the predicted proportion of rat pups that survive in the treated group, is $\widehat{\text{treat}}_1 \pm 1.96\hat{\gamma}$ from which the likely range of values of $\hat{p}_1$ is from 0.50 to 0.99. Similarly, the estimated value of the survival probability for the rats in the control group that survive could range from 0.25 to 0.98. These estimates are in line with the observed proportions which were given in Table 1.4.

When the model

$$\text{logit}(\vartheta_{jk}) = \beta_0 + \gamma z_{jk} \tag{6.12}$$

is fitted using the same procedure, the deviance is 66.31 on 30 d.f., and so the change in deviance on including the term treat$_j$, allowing for overdispersion, is 2.87 on 1 d.f. This is significant at the 10% level of significance ($P = 0.094$), and so there is some evidence of a difference in the survival probabilities of the rat pups in the two groups, after allowing for overdispersion. If no allowance is made for overdispersion, the change in deviance on introducing treat$_j$ into a linear logistic model that only includes a constant term is 9.02 on 1 d.f., which is significant at the 1% level ($P = 0.003$). We therefore see that when the overdispersion in the data is ignored, the significance of the treatment difference is substantially overestimated.

When the random effect is not included in the model that contains treat$_j$, we are back to the model fitted in Example 5.20, where

$$\text{logit}(p_j) = \text{treat}_j \qquad (6.13)$$

and p_j is the response probability for rats in litters exposed to the jth treatment. The fitted response probabilities under this model are 0.90 and 0.77 respectively which are similar to those for the model described by equation (6.11). However, the standard errors of the estimated values of treat$_1$ and treat$_2$ under the random effects model are 0.483 and 0.434, respectively, which are larger than their corresponding values of 0.264 and 0.198 in the model underlying equation (6.13). The precision of the estimated survival probabilities will therefore be overestimated if no allowance is made for the overdispersion. The deviance for the model in (6.13) is 86.19 on 30 d.f., and so we see that the random effect which models within-treatment variation does reduce the deviance by a considerable amount. However, the residual deviance for the model in (6.11), 63.44, is still somewhat larger than the corresponding number of degrees of freedom, 29, suggesting that some unexplained overdispersion remains.

Example 6.9 Germination of Orobanche
In this example, a random effect will be used to model variation in the germination probabilities of batches of seed of a given species of *Orobanche*, germinated in a particular root extract. The full model will therefore be

$$\text{logit}(\vartheta_{jkl}) = \beta_0 + \text{species}_j + \text{extract}_k + (\text{species} \times \text{extract})_{jk} + \gamma z_{jkl}$$

where $z_{jkl} \sim N(0, 1)$, and so the variation in the lth batch of seeds of the jth species germinated in the kth root extract is modelled by a normally distributed random effect with zero mean and variance γ^2. The deviances on fitting different models to the data, all of which contain a random effect, were computed using EGRET and are given in Table 6.8.

The models in Table 6.8 can be compared by examining changes in the deviance relative to the percentage points of a χ^2-distribution. The first step is to look at the significance of the interaction between species and root extract. The change in deviance on adding the term (species $\times$ extract)$_{jk}$ into the model that contains the main effects of species and extract, as well as the random effect, is 4.15 on 1 d.f.,

Table 6.8 Deviances on fitting linear logistic models with a random effect to the data
on the germination of *Orobanche*

Terms fitted in model	Deviance	d.f.
$\beta_0 + \gamma z_{jkl}$	52.32	19
$\beta_0 + \text{species}_j + \gamma z_{jkl}$	50.33	18
$\beta_0 + \text{extract}_k + \gamma z_{jkl}$	37.77	18
$\beta_0 + \text{species}_j + \text{extract}_k + \gamma z_{jkl}$	35.07	17
$\beta_0 + \text{species}_j + \text{extract}_k + (\text{species} \times \text{extract})_{jk} + \gamma z_{jkl}$	30.92	16

which is significant at the 5% level ($P = 0.042$). Using a random effect to account
for overdispersion, essentially the same conclusion is reached about the size of the
interaction as in Example 6.4, where the overdispersion was modelled in a different
way. The parameter estimates and their standard errors, obtained using EGRET,
are as shown in Table 6.9.

Table 6.9 Parameter estimates and their standard errors for
the model that contains the terms species$_j$, extract$_k$, (species $\times$
extract)$_{jk}$ and a random effect

Parameter	Estimate	s.e.
constant	-0.548	0.167
species$_2$	0.097	0.278
extract$_2$	1.337	0.237
(species $\times$ extract)$_{22}$	-0.810	0.385
γ	0.236	0.110

In this table, the effect due to the first level of the two factors is zero, and so the
estimates can be interpreted in the same way as estimates obtained using GLIM.
The parameter estimates in Table 6.9 are very similar to those reported in Example
6.4, and lead to fitted probabilities which, at most, differ by 0.01 from those given
in Table 6.3. The residual mean deviance for the full model is 30.92/16, which is
much greater than unity. This suggests that the introduction of a single random
effect to represent the between-batch variability of the seeds has not been entirely
successful.

6.7.2 An approximate method for fitting a model with a random effect

The parameters $\boldsymbol{\beta}$ and γ in equation (6.9) can also be estimated using an
approximate method proposed by Williams (1982a). According to the model
depicted in equation (6.9), logit(ϑ_i) has a normal distribution with mean η_i and
variance γ^2, that is $E[\text{logit}(\vartheta_i)] = \eta_i$ and Var $[\text{logit}(\vartheta_i)] = \gamma^2$. If γ is sufficiently small
for the expression (4.16) for the approximate variance of a function of a random

variable to be valid,

$$\text{Var}\left[\text{logit}(\vartheta_i)\right] \approx \left\{\frac{\text{d logit}(\vartheta_i)}{\text{d}\vartheta_i}\right\}^2\Bigg|_{p_i} \text{Var}(\vartheta_i)$$

$$= \frac{1}{p_i^2(1-p_i)^2} \text{Var}(\vartheta_i)$$

where p_i is the expected response probability for the ith observation, that is $p_i = E(\vartheta_i)$. Consequently,

$$\text{Var}(\vartheta_i) \approx p_i^2(1-p_i)^2 \text{Var}\left[\text{logit}(\vartheta_i)\right]$$
$$= \gamma^2 p_i^2(1-p_i)^2$$

This is similar to the model for variation in the response probabilities described in Section 6.2, for which the variance of ϑ_i was given in expression (6.1). Following an argument similar to that used in Section 6.2, the unconditional mean and variance of y_i are given by $E(y_i) \approx n_i p_i$ and

$$\text{Var}(y_i) \approx n_i p_i (1-p_i)[1 + \gamma(n_i - 1)p_i(1 - p_i)] \qquad (6.14)$$

respectively. This variance is the same as that in equation (6.2), after replacing ϕ by $\gamma p_i(1 - p_i)$. The method of fitting the model of Section 6.2, outlined in Section 6.4, can therefore be used to fit the model where the variance of y_i is given in equation (6.14). The only change that has to be made to the algorithm is that the quantity σ_i^2 is replaced by $1 + \gamma(n_i - 1)p_i(1 - p_i)$. As in Section 6.4, γ can be estimated iteratively by equating the value of Pearson's X^2-statistic to its approximate expected value. Again, this procedure cannot be used when the observations are binary, since then the variance in equation (6.14) reduces to $p_i(1 - p_i)$, the variance of a Bernoulli response variable, and the weights are all equal to unity.

The GLIM macro that implements the algorithm in Section 6.4 can easily be modified to fit this model. A note on the changes required is included in the listing of the macro in Appendix C.7.

The method of estimating σ^2 used in this approximate procedure does force the deviance for the model that includes the random effect to be very similar to its number of degrees of freedom, and so the model is forced to explain all of the overdispersion. The residual mean deviance cannot therefore be used as a measure of model adequacy. On the other hand, if the model underlying equation (6.9) is fitted by maximizing the marginal likelihood function, as in Section 6.7.1, the deviance for the full model will not necessarily be as small as its number of degrees of freedom. For this reason, the estimated variance of the random effect, $\hat{\gamma}^2$, obtained using the approximate approach, can be considerably smaller than that found using the marginal likelihood method.

When the approximate method of estimating γ is used to fit the model described in equation (6.9), models with different linear systematic components can be compared by examining the difference in deviance between two models, after fitting these models with weights based on the full model. Specifically, suppose that model (1) contains a subset of terms in model (2), and it is desired to look at the explanatory power of the extra terms in model (2), after allowing for overdispersion. The full model is first fitted, given an estimate $\hat{\gamma}$ of γ, and fitted probabilities, $\hat{p}_i$. Models (1) and (2) are then fitted using a weighted analysis, with weights $w_i = 1/[1 + \hat{\gamma}(n_i - 1)\hat{p}_i(1 - \hat{p}_i)]$. This is essentially the same procedure as that described in Section 6.4. Indeed, since the values of $\hat{p}_i(1 - \hat{p}_i)$ will not differ much when $0.2 \leqslant \hat{p}_i \leqslant 0.8$, this approximate procedure for fitting a linear logistic model with a random effect can only be expected to give different results to the approach described in Section 6.4 when some of the fitted probabilities under the full model are close to zero or one. Because of this, and the fact that the approximation is only valid when γ is small, this procedure is not recommended for general use.

Example 6.10 A toxicological study
When the approximate method is used to fit the model depicted in equation (6.11), the full model, the estimated value of γ is 1.293. This estimate is slightly smaller than 1.345, the estimate based on the maximization of the marginal likelihood function in Example 6.8. The reason for this is that the approximate method forces γ to explain all of the overdispersion in the data, whereas when the exact method is used, as in Example 6.8, some overdispersion remains unaccounted for by the random effect. When the estimated value of γ, and the fitted probabilities, $\hat{p}_{jk}$, for the full model, are used to compute weights, w_{jk}, for the kth litter of rats in the jth treatment group, the deviance obtained on fitting the model underlying equation (6.11), using weights w_{jk}, is 32.99 on 30 d.f. The fact that the mean deviance is very close to unity is a by-product of the iterative procedure used to estimate γ in equation (6.11), and does not imply that the model is adequate.

To fit a model that contains a constant and a random effect only, using the approximate method, the same weights, w_{jk}, are used to fit the model described in equation (6.12). The resulting deviance is 37.36 on 31 d.f., and so the change in deviance due to the treatment effect is 4.37 on 1 d.f., which, when compared to percentage points of the χ^2-distribution, is significant at the 5% level. The corresponding P-value is 0.037, which is considerably smaller than that found using the exact method in Example 6.8. The fitted survival probabilities are

$$\frac{\exp(\widehat{\text{treat}}_j)}{1 + \exp(\widehat{\text{treat}}_j)}$$

as in Example 6.8. The estimates of the parameters treat_1 and treat_2 when the model in (6.11) is fitted using the approximate method are 2.153 and 1.117, respectively, and so the corresponding survival probabilities for the rat pups in the two treatment groups are 0.90 and 0.75. These are quite close to those found in Example 6.8 when the exact method was used.

Example 6.11 Germination of Orobanche
The estimated value of γ obtained when the full model,

$$\text{logit} (\vartheta_{jkl}) = \beta_0 + \text{species}_j + \text{extract}_k + (\text{species} \times \text{extract})_{jk} + \gamma z_{jkl}$$

is fitted using the approximate method is 0.107, which is much less than the exact value found in Example 6.9. The deviance for this model is 18.44 on 17 d.f. When the weights $w_{jkl} = [1 + \hat{\gamma}(n_{jkl} - 1)\hat{p}_{jk}(1 - \hat{p}_{jk})]^{-1}$, where $\hat{\gamma}$ and $\hat{p}_{jk}$ are found for the full model, are used to fit the model that does not include the term (species $\times$ extract)$_{jk}$, the deviance is 21.98 on 18 d.f. Hence the deviance due to the interaction term is 3.54 on 1 d.f., and by comparing this change in deviance with percentage points of the χ^2-distribution on 1 d.f., we find that this change is significant at the 6% level ($P = 0.060$). This is in close agreement with that found using the exact approach in Example 6.9. When the parameter estimates for the full model are used to calculate the fitted germination probabilities, they turn out to be very close to those in Table 6.3.

6.8 COMPARISON OF METHODS

In this chapter, we have considered a number of procedures that can be used to model overdispersion. For convenience, these are catalogued below.

Method 1: Williams procedure (section 6.4)
Method 2: Use of a heterogeneity factor (section 6.5)
Method 3: Beta-binomial modelling (section 6.6)
Method 4: Inclusion of a random effect (section 6.7)
Method 5: Approximation to method 4 (section 6.7.2)

We have seen that method 3 can be less flexible than the procedure which forms the basis of method 1, and so this procedure need not be considered further. Method 5 will often give very similar results to method 1, and since the approximation is only valid when the variance of the random effect is small, method 5 cannot be generally recommended. Moreover, computer software that can be used to fit models that contain a random effect is becoming more widely available, and so it is no longer necessary to rely on approximate methods that are more easy to implement than the exact approach. A similar argument can be applied to method 2, which is again simple to apply, but only valid when the binomial denominators are all equal.

Method 4 is a natural way of modelling overdispersion when it is believed that values of relevant explanatory variables have not been recorded, or when variables in the model have not been adequately measured or controlled. This is also the only method of those being considered that can be used to model overdispersion in binary response data. Unfortunately, after fitting a model that includes a random effect to overdispersed data, it will not be clear whether the overdispersion has actually been accommodated. Method 1 is a general procedure for modelling overdispersion in grouped binary data when there is variation between the response probabilities, and includes method 2 as a special case. Method 1 and method 4 are therefore the procedures that are recommended for general use.

6.9 A FURTHER EXAMPLE ON MODELLING OVERDISPERSION

A rotifer, or wheel animalcule, is a microscopic aquatic invertebrate that has a ciliated wheel-like organ which is used in feeding and locomotion. Rotifers are common constituents of freshwater zooplankton, and in spite of their small size, account for a considerable proportion of the zooplankton biomass. The aim of an experimental procedure described by Saunders-Davies and Pontin (1987) was to determine the relative density (or specific gravity) of two species of rotifer, *Polyarthra major* and *Keratella cochlearis*, and to make comparisons between them. The relative densities of the animals were obtained using an indirect method that involved centrifugation of rotifers in solutions that have different relative densities. When rotifers are centrifuged in a solution that has a density less than their own, they will settle to the bottom of the tube. On the other hand, when they are centrifuged in a solution that has a relative density that is equal to or greater than their own, they will remain in suspension. By using a series of solutions of increasing relative density, it is possible to estimate the density at which the animals just fail to settle out. This density will be the median relative density of the rotifers. In the experiment, stock solutions of Ficoll, a sugar polymer which has negligible osmotic effects, were prepared with known densities. For each stock solution of Ficoll that was prepared, two tubes of the solute were taken. Rotifers of one of the two species were added to one of these tubes, and rotifers of the other species were added to the other. Each tube was then centrifuged, and the solution containing the rotifers drawn off. The number of rotifers in this supernatant was then recorded and expressed as a proportion of those introduced into the solute. Data from this experiment are given in Table 6.10.

The median relative density of the rotifers is the density of the solute for which the proportion of rotifers in the supernatant is 0.5. This is equivalent to the *ED50* value used in bioassay, and discussed in Section 4.2. The methods of that section can therefore be used to estimate the median relative density and to provide a confidence interval for the true median density of the rotifers of each species.

In order to estimate and compare the relative densities of the two species of rotifer, a linear logistic model is adopted for the probability that a rotifer in a solution of a given relative density remains in the supernatant after centrifugation. Let p_{jk} be the probability that a rotifer of the jth species remains in the supernatant of a solution of Ficoll with a relative density of d_k. The relationship between p_{jk} and d_k can then be modelled by assuming that

$$\text{logit}(p_{jk}) = \alpha_j + \beta_j d_k \tag{6.15}$$

for $j = 1, 2$, where α_j and β_j are the intercepts and slopes in the relationship between

Table 6.10 Numbers of rotifers of two species that remain in suspension in tubes containing solutions of different relative densities of Ficoll, out of the number introduced into each of the tubes

Density of solution	Polyarthra major		Keratella cochlearis	
	Number in suspension	Number introduced	Number in suspension	Number introduced
1.019	11	58	13	161
1.020	7	86	14	248
1.021	10	76	30	234
1.030	19	83	10	283
1.030	9	56	14	129
1.030	21	73	35	161
1.031	13	29	26	167
1.040	34	44	32	286
1.040	10	31	22	117
1.041	36	56	23	162
1.048	20	27	7	42
1.049	54	59	22	48
1.050	20	22	9	49
1.050	9	14	34	160
1.060	14	17	71	74
1.061	10	22	25	45
1.063	64	66	94	101
1.070	68	86	63	68
1.070	488	492	178	190
1.070	88	89	154	154

logit (p_{jk}) and relative density, for the jth species. The deviance on fitting this model is 434.02 on 36 d.f., and the fitted logistic regression models for each species are:

Species 1 (*Polyarthra*): $\quad$ logit ($\hat{p}_{jk}$) = $-109.72 + 105.66 d_k$

Species 2 (*Keratella*): $\quad$ logit ($\hat{p}_{jk}$) = $-114.35 + 108.74 d_k$

A plot of the data, together with the fitted logistic regression models, is given in Fig. 6.2. This figure does not suggest that the model is inappropriate, although the curve fitted to the data from *Keratella* is not a particularly good fit at relative densities of 1.04 and 1.05. An index plot of the standardised deviance residuals, shown in Fig. 6.3, indicates that the model is satisfactory, in that there is no systematic pattern in the plot and there are no observations that have unusually large positive or negative residuals. Using the logarithm of the relative density, rather than the relative density itself, does not appear to improve the fit of the model to the observed proportions. We therefore conclude that the large deviance for the model in equation (6.15) is due to overdispersion. Indeed, Figs. 6.2 and 6.3

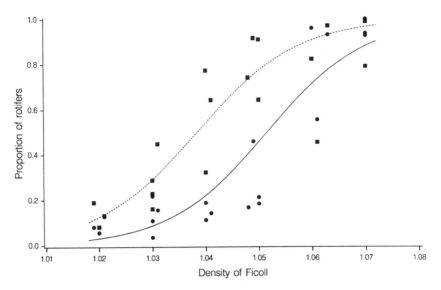

Fig. 6.2 The observed proportion of rotifers of the two species, *Polyarthra* (⋯⋯■⋯⋯) and *Keratella* (──●──), plotted against the relative density of the solute, Ficoll. The fitted logistic regression curves have been superimposed.

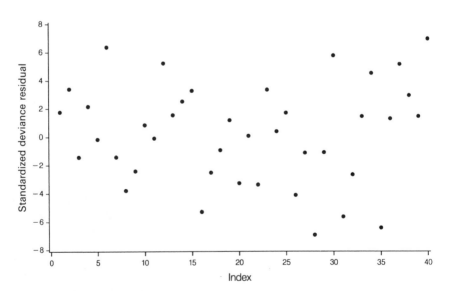

Fig. 6.3 Index plot of the standardized deviance residuals on fitting separate logistic regression lines to the data from the two species of rotifer.

show that at the different densities of Ficoll, there is a considerable amount of variation in the observed proportions of rotifers remaining in suspension.

A possible explanation of this overdispersion is contained in the article by Saunders-Davies and Pontin. They point out that immature rotifers tend to be less dense than the more mature ones, and that there were very likely to have been differences in the proportions of immature rotifers in the different batches used. This would lead to excess variability in the data. Had the experimenters been able to use rotifers of similar ages, or had they been able to determine the ages of those actually used in their experiment and allowed for this in the subsequent modelling, there would have been the potential for reducing the variability in the observed proportions.

6.9.1 Fitting a linear logistic model with a random effect

Since the most likely cause of the overdispersion in the data given in Table 6.10 is that the age of the rotifers used in the experiment was not controlled, a natural approach to modelling this overdispersion is to include a random effect in the model. The model with separate logistic regression lines for each species can then be written in the form

$$\text{logit}\,(\vartheta_{jk}) = \alpha_j + \beta_j d_k + \gamma z_{jk}$$

where ϑ_{jk} is the (random) response probability for the jth species at density d_k and z_{jk} is a standard normal random variable. Under this model, the logistic transform of ϑ_{jk} has a normal distribution with zero mean and variance γ^2. When the package EGRET is used to fit this model, the deviance is 132.80 on 35 d.f. This is substantially less than the deviance for the corresponding model that does not include the random effect, indicating that the random effect explains a large proportion of the overdispersion. We can now go on to investigate whether the model can be simplified, and so the model

$$\text{logit}\,(\vartheta_{jk}) = \alpha_j + \beta d_k + \gamma z_{jk} \qquad (6.16)$$

under which the coefficient of d_k does not depend on species, is fitted. The deviance for this model is 133.30 on 36 d.f., and so the increase in deviance brought about by constraining the individual logistic regression lines to be parallel is 0.50 on 1 d.f. This is not significant when compared to percentage points of the χ^2-distribution on 1 d.f., and so we may conclude that the parallel line model fits the data as well as two separate lines. To determine whether the model can be further simplified, a single logistic regression line with a random effect, where

$$\text{logit}\,(\vartheta_{jk}) = \alpha + \beta d_k + \gamma z_{jk}$$

is fitted to the data from the two species. The deviance for this model is 147.41 on 37 d.f., an increase of 14.11 from that for the parallel line model. This increase in deviance is highly significant, and so the same relationship between the logistic transform of the probability that a rotifer remains in the supernatant, and relative density, cannot be used for the two species. The fitted response probabilities under the parallel line model are given by

Species 1 (*Polyarthra*): $\text{logit}(\hat{p}_{jk}) = -113.32 + 109.02d_k$

Species 2 (*Keratella*): $\text{logit}(\hat{p}_{jk}) = -114.46 + 109.02d_k$

The estimated standard deviation of the random effect term is given by $\hat{\gamma} = 1.04$.

The fitted model can now be used to estimate the relative density of each species of rotifer. The median relative density of the jth species of rotifer can be estimated by $-\hat{\alpha}_j/\hat{\beta}$, where $\hat{\alpha}_j$ is the estimated intercept in the logistic regression model for the jth species, and $\hat{\beta}$ is the estimated slope. The estimated relative density of the first species, *Polyarthra major*, is 1.039, while that for *Keratella cochlearis* is 1.050. Confidence intervals for the corresponding true relative densities can be obtained using Fieller's theorem given in section 4.2.2. The variance–covariance matrix of the parameter estimates in the model of equation (6.16) is

$$
\begin{array}{c}
\hat{\alpha}_1 \\
\hat{\alpha}_2 \\
\hat{\beta}
\end{array}
\begin{bmatrix}
29.65 & & \\
29.11 & 28.60 & \\
-28.08 & -27.59 & 26.61
\end{bmatrix}
$$
$$
\qquad \hat{\alpha}_1 \qquad \hat{\alpha}_2 \qquad \hat{\beta}
$$

Using expression (4.7) of Chapter 4 to obtain 95% confidence limits, $z_{\alpha/2}$ is replaced by the $\alpha/2$ point of a t-distribution with 36 d.f., and $\hat{v}_{00}$, $\hat{v}_{01}$, and $\hat{v}_{11}$ are 29.65, -28.08, 26.61, respectively for the first species, and 28.60, -27.59, 26.61, respectively for the second species. A 95% confidence interval for the true relative density of the species *Polyarthra major* is found to be (1.036, 1.042), and that for *Keratella cochlearis* is the interval (1.049, 1.051). These intervals are slightly wider than those obtained when no allowance is made for the ovedispersion.

In this example, the deviance on fitting a model that contains a random effect, 133.30, is still large relative to the number of degrees of freedom, 36. This indicates that there may be additional overdispersion that is not accounted for by the inclusion of a random effect. In the absence of theoretical results on the distribution of the deviance for such models, it is not clear how to proceed with the analysis.

6.9.2 Modelling the overdispersion using the Williams procedure

The method of modelling the overdispersion presented in section 6.4 will now be used in an alternative analysis of the data on rotifers. Using the procedure for fitting the model outlined in Section 6.4, the parameter ϕ is estimated by 0.136. When the full model described in equation (6.15) is fitted, using weights $w_{jk} = 1/[1 + (n_{jk} - 1)\hat{\phi}]$, the deviance is 34.24 on 36 d.f. The next step is to fit parallel logistic regression lines for the two species, using the same weights. The

868

888866866677888

deviance for the model

$$\text{logit}(p_{jk}) = \alpha_j + \beta d_k$$

is 35.40 on 37 d.f. The change in deviance on constraining the parameters β_1 and β_2 to be equal is therefore 1.16 on 1 d.f., which is not significant. A pair of parallel logistic regression lines therefore fit the observed proportions as well as separate lines, after allowing for the overdispersion. When the single logistic regression model

$$\text{logit}(p_{jk}) = \alpha + \beta d_k$$

is fitted to the combined data from each of the two species, again using weights w_{jk}, the deviance is 51.47 on 38 d.f. The increase in deviance from that of the parallel line model is 16.07 on 1 d.f., which is highly significant, and so the model with parallel logistic regression lines is again the most satisfactory one. The fitted logistic regression equations under this model are as follows:

Species 1 (*Polyarthra*): $\text{logit}(\hat{p}_{jk}) = -101.90 + 98.04 d_k$

Species 2 (*Keratella*): $\text{logit}(\hat{p}_{jk}) = -103.18 + 98.04 d_k$

From these equations, the estimated relative density of the first species, *Polyarthra major*, is 1.039, while that for *Keratella cochlearis* is 1.052. These estimates are practically identical to those found using the random effects model. Confidence intervals for these estimates can be obtained using Fieller's theorem given in Section 4.2.2, modified to accommodate overdispersion in the manner illustrated in Example 6.5. Details will be omitted, but the resulting 95% confidence intervals are (1.034, 1.044) for *Polyarthra major* and (1.048, 1.057) for *Keratella cochlearis*. These intervals are wider than those obtained when a random effect is included in the model, because the model used in this section accounts for all of the overdispersion in the observed data.

FURTHER READING

Much of the earlier work on the phenomenon of overdispersion in data from toxicological experiments is reviewed by Haseman and Kupper (1979). Anderson (1988) gives a summary of models for overdispersion, and illustrates their use in the analysis of two data sets. The books of Aitkin et al. (1989), Cox and Snell (1989) and McCullagh and Nelder (1989) also contain sections on overdispersion.

A key paper in the literature on overdispersion is Williams (1982a), which includes the material on which Sections 6.2, 6.4 and 6.7.2 of this chapter are based. The use of the beta-binomial model for overdispersed proportions was proposed by Williams (1975) and Crowder (1978). Pierce and Sands (1975) proposed linear logistic models with a random effect for the analysis of binary data. Follmann and

Lambert (1989) describe a non-parametric method for fitting models with random effects, in which the distribution of the random term remains unspecified.

An alternative way of obtaining the estimates $\hat{\beta}, \hat{\gamma}$ which minimise the marginal likelihood function in (6.10) is to use what is known as the **E–M algorithm**, due to Dempster, Laird and Rubin (1977). This algorithm can be implemented in a GLIM macro, following the method outlined by Anderson and Aitkin (1985). Hinde (1982) discusses the use of this algorithm in the analysis of overdispersed Poisson data, and gives GLIM code which can be easily modified for use with binary data. However, the algorithm does tend to be rather slow, which is why it has not been discussed in this chapter.

The model in Section 6.5 for the variance of the number of successes for the ith observation can be generalized to give $\mathrm{Var}(y_i) = \sigma^2 n_i p_i (1 - p_i)$, where the n_i are no longer assumed constant. Moreover, by allowing σ^2 to lie in the range $(0, \infty)$, underdispersion can be modelled by values of σ^2 less than unity. This model can be fitted using the notion of **quasi-likelihood**, which is discussed in detail in Chapter 9 of McCullagh and Nelder (1989). In fact, the quasi-likelihood approach leads to the same procedure for adjusting for overdispersion as that described in Section 6.4.

7

Modelling data from epidemiological studies

Epidemiology is the study of patterns of disease occurrence and the factors that influence such patterns. Historically, epidemiology was concerned with epidemics of those infectious diseases that were responsible for much suffering during the eighteenth and nineteenth centuries, such as cholera and typhoid. Nowadays, the scope of epidemiology has been extended to include acute conditions, such as the occurrence of episodes of food poisoning, and chronic diseases, such as various types of cancer.

The aim of an epidemiological study is to investigate the cause, or **aetiology**, of a disease or physiological condition. Many diseases are not caused by one particular agent, but through a combination of different circumstances, termed **exposure factors** or **risk factors**. These exposure factors may relate to the physical characteristics of an individual, such as age and body mass index; physiological variables, such as serum cholesterol level and measures of lung function; personal habits, such as history of tobacco smoking and diet; socio-economic factors, such as social class and occupational group; and environmental effects, such as proximity to a nuclear power station and the extent of industrialization in the area in which the individual lives. An aetiological study seeks to determine whether there is an association between one or more exposure factors and the occurrence of a particular disease state.

Typical examples of epidemiological studies include investigations into the relation between the use of combined oral contraceptives and the occurrence of gall bladder cancer in women, the extent to which the risk of myocardial infarction is related to factors such as age, smoking history and dietary habits, and the association between high alcohol consumption and the incidence of oesophageal cancer. In each of these studies, it would not be feasible to use an experimental approach to determine whether or not particular factors are indeed risk factors for the disease in question. Instead, one of two basic types of survey design would be used. These two designs are outlined in the next section and measures of association between the occurrence of a disease and exposure factors are described in Section 7.2. However, the main purpose of this chapter is to show how the linear logistic model can be used to analyse data from aetiological studies.

7.1. BASIC DESIGNS FOR AETIOLOGICAL STUDIES

Two basic designs used in epidemiological surveys into the aetiology of a disease are the **cohort study** and the **case-control study**. Each of these designs is described in the following sub-sections. Fuller details on the design of aetiological studies can be found in Breslow and Day (1980, 1987).

7.1.1 The cohort study

In a cohort study to determine if there is an association between certain exposure factors and the occurrence of a particular disease, a sample consisting of individuals who are free of the disease in question is selected. The individuals in this sample are then stratified according to the exposure factors of interest, and followed-up for a given period of time. Each individual is then classified according to whether or not he or she has developed the disease that is being studied. The relationship between the probability of disease occurrence and the exposure factors can then be investigated.

Example 7.1 The Framingham study
Framingham is an industrial town located some twenty miles west of Boston, Massachusetts, USA. In 1948, a cohort study was begun with the broad aim of determining which of a number of potential risk factors are related to the occurrence of coronary heart disease (CHD). At the start of the study, a large proportion of the town's inhabitants were examined for the presence of CHD. Measurements were also made on a number of other variables, including age, serum cholesterol level, systolic blood pressure, smoking history and the result of an electrocardiogram. Those individuals found to be free of CHD at that time were followed-up for twelve years and those who developed CHD during that period were identified. The resulting data set consisted of this binary response variable and information on the risk factors for 2187 men and 2669 women aged between 30 and 62. The summary data in Table 7.1 are adapted from Truett, Cornfield and

Table 7.1 Proportions of cases of CHD, cross-classified by age and sex, and initial serum cholesterol level

Sex	Age group	Serum Cholesterol level			
		< 190	190–219	220–249	⩾ 250
Male	30–49	13/340	18/408	40/421	57/362
	50–62	13/123	33/176	35/174	49/183
Female	30–49	6/542	5/552	10/412	18/357
	50–62	9/58	12/135	21/218	48/395

Kannel (1967) and relate to the initial serum cholesterol level (in units of mg/100 ml) of these individuals, cross-classified according to their age and sex. In analysing these data, the epidemiologist would be interested in the extent to which the occurrence of CHD is associated with initial serum cholesterol level, after age and sex effects have been allowed for, and whether or not the degree of association is similar for each sex and each age group.

A cohort study is often termed a **prospective study**, since the individuals are followed prospectively in time. However, a modification of the design that is useful in studies of occupational health would involve workers who have been in that occupation for a relatively long period of time, ten or twenty years, for example. The current health status of these employees would then be compared with historical records, and comparisons made between different occupational groups on the basis of the probability of disease occurrence. Investigations of this type, where the outcome of interest has occurred before the study has begun, are known as historical cohort studies.

If the follow-up time differed for each individual, if they had been at risk for different periods before the study began, or at risk intermittently through the duration of the study, it would be sensible to take account of this in the analysis. One possibility is to compute the number of **person-years** of exposure for individuals who have been cross-classified according to factors such as age group. The number of individuals who develop the disease in a particular group is then expressed as a proportion of the person-years of exposure. This rate of occurrence of the disease can then be modelled using the **Poisson regression model**, described briefly in Section 8.2. When the time from entry into the study until the occurrence of a particular disease is of interest, models developed for the analysis of survival data, particularly the **proportional hazards model** due to Cox (1972), become appropriate. These techniques for analysing data from occupational health studies can similarly be used in the analysis of occupational mortality data. The methods will not be considered in this book, but are described by Breslow and Day (1987), for example.

7.1.2 The case-control study

A case-control study is an aetiological study in which comparisons are made between individuals who have a particular disease or condition, known as **cases**, and individuals who do not have the disease, known as **controls**. A sample of cases is selected from a population of individuals who have the disease being studied and a sample of controls is selected from individuals that do not have the disease. Information about factors which might be associated with the disease is then obtained retrospectively for each person in the study. For this reason, a case-control study is also known as a **retrospective study**.

Example 7.2 Age at first coitus and cervical cancer
Adelusi (1977) describes a case-control study to investigate whether coital characteristics are associated with the subsequent development of cervical cancer. In this

study, the cases were married Nigerian women with a histologic diagnosis of invasive cancer of the cervix, presenting at University College Hospital, Ibadan, Nigeria, between October, 1972 and September, 1974. The control group consisted of healthy married women of child bearing age who attended the hospital's family planning clinic during the same period. A total of 47 cases were selected and nearly four times as many controls were used. A questionnaire was administered to each of the 220 women in the study, in which they were asked about their sexual habits. One particular question asked the age at which they first had sexual intercourse, and the women were subsequently classified according to whether or not their reported age at first coitus was less than or equal to fifteen years. The data recorded are given in Table 7.2. In Section 7.6, we will see how the risk of cervical cancer

Table 7.2 Data from a study to examine the association between the occurrence of cervical cancer and age at first coitus

Age at first coitus	Cases	Controls
≤ 15 years	36	78
> 15 years	11	95
Total	47	173

occurring in women who were aged 15 or less at first coitus, relative to those who were older than 15, can be estimated. However, these data cannot be used to estimate the risk of cervical cancer occurring in, say, women who were 15 or less at first intercourse. This is because the proportion of women with cervical cancer in the case-control study is much greater than the proportion of women with cervical cancer in the female population of Ibadan. A similar point was made in the discussion of Example 1.10, and will be discussed in greater detail in section 7.6.

Had the investigation described in Example 7.2 been conducted as a cohort study, a large number of women without cervical cancer would have been stratified according to age at first intercourse, and followed prospectively, or perhaps retrospectively, for a long period of time. The main advantage of this design would be that the risk of developing cervical cancer for the women in each group could be estimated directly from the corresponding proportion who develop the cancer during the follow-up period. However, the cohort study would take a much greater length of time to complete than the case-control study.

In general, it is much quicker to carry out a case-control than a cohort study, since cases of the disease are selected at the outset. When the disease under study is rare, a cohort study will also need a much greater number of subjects in order to ensure that there is a sufficient number of cases at the end of the study. It may be difficult and costly to follow up the patients over the length of time required for a cohort study and so it will tend to be much cheaper to conduct a case-control study than a cohort study. One disadvantage of the case-control design is that it relies on a subject's recall of past experiences or on suitable documentary evidence of past exposure being available. This could lead to unreliable data. Also, the case-control study is restricted to one disease, whereas information on the incidence of a number of different diseases can be collected in a cohort study. Finally, in the case-control design, considerable care must be taken to ensure that the cases and controls used in the study are representative of the underlying populations of individuals. In particular, the probability that an individual is included in the study as either a case or a control must not be associated with exposure factors of interest.

7.2 MEASURES OF ASSOCIATION BETWEEN DISEASE AND EXPOSURE

The probability of a disease occurring during a given period of time, also known as the **risk** of disease occurrence, is the number of new cases of the disease that occur in that time period, expressed as a proportion of the population at risk. In a cohort study, where a sample of the population at risk is monitored to determine the incidence of a disease, the risk can be estimated directly from the proportion of individuals in the sample who develop the disease during the follow-up period. Risks are most conveniently expressed as incidence rates per unit time. For example, in Example 7.1, the estimated risk of coronary heart disease occurring during a twelve year perod, in males aged 30–49 with an initial serum cholesterol level that exceeds $250\,mg/100\,ml$, is $57/362$, that is, 0.157. The incidence rate is therefore 0.0131 per person per year, or 13.1 per 1000 persons per year.

The main purpose of estimating risks is to be able to make comparisons between the risk of disease at different levels of an exposure factor, and thereby assess the association between the disease and that exposure factor. For this, a measure of the difference in risk of disease occurrence for one level of an exposure factor relative to another is required. Suppose that a particular exposure factor has two distinct values or levels, labelled exposed and unexposed. For example, if the exposure factor were smoking history, a person may be classified as a smoker (exposed) or non-smoker (unexposed). Denote the risk of the disease occurring during a given time period by p_e for the exposed persons and p_u for the unexposed persons. The risk of the disease occurring in an exposed person relative to that for an unexposed person is then

$$\rho = \frac{p_e}{p_u}$$

This is called the **relative risk** of disease, and is a measure of the extent to which an individual who is exposed to a particular factor is more or less likely to develop a disease than someone who is unexposed. In particular, an exposed person is ρ times more likely to contract the disease in a given period of time than an unexposed person. When the risk of the disease is similar for the exposed and unexposed groups, the relative risk will be close to unity. Values of ρ less than one indicate that an unexposed person is more at risk than an exposed person, while values of ρ greater than one indicate that an exposed person is at greater risk.

Example 7.3 The Framingham study
The data in Table 7.1 give the twelve year incidence rates of coronary heart disease amongst the inhabitants of Framingham. From this table, the risk of CHD for males aged 30–49 with an initial serum cholesterol level that is greater than or equal to $250 \, mg/100 \, ml$ is $57/362 = 0.157$. The corresponding risk for males aged 30–49 with an initial serum cholesterol level less than $250 \, mg/100 \, ml$ is $(13 + 18 + 40)/(340 + 408 + 421) = 0.061$. The risk of CHD in a male aged 30–49 with cholesterol level greater than or equal to 250 relative to someone with a cholesterol level less than 250 is therefore given by $\rho = 0.157/0.061 = 2.58$. In other words, a man aged between 30 and 49 with an initial serum cholesterol level of $250 \, mg/100 \, ml$ is more than two and a half times more likely to develop CHD than someone whose serum cholesterol level is less than $250 \, mg/100 \, ml$.

Another useful measure of association between the occurrence of a particular disease state or condition and an exposure factor is the **odds ratio**, described in Section 2.3.4. Again suppose that the exposure factor is dichotomous with levels labelled exposed and unexposed. Let p_e and p_u be the corresponding risks of disease in the exposed and unexposed groups, respectively. The odds of a person having the disease in the exposed group are then $p_e/(1 - p_e)$, and similarly, the odds of disease for someone in the unexposed group are $p_u/(1 - p_u)$. The odds ratio is the ratio of the odds of disease for an exposed relative to an unexposed person, given by

$$\psi = \frac{p_e/(1 - p_e)}{p_u/(1 - p_u)}$$

When a disease is rare, p_e and p_u are both small and the odds ratio, ψ, is approximatey equal to the relative risk, ρ. To show this, consider the odds of disease for an individual in the exposed group, $p_e/(1 - p_e)$. This can be written as

$$p_e(1 - p_e)^{-1} = p_e(1 + p_e + p_e^2 + \cdots)$$

If p_e is small, p_e^2 and higher-order powers of p_e will be negligible and so $p_e/(1 - p_e)$ will be approximately equal to p_e. By the same token, $p_u/(1 - p_u)$ is approximately equal to p_u, when p_u is small. Consequently, the relative risk, ρ, can be approximated by the odds ratio, ψ, so long as the disease is sufficiently rare. This approximation is likely to be valid in most epidemiological studies of the association between disease occurrence and exposure factors.

Since the odds ratio is approximately equal to the relative risk when the disease under study is rare, the odds ratio is widely used as a measure of association between a disease and an exposure factor. In addition, the odds ratio can easily be estimated after fitting a linear logistic model to cohort data, as we will see in Section 7.5. But the main reason why the odds ratio is so important in quantitative epidemiology is that it is the only quantity that can be estimated from the results of a case-control study, to be shown in Section 7.6. For these reasons, ψ is generally used as a measure of the relative risk of disease occurrence in an exposed relative to an unexposed person, in both cohort and case-control studies.

7.2.1 Inference about the odds ratio

The odds ratio ψ can easily be estimated when the exposure factor is dichotomous. Suppose that we have data from a cohort study where the individuals have all been observed for the same period of time, and where the numbers of individuals in the four combinations of disease state and exposure level are a, b, c and d, as in the following table.

	Diseased	Not diseased
Exposed	a	b
Unexposed	c	d

The estimated risk of disease for individuals in the exposed and unexposed groups are given by $\hat{p}_e = a/(a + b)$ and $\hat{p}_u = c/(c + d)$, respectively, and so the relative risk is estimated by

$$\hat{\rho} = \frac{\hat{p}_e}{\hat{p}_u} = \frac{a(c + d)}{c(a + b)}$$

The estimated odds ratio is

$$\hat{\psi} = \frac{\hat{p}_e/(1 - \hat{p}_e)}{\hat{p}_u/(1 - \hat{p}_u)} = \frac{ad}{bc} \tag{7.1}$$

which is the ratio of the products of the two pairs of diagonal elements in the above 2×2 table. As stated in Section 2.3.4, the approximate standard error of $\log \hat{\psi}$ is given by

$$\text{s.e.} (\log \hat{\psi}) \approx \sqrt{\left(\frac{1}{a} + \frac{1}{b} + \frac{1}{c} + \frac{1}{d}\right)} \tag{7.2}$$

while the standard error of $\hat{\psi}$ itself is approximately $\hat{\psi}$ s.e. $(\log \hat{\psi})$. For reasons given in Section 2.3.4, confidence intervals for the true odds ratio, ψ, are best obtained from exponentiating the confidence limits for $\log \psi$, so that approximate

$100(1 - \alpha)\%$ confidence limits for ψ are $\exp[\log \hat{\psi} \pm z_{\alpha/2} \text{ s.e. } (\log \hat{\psi})]$, where $z_{\alpha/2}$ is the upper $(100\alpha/2)\%$ point of the standard normal distribution.

Example 7.4 The Framingham study
The data used in Example 7.3 to illustrate the computation of an estimated relative risk are presented in the following table.

Cholesterol level	CHD	No CHD
$\geqslant 250$	57	305
< 250	71	1098

The estimated odds ratio is given by

$$\hat{\psi} = \frac{57 \times 1098}{305 \times 71} = 2.890$$

Thus, the odds of developing CHD for a person with cholesterol level greater than or equal to 250 mg/100 ml, relative to someone with a cholesterol level less than 250 mg/100 ml, is 2.89. This is quite close to the estimated value of the relative risk obtained in Example 7.3. The approximate standard error of the estimated odds ratio is given by

$$\text{s.e. } (\log \hat{\psi}) \approx \sqrt{\left(\frac{1}{57} + \frac{1}{305} + \frac{1}{71} + \frac{1}{1098}\right)} = 0.189$$

and the standard error of $\hat{\psi}$ itself is 2.890×0.189, that is, 0.547. An approximate 95% confidence interval for $\log \psi$ is the interval from 0.690 to 1.432, and the corresponding interval for the true odds ratio ranges from 1.99 to 4.19. This interval does not include unity and so we may conclude that there is real evidence to suggest that a raised cholesterol level increases the risk of coronary heart disease.

The odds ratio can similarly be estimated from data from a case-control study, as in the following example.

Example 7.5 Age at first coitus and cervical cancer
From the data in Table 7.2 concerning the results of a case-control study to investigate the association between age at first sexual intercourse and the development of cervical cancer in Nigerian women, the estimated odds ratio is $\hat{\psi} = (36 \times 95)/(78 \times 11) = 3.99$. This odds ratio can be interpreted as the approximate relative risk of cervical cancer in women aged 15 years or less at first coitus, relative to those who are older than 15. The risk of cervical cancer occurring in women who have first intercourse when they are relatively young is therefore about

four times that of women who have first intercourse when they are older than 15 years.

Using relation (7.2), s.e. $(\log \hat{\psi}) \approx 0.377$, and the approximate standard error of the odds ratio itself is 1.504. An approximate 95% confidence interval for $\log \psi$ is the interval from 0.644 to 2.122. Consequently, a 95% confidence for the true odds ratio is the interval from 1.90 to 8.35. This interval clearly shows that early age at first coitus is a risk factor for cervical cancer.

In this section, we have seen how the association between a particular disease and a dichotomous exposure factor can be estimated. In practice, we would rarely want to consider a particular exposure factor in isolation, and account would need to be taken of other variables measured on the individuals in the study. Some implications of this are discussed in general terms in the next section.

7.3 CONFOUNDING AND INTERACTION

In an aetiological study designed to investigate whether a disease or condition is associated with particular exposure factors, information will usually be obtained on a number of factors that are expected to be associated with the occurrence of the disease. Although some of these will be of interest as possible causes of the disease, others may simply disguise or exaggerate the true effect of an exposure factor of direct interest. This may mean that an apparent association between an exposure factor and disease occurrence is entirely due to another factor. Alternatively, a lack of association between an exposure factor and the disease may be due to the fact that proper account has not been taken of that other factor in the analysis of the data.

7.3.1 Confounding

A variable that completely or partially accounts for the apparent association between a disease and an exposure factor is called a **confounding variable** or **confounder**. Specifically, a confounding variable is defined to be a variable that is a risk factor for the disease under study, and is associated with, but not a consequence of, the exposure factor. A variable which leads to exposure which in turn causes the disease will not usually be regarded as a confounding variable.

In practice, information will often be collected on a number of variables that might turn out to be confounding variables. Whether or not they are confounding variables will only be apparent after the data have been analysed. For this reason, much of the ensuing discussion is phrased in terms of potential confounding variables, although this distinction will not be made when it is clear from the context.

Example 7.6 The Framingham study
In the Framingham study outlined in Example 7.1, interest centred on the extent of association between the occurrence of coronary heart disease and serum cholesterol level. However, information was also collected on the age and sex of the

individuals in the survey, as well as on a number of other variables. Age and sex both tend to be associated with serum cholesterol level, in that raised cholesterol levels more commonly occur in males and amongst older persons. Neither age nor sex are a consequence of the exposure factor, in the sense that a high serum cholesterol level does not predispose a person to be in a particular age group or of particular sex! Both age and sex are also risk factors for CHD, in that CHD is predominantly a male disease whose incidence increases with age. Consequently, age and sex are possible confounding variables, and any assessment of the overall effect of elevated serum cholesterol levels on the occurrence of CHD should take account of the age and sex distribution of the individuals in the cohort study.

The consequence of ignoring an important confounding variable in the analysis is that incorrect estimates of an odds ratio can be obtained. The following example illustrates this in the context of a case-control study, but the same phenomenon can also occur in cohort studies.

Example 7.7 CHD, alcohol consumption and smoking history
Consider a case-control study to investigate whether coronary heart disease is associated with high alcohol consumption, in which the cases are individuals with CHD and a dichotomous exposure factor corresponds to high and low alcohol consumption. Suppose that this study yields the artificial data shown in Table 7.3.

Table 7.3 Artificial data from a case-control study on the association between CHD and alcohol consumption

Alcohol consumption	Cases	Controls
High	68	33
Low	32	72

From these data, the ratio of the odds of CHD occurring in persons whose alcohol consumption is high relative to those for whom it is low is 4.64. This suggests that the risk of CHD in persons who consume large amounts of alcohol is nearly five times that of those who consume small amounts. However, we also know that tobacco smoking is a risk factor for CHD and tends to be associated with alcohol consumption: heavy smokers are often heavy drinkers. Might the observed association between CHD and alcohol consumption be due to smoking history? Suppose that information is available on the smoking habits of the 205 individuals in this case-control study and that they are stratified according to whether they are smokers or non-smokers. This leads to the data in Table 7.4. From the information in this table, the odds ratios for non-smokers and smokers are 1.11 and 1.12, respectively. Both of these are very close to unity and suggest that there is no association between CHD and high alcohol consumption in both smokers and non-smokers. The reason why this result contradicts that based on the unstratified

Table 7.4 Data from Table 7.3 stratified according to smoking history

Alcohol consumption	Non-smokers		Smokers	
	Cases	Controls	Cases	Controls
High	5	18	63	15
Low	17	68	15	4

data in Table 7.3 is that there is a much higher proportion of smokers amongst the cases. It is this that accounts for the apparent association between alcohol consumption and CHD when smoking habit is ignored. When due account is taken of smoking habit, we would conclude that the occurrence of CHD is unrelated to alcohol consumption. Smoking habit is therefore a confounding variable.

Example 7.7 stressed the importance of allowing for potential confounding variables by stratification. It is also important to avoid stratifying by variables that do not satisfy the criteria for a confounding variable. For example, suppose that the disease being studied is oesophageal cancer and the exposure factor of interest is high alcohol consumption. Suppose also that information is available on whether or not each person in the study has cirrhosis of the liver. One might expect that the pattern of alcohol consumption amongst oesophageal cancer patients would be similar to that for cirrhotic patients. Accordingly, if the individuals in the study are first stratified according to whether or not they have cirrhosis of the liver, the association between high alcohol consumption and oesophageal cancer may disappear. In this illustraton, the presence or absence of cirrhosis of the liver should not be regarded as a confounding variable, because there is a causal association between the exposure factor and this variable: heavy drinking leads to cirrhosis of the liver.

In some circumstances, it might be legitimate and informative, to carry out two analyses, one in which the confounding variable is ignored, and one where account is taken of the confounder. For example, consider an aetiological study in which the exposure factor is the occurrence of sickle-cell anaemia in an infant, and the disease is low birth weight. A possible confounding variable is the period of gestation, since this variable is a risk factor for low birth weight. It could be argued that a short gestation period is a consequence of sickle-cell anaemia, and so this variable should not be regarded as a confounder when studying the association between the occurrence of sickle-cell anaemia and low birth weight. However, another hypothesis of interest is whether or not there is an alternative route to low birth weight, other than through being premature. An analysis in which gestational age is taken into account would then be used to investigate this hypothesis.

7.3.2 Interaction

In Example 7.7, the odds ratios for smokers and non-smokers were very similar, showing that the association between CHD and alcohol consumption was consistent between smokers and non-smokers. When the degree of association between a disease and an exposure factor is different for each level of a confounding variable, there is said to be **interaction** between the exposure factor and the confounding variable. Another way of putting it is to say that the confounding variable modifies the effect of the exposure factor on the disease. A confounding variable that interacts with an exposure factor may therefore be called an **effect modifier**. This concept is illustrated in the following example.

Example 7.8 CHD, alcohol consumption and smoking history.
Suppose that the 205 individuals in the case-control study described in Example 7.7 are again stratified according to smoking habit, but that now the data in Table 7.5 are obtained. The odds ratios for the non-smokers and smokers are now 1.02

Table 7.5 Data from Table 7.3 stratified according to smoking habit

Alcohol consumption	Non-smokers		Smokers	
	Cases	Controls	Cases	Controls
High	24	18	44	15
Low	17	13	15	59

and 11.54, respectively. This suggests that there is no association between CHD and high alcohol consumption in non-smokers, but that the risk of CHD in smokers is increased by a factor of 11.5 if they also consume large amounts of alcohol. Smoking habit and alcohol consumption are said to interact in terms of their effect on the occurrence of CHD. Alternatively, we can say that smoking modifies the effect of high alcohol consumption on the occurrence of CHD.

When the ratios of odds for exposed relative to unexposed individuals at different levels of a confounding variable are comparable, they can be combined to give an overall odds ratio. This overall value is then said to be **adjusted** for the confounding variable. However, when the odds ratios differ across the levels of a confounding variable, that is, when there is interaction, it is not appropriate to combine them, and the individual odds ratios for each level of the confounding variable must then be reported. Linear logistic modelling provides a straightforward way of adjusting odds ratios to take account of confounding variables, and for studying the extent of interaction between an exposure factor and a confounding variable in aetiological studies.

7.4 THE LINEAR LOGISTIC MODEL FOR DATA FROM COHORT STUDIES

The first published applications of linear logistic modelling to data from cohort studies appeared during the early 1960s and were the result of work by Cornfield and his associates on the analysis of data from the Framingham study on coronary heart disease. For a number of reasons, this approach is now widely used in the analysis of data from aetiological studies. In particular, it provides a systematic way of exploring the relationship between the probability of disease occurrence and exposure factors of interest, after suitable allowance has been made for any confounding variables. It is straightforward to determine whether there are inter-actions between exposure factors and confounding variables and estimates of the relative risk of disease occurrence can easily be obtained from the fitted model.

Suppose that in a cohort study that involves n individuals, information is collected on exposure factors and potential confounding variables for each person in the study. These variables may be continuous variates such as weight and systolic blood pressure; they may be factors such as occupational group and parity, in which case they are represented as sets of indicator variables constructed as in Section 3.2.1, or they may correspond to interactions between two or more factors. The values of k such explanatory variables measured on the ith individual will be denoted by $x_{1i}, x_{2i}, \ldots, x_{ki}, i = 1, 2, \ldots, n$. Whether or not the ith individual in the study develops the disease during the follow-up period will be represented by the value y_i of a binary response variable, where y_i takes the value unity if that individual develops the disease and zero otherwise. The corresponding probability that the ith individual develops the disease during the follow-up period will be denoted by p_i, so that the dependence of p_i on the k explanatory variables can be modelled using a linear logistic model, where

$$\operatorname{logit}(p_i) = \log\left(\frac{p}{1 - p_i}\right) = \beta_0 + \beta_1 x_{1i} + \beta_2 x_{2i} + \cdots + \beta_k x_{ki}$$

In order to investigate the association between particular exposure factors and the occurrence of the disease, after allowing for the effects of potential confounding variables, an analysis of deviance is used. In summary, relevant confounding variables are first included in the linear logistic model. Terms associated with the exposure variables are then added to the model, and the extent to which they are significant is assessed. The effect of the exposure factors on the logistic transform-ation of the probability of disease occurrence is then said to be adjusted for the confounding variables, a concept which was first discussed in section 3.9. The general strategy for modelling data from cohort studies is described in greater detail in the following sub-section, and this is followed by an illustrative example.

7.4.1 Model selection

Before the effects of particular exposure factors on the occurrence of the disease can be studied, confounding variables of potential importance must be identified. The decision on which variables to use as confounding variables is best made on the basis of epidemiological considerations. Thus, those variables that are expected to be related to the disease being studied, and are not a consequence of exposure factors of interest, will be the ones that are adopted as potential confounding variables in the analysis. However, in some circumstances, such as when it is unclear whether to allow for an interaction between two confounding variables, statistical arguments may be needed to identify which confounding variables should be adjusted for.

There are two possible statistical procedures that can be used to determine whether a potential confounding variable should be included in the linear logistic model. In the first approach, the effect of possible confounding variables on the probability of disease occurrence is assessed by adding these variables to a linear logistic model that contains a constant term. The decision on whether or not to include a particular confounding variable could then be based on the difference in deviance that results from including it in the model. Since it is important to include all confounding variables that are likely to have an effect on the inference to be drawn about the relevance of the exposure factors, potential confounding variables should be included even if the resulting change in deviance is not significant at conventional levels of significance. When there are a number of potential confounding variables, each of them may be added on their own to the model that includes a constant term to determine which are the most important. Combinations of these variables, as well as interactions amongst them, can then be included in order to identify a set of confounding variables that the exposure factors will be adjusted for.

A more direct procedure for assessing whether or not a potential confounding variable should be adjusted for is to examine the effect of the confounder on estimates of odds ratios of interest. For example, if adjusting for the confounder has a negligible effect on an estimated odds ratio, there is no need to take account of the confounder, and it need not be included in the model. On the other hand, if the effect of including the confounder in the model is to substantially change the estimated odds ratio, it will be important to include that confounding variable in the model.

Epidemiological conclusions are generally based on the magnitude of estimated odds ratios, and so it will usually be preferable to assess the impact of a confounding variable on the estimated values of relevant odds ratios. However, since estimates of odds ratios need to be obtained from the fitted linear logistic model in order to implement this procedure, which will be discussed in the next section, this approach will not be illustrated until Examples 7.13 and 7.15.

Once relevant confounding variables have been identified, the next step is to see which exposure factors are important, after adjusting for the effects of the confounding variables. To do this, a model that contains terms that correspond to each of the relevant confounding variables is fitted. Exposure factors are then added to this model, both on their own and in combination with one another, until a set of important exposure factors has been identified. The investigator will now be interested in the size of the effects due to each exposure factor, and so the extent to which these effects are significant will need to be judged. The significance of an exposure factor can be assessed in the usual way by comparing the change in deviance on including that term with percentage points of the relevant χ^2-distribution.

To determine whether or not terms that correspond to interactions between exposure factors and confounding variables are needed in the model, the effect of adding such terms to the model that includes the important confounding variables and exposure factors is examined. In particular, if the reduction in deviance due to including a particular interaction is significantly large, we would conclude that the ratio of the odds of disease, for exposed relative to unexposed persons, is not consistent over the different levels of the confounding variable.

The model fitting strategy described in this section is not specific to the analysis of epidemiological data. For example, a similar procedure would be adopted in the analysis of binary response data from individuals in a clinical trial to compare two or more treatments. In this situation, it would be important to estimate the magnitude of the treatment effect after allowing for the effects of other explanatory variables that have been recorded for each individual. These explanatory variables then play the role of confounding variables, while treatment corresponds to exposure.

Example 7.9 The Framingham study
In this example, the application of linear logistic modelling in the analysis of the data from Example 7.1 is described. There are two potential confounding variables in this data set, age group and sex, and one exposure factor, serum cholesterol level. These three variables will be labelled age, sex and chol in the following discussion.

The deviances obtained on fitting a sequence of models to the data in Table 7.1 are summarized in Table 7.6. In most aetiological studies, and this one in

Table 7.6 Deviances on fitting a sequence of models to the data from Example 7.1

Terms fitted in model	Deviance	d.f.
Constant	301.21	15
Age	171.05	14
Sex	221.50	14
Age + sex	88.38	13
Age + sex + age × sex	77.44	12
Age + sex + age × sex + chol	22.39	9
Age + sex + age × sex + chol + chol × age	7.58	6
Age + sex + age × sex + chol + chol × sex	18.28	6
Age + sex + age × sex + chol + chol × age + chol × sex	3.47	3

particular, age and sex will be regarded as confounding variables. To determine whether both age and sex are associated with the probability of CHD, a linear logistic model that contains a constant term is fitted to the data in Table 7.1. The deviance for this model is then compared with that for the model with age, and the model with sex. Both changes in deviance are highly significant, and so both of these potential confounding variables need to be included in the model. A term corresponding to the interaction between age and sex is also needed, since the change in deviance on adding age × sex to the model that contains age and sex is 10.9 in 1 d.f., which is large enough to be significant at the 0.1% level.

When chol is added to the model that contains the confounding variables age, sex and age × sex, the reduction in deviance is 55.0 on 3 d.f. This highly significant change in deviance means that chol is a risk factor for CHD, after allowing for the effects of age and sex.

To see if there is an interaction between chol and age and between chol and sex, the terms chol × age and chol × sex are each added to the model that contains chol, age, sex and age × sex. The reduction in deviance on including chol × age is 14.8 on 3 d.f., while that on adding chol × sex is 4.1 on 3 d.f. These results suggest that there is an interaction between chol and age but not between chol and sex. This is confirmed by including chol × sex in the model with chol, age, sex, age × sex and chol × age, when the reduction in deviance is also 4.1 on 3 d.f. In view of the interaction between chol and age, the estimated relative risks for different initial serum cholesterol levels, relative to a suitable baseline level, such as less than 190 mg/100 ml, will be different for each age group. However they will be similar for persons of each sex. The deviance for the model that seems to be appropriate is 7.6 on 6 d.f., and since the mean deviance is very close to unity, there is no evidence of overdispersion.

Having found what appears to be a suitable model for these data, it will need to be checked using some of the procedures described in Chapter 5. Tabular displays or index plots of the standardised deviance residuals and the values of the D-statistic may be useful, but the most relevant plot is that of the $\Delta\hat{\beta}$-statistic for those parameter estimates associated with the exposure factor. One reason for this is that these parameter estimates are directly related to the odds ratio for the occurrence of disease in exposed relative to unexposed persons. This is discussed in the next section. The role of diagnostic methods in analysing data from aetiological studies will be illustrated later in Example 7.15, in the context of a case-control study.

7.5 INTERPRETING THE PARAMETERS IN A LINEAR LOGISTIC MODEL

One of the reasons for using the linear logistic model in the analysis of data from aetiological studies is that the coefficients of the explanatory variables in the model can be interpreted as logarithms of odds ratios. This means that estimates of the relative risk of a disease, and corresponding standard errors, can easily be obtained from a fitted model. In the following sub-sections, the interpretation of parameters in a linear logistic model will be examined for a number of different situations.

7.5.1 Single dichotomous exposure factor

The simplest model is that for which the probability of disease occurrence is related to a single dichotomous exposure factor. We will suppose that the two levels of this factor are labelled unexposed and exposed, respectively. Suppose that the binary responses that have been recorded for each individual are grouped so that the basic data consist of proportions of individuals who develop the disease in the unexposed and exposed groups, respectively. A linear logistic model for p_j, the probability that the disease occurs in an individual in the jth exposure group, $j = 1, 2$, can be written as

$$\text{logit}\,(p_j) = \beta_0 + \gamma_j$$

where γ_j is the effect due to the jth level of the exposure factor. This model is overparameterized and so a constraint must be placed on the parameters in the model. If the constraint $\gamma_1 = 0$ is adopted, the model can be written in the form

$$\text{logit}\,(p_j) = \beta_0 + \beta_1 x_j \tag{7.3}$$

where x is an indicator variable that takes the value zero for individuals in the unexposed group (level 1 of the exposure factor) and unity for those in the exposed group (level 2 of the exposure factor). In other words, $x_1 = 0$ and $x_2 = 1$, and the parameter β_1 is equivalent to γ_2.

Next, p_1 is the probability that the disease occurs in an individual who is unexposed, and the odds of this event are $p_1/(1 - p_1)$. From equation (7.3), $\text{logit}\,(p_1) = \beta_0$, since $x_1 = 0$, and so the odds of disease in an unexposed person are given by

$$\frac{p_1}{1 - p_1} = e^{\beta_0}$$

Similarly, the odds of disease in an exposed person are given by

$$\frac{p_2}{1 - p_2} = e^{\beta_0 + \beta_1}$$

The ratio of the odds of disease for an exposed relative to an unexposed person is therefore

$$\psi = \frac{p_2/(1 - p_2)}{p_1/(1 - p_1)} = e^{\beta_1}$$

and consequently $\beta_1 = \log \psi$ is the log odds ratio.

When the model described by equation (7.3) is fitted to an observed set of data, maximum likelihood estimates of the unknown parameters β_0 and β_1 are obtained. Denoting these estimates by $\hat{\beta}_0$ and $\hat{\beta}_1$, respectively, the corresponding maximum likelihood estimate of the odds ratio is $\hat{\psi} = \exp(\hat{\beta}_1)$. When a statistical package is used to carry out the fitting procedure, the standard errors of the parameter estimates are usually provided. For the model described by equation (7.3), the standard error of

$\hat{\beta}_1$ will be the standard error of the estimated log odds ratio, that is s.e. $(\hat{\beta}_1)$ = s.e. $\{\log \hat{\psi}\}$. Using this result, an approximate $100(1 - \alpha)\%$ confidence interval for the true log odds ratio is the interval with limits

$$\log \hat{\psi} \pm z_{\alpha/2} \text{ s.e. } \{\log \hat{\psi}\}$$

where $z_{\alpha/2}$ is the upper (one sided) $\alpha/2$ point of the standard normal distribution. To obtain a confidence interval for the true odds ratio, ψ, these limits are exponentiated.

The standard error of the parameter estimate $\hat{\beta}_1$ can also be used to obtain the standard error of the estimated odds ratio $\hat{\psi}$. Using the result for the approximate variance of a function of a parameter estimate, expressed in equation (4.16), the approximate variance of $\log \hat{\psi}$ is given by

$$\text{Var} \{\log \hat{\psi}\} \approx \frac{1}{\hat{\psi}^2} \text{Var} (\hat{\psi})$$

It then follows that

$$\text{s.e. } (\hat{\psi}) \approx \hat{\psi} \text{ s.e. } \{\log \hat{\psi}\}$$

where s.e. $\{\log \hat{\psi}\}$ = s.e. $(\hat{\beta}_1)$.

Example 7.10 The Framingham study
Consider the summary data in Example 7.4, which relate to the proportions of males aged 30–49, with initial serum cholesterol levels that are classified as less than 250 or greater than or equal to 250 mg/100 ml, who develop CHD during a twelve year follow-up period. In this example, we look at the results of fitting linear logistic models to these data. Let x be an indicator variable that takes the value zero if the initial serum cholesterol level of an individual is less than 250 mg/100 ml, and unity otherwise. The raw data are given in Table 7.7. Let p_j be the probability that an individual with the jth level of serum cholesterol, $j = 1, 2$, develops CHD. On fitting the linear logistic model logit $(p_j) = \beta_0$ to these two proportions, the deviance is 29.66 on 1 d.f. When the term x_j is included in the model, the model becomes a perfect fit to the data, and the deviance for this model is zero. The reduction in deviance due to including x_j in the model is therefore 29.66 on 1 d.f.,

Table 7.7 Summary data for males aged 30–49 in the Framingham study

Serum cholesterol level	Proportion with CHD
< 250 ($x = 0$)	71/1169
≥ 250 ($x = 1$)	57/362

which is highly significant. On the basis of these data, we may therefore conclude that there is a significant difference between the proportions of men who develop CHD in two groups defined by their initial serum cholesterol levels.

The fitted model for the estimated probability of developing CHD, with the standard errors of the parameter estimates given in parentheses, is as follows:

$$\text{logit}(\hat{p}_j) = -2.739 + 1.061 x_j$$
$$(0.122) \quad (0.189)$$

From this, the estimated ratio of the odds of developing CHD for persons with initial serum cholesterol levels greater than or equal to 250 mg/100 ml, relative to those whose initial level is less than 250 mg/100 ml, is given by $\hat{\psi} = \exp(1.061) = 2.89$. The standard error of $\log \hat{\psi}$ is 0.189 and so a 95% confidence interval for the true log odds ratio ranges from $1.061 - (1.96 \times 0.189)$ to $1.061 + (1.96 \times 0.189)$, that is, the interval (0.69, 1.43). The corresponding 95% confidence interval for the true odds ratio is therefore the interval from $e^{0.69}$ to $e^{1.43}$, that is, from 1.99 to 4.19. Finally, the standard error of the estimated odds ratio is approximately $\hat{\psi}$ s.e. $\{\log \hat{\psi}\}$, which is 0.547.

In this example, it would have been simpler to estimate the odds ratio directly using equation (7.1), as in Example 7.4. Equation (7.2) could then be used to give the standard error of the estimate. However, the values of $\hat{\psi}$ and s.e. $(\hat{\psi})$ in Example 7.4 are, to all intents and purposes, the same as those obtained using linear logistic modelling. Indeed, the two methods for obtaining $\hat{\psi}$ and its standard error are asymptotically equivalent, and since the proportions of men with CHD given in Table 7.7 are based on large numbers of individuals, it is not surprising that the results are so similar.

Using the statistical package GLIM, the same results would be obtained if the initial serum cholesterol level were denoted by a two-level factor, where the first level of the factor corresponds to a cholesterol level less than 250 mg/100 ml. This is because the coding used for indicator variables in this package is such that the value of the indicator variable that corresponds to the factor is zero at the first level of the factor. Other packages may use different codings and different parameter estimates will then be obtained. For example, suppose that the single indicator variable x, corresponding to a factor with two levels, takes the value -1 at the first level of the factor and $+1$ at the second. Then, under the model $\text{logit}(p_j) = \beta_0 + \beta_1 x_j$, the odds of disease in the two exposure groups are, respectively,

$$\frac{p_1}{1 - p_1} = e^{\beta_0 - \beta_1} \quad \text{and} \quad \frac{p_2}{1 - p_2} = e^{\beta_0 + \beta_1}$$

so that the odds ratio for exposed relative to unexposed persons is now $\exp(2\beta_1)$. After fitting this model, the parameter estimate $\hat{\beta}_1$ would be obtained and so now $\log \hat{\psi} = 2\hat{\beta}_1$. The standard error of $\log \hat{\psi}$ would then be given by s.e. $(\log \hat{\psi}) = 2$ s.e. $(\hat{\beta}_1)$. A similar argument shows that if the coding used for the indicator variable were such that the second level of the factor corresponded to zero and the first level to one, the odds ratio would be $\exp(-\beta_1)$. So even in this simple case, it is essential to know what coding of the indicator variable has been used in order to be able to interpret the parameter estimates correctly.

7.5.2 Polychotomous exposure factor

Now consider the situation where the probability of disease occurrence is to be related to a single exposure factor that has m levels. We will suppose that the data are grouped so that the linear logistic model for the probability of disease occurrence in an individual in the jth exposure group, $j = 1, 2, \ldots, m$, can be written

$$\text{logit}(p_j) = \beta_0 + \gamma_j$$

where γ_j is the effect due to the jth level of the exposure factor. Again this model is overparameterised, and a constraint needs to be imposed on the parameters in the model. The constraint $\gamma_1 = 0$ is particularly convenient. In order to fit the model with this constraint, $m - 1$ indicator variables, $x_2, x_3, \ldots, x_m$, are defined, with values shown in the following table.

Level of exposure factor	x_2	x_3	x_4	$\cdots$	x_m
1	0	0	0	$\cdots$	0
2	1	0	0	$\cdots$	0
3	0	1	0	$\cdots$	0
$\cdots$					
m	0	0	0	$\cdots$	1

The model may therefore be expressed as

$$\text{logit}(p_j) = \beta_0 + \beta_2 x_{2j} + \beta_3 x_{3j} + \cdots + \beta_m x_{mj} \tag{7.4}$$

The parameters $\beta_2, \beta_3, \ldots, \beta_m$ in equation (7.4) can be interpreted as log odds ratios for individuals exposed to level j of the exposure factor, $j = 2, 3, \ldots, m$, relative to those exposed to the first level. To see this, first consider the odds of disease in an individual exposed to the first level of the exposure factor, for whom the probability of disease occurrence is p_1. The values of each of the $m - 1$ indicator variables that correspond to the first level of the factor are zero and so $\text{logit}(p_1) = \beta_0$. At the jth level of the exposure factor, $j = 2, 3, \ldots, m$, the only indicator variable to have a non-zero value is x_j, and so the odds of the disease occurring in an individual in the jth exposure group will be such that $\text{logit}(p_j) = \beta_0 + \beta_j$. The ratio of the odds of disease occurrence for individuals in the jth exposure group, relative to those in the first, will therefore be given by

$$\psi_{j1} = \frac{p_j/(1 - p_j)}{p_1/(1 - p_1)} = e^{\beta_j}$$

When the model described by equation (7.4) is fitted to observed data, parameter estimates $\hat{\beta}_2, \hat{\beta}_3, \ldots, \hat{\beta}_m$ are obtained, from which estimates of the odds ratios $\hat{\psi}_{j1}$ can be calculated. As in Section 7.5.1, the standard errors of the estimates $\hat{\beta}_j$ will be the standard errors of these estimated log odds ratios, from which confidence intervals for the corresponding true odds ratios can be derived.

Example 7.11 The Framingham study
The proportion of males aged between 30 and 49 who develop CHD during a twelve-year period, stratified according to initial serum cholesterol level, form the first row of Table 7.1. An appropriate linear logistic model for these data is such that

$$\text{logit}\,(p_j) = \beta_0 + \text{chol}_j$$

where p_j is the probability of CHD occurring in an individual whose serum cholesterol level falls in the jth group, $j = 1, 2, 3, 4$, and chol_j refers to the effect of the jth level of this exposure factor. Parameter estimates and their standard errors on fitting this model, reproduced from GLIM output, are given below.

	estimate	s.e.	parameter
1	−3.225	0.2828	1
2	0.1492	0.3716	CHOL(2)
3	0.9711	0.3280	CHOL(3)
4	1.548	0.3175	CHOL(4)

When GLIM is used, the indicator variables corresponding to a factor are such that they are zero at the first level of the factor, as in the above discussion, and so the estimates given in the output correspond to $\hat{\beta}_0, \hat{\beta}_2, \hat{\beta}_3$ and $\hat{\beta}_4$ in equation (7.4). Ratios of the odds of developing CHD for individuals in the jth cholesterol group, $j = 2, 3, 4$, relative to those in the first, are given by

$$\hat{\psi}_{21} = e^{0.149} = 1.16; \quad \hat{\psi}_{31} = e^{0.971} = 2.64; \quad \hat{\psi}_{41} = e^{1.548} = 4.70$$

These results show that the risk of developing CHD during a twelve-year period increases with initial serum cholesterol level, so that, for example, a person whose serum cholesterol level is greater than or equal to 250 mg/100 ml is nearly five times more likely to develop CHD than someone whose initial cholesterol level is less than 190 mg/100 ml.

The standard error of the parameter estimates given in the GLIM output can be used to obtain confidence intervals for the true odds ratios. For example, an approximate 95% confidence interval for the log odds ratio $\log \psi_{21}$ is found from the standard error of $\log \hat{\psi}_{21}$, which is 0.372. The interval for $\log \psi_{21}$ becomes $0.149 \pm (1.96 \times 0.372)$, which gives the interval $(-0.58, 0.88)$. The corresponding 95% confidence interval for ψ_{21} itself is $(0.56, 2.41)$. Similarly, confidence intervals for the other two odds ratios are $(1.39, 5.02)$ and $(2.53, 8.75)$, respectively. The confidence interval for ψ_{21} includes unity, which suggests that the risk of CHD' occurring in men aged 30–49 is not significantly greater if their initial serum cholesterol level lies between 190 and 219 mg/100 ml, compared to those for whom it is less than 190 mg/100 ml. The confidence intervals for both ψ_{31} and ψ_{41} exclude unity, from which we may conclude that those with initial cholesterol levels in excess of 220 mg/100 ml run a significantly increased risk of developing CHD.

The reason why the parameter estimates in the model labelled (7.4) are log odds ratios for the jth exposure level relative to the first is that the corresponding indicator variables are all zero for this reference exposure level. In some applications, odds ratios that use a level of the exposure factor other than the first as the reference level may be required, although the indicator variables used for the factor are coded as zero for the first level. For example, consider the odds of disease for an individual with an initial cholesterol concentration greater than or equal to 250 (level 4 of the exposure factor) relative to someone whose initial cholesterol concentration is between 220 and 249 (level 3 of the exposure factor). The odds of disease for persons in these two groups are given by

$$\frac{p_3}{1-p_3} = e^{\beta_0 + \beta_3} \quad \text{and} \quad \frac{p_4}{1-p_4} = e^{\beta_0 + \beta_4}$$

Consequently, the ratio of these odds is $\psi_{43} = \exp(\beta_4 - \beta_3)$ and the corresponding estimate is $\hat{\psi}_{43} = \exp(\hat{\beta}_4 - \hat{\beta}_3)$. To obtain the standard error of $\log(\hat{\psi}_{43})$, used in the construction of a confidence interval for ψ_{43}, we proceed as follows. Since $\log(\hat{\psi}_{43}) = \hat{\beta}_4 - \hat{\beta}_3$, the variance of the estimated log odds ratio is given by

$$\text{Var}\{\log(\hat{\psi}_{43})\} = \text{Var}(\hat{\beta}_4) + \text{Var}(\hat{\beta}_3) - 2\,\text{Cov}(\hat{\beta}_4, \hat{\beta}_3)$$

Since this expression involves covariance terms as well as variances, the full variance–covariance matrix of the parameter estimates will now be needed. The calculations involved are illustrated in Example 7.12.

Example 7.12 The Framingham study
Consider again the data on the proportion of males aged 30–49 who develop CHD, whose initial serum cholesterol levels fall in one of four groups. In Example 7.11, parameter estimates on fitting a linear logistic model to these data were given. The estimated ratio of the odds of disease for an individual in the third group of the exposure factor relative to someone in the fourth is $\hat{\psi}_{43} = \exp(\hat{\text{chol}}_4 - \hat{\text{chol}}_3) = \exp(1.548 - 0.971) = 1.78$. This suggests that the risk of CHD is greater for those with an initial serum cholesterol level greater than or equal to 250 mg/100 ml, compared to those with initial cholesterol values between 220 and 249 mg/100 ml. The variance of $\log(\hat{\psi}_{43})$ is given by

$$\text{Var}\{\log(\psi_{43})\} = \text{Var}(\hat{\text{chol}}_4) + \text{Var}(\hat{\text{chol}}_3) - 2\text{Cov}(\hat{\text{chol}}_4, \hat{\text{chol}}_3)$$

and the required variances and covariance are found from the variance–covariance matrix of the four parameter estimates. Using GLIM, this matrix is found to be

$$
\begin{array}{c}
\hat{\beta}_0 \\
\text{chol}_2 \\
\text{chol}_3 \\
\text{chol}_4
\end{array}
\left[
\begin{array}{cccc}
0.07998 & & & \\
-0.07998 & 0.13810 & & \\
-0.07998 & 0.07998 & 0.10761 & \\
-0.07998 & 0.07998 & 0.07998 & 0.10080
\end{array}
\right]
$$
$$
\quad\ \hat{\beta}_0 \qquad\quad \text{chol}_2 \qquad\ \ \text{chol}_3 \qquad\ \ \text{chol}_4
$$

from which $\text{Var}(\hat{\text{chol}}_4) = 0.1008$, $\text{Var}(\hat{\text{chol}}_3) = 0.1076$ and $\text{Cov}(\hat{\text{chol}}_4, \hat{\text{chol}}_3) = 0.0800$. It then follows that

$$
\text{Var}\{\log(\hat{\psi}_{43})\} = 0.1008 + 0.1076 - (2 \times 0.0800) = 0.0484
$$

and so the standard error of $\log \hat{\psi}_{43}$ is 0.220. Consequently, a 95% confidence interval for $\log(\psi_{43})$ is the interval from 0.15 to 1.01, and so a 95% confidence interval for ψ_{43} itself is $(1.16, 2.74)$.

As for the case where the exposure factor is dichotomous, the way in which coefficients of the indicator variables are interpreted is crucially dependent upon the coding that has been used for them. For example, suppose that the coding used for the $m - 1$ indicator variables corresponding to a factor with m levels is as shown in the following table.

Level of exposure factor	x_2	x_3	x_4 ...	x_m
1	-1	-1	-1 ...	-1
2	1	0	0 ...	0
3	0	1	0 ...	0
. . .				
m	0	0	0 ...	1

A linear logistic model that includes the factor can then be expressed as

$$
\text{logit}(p_j) = \beta_0 + \beta_2 x_{2j} + \beta_3 x_{3j} + \cdots + \beta_m x_{mj}
$$

The odds of disease occurrence in an individual exposed to the first level of the exposure factor is

$$
\frac{p_1}{1 - p_1} = \exp(\beta_0 - \beta_2 - \beta_3 - \cdots - \beta_m)
$$

while that for an individual exposed to the jth level of the factor is

$$\frac{p_j}{1 - p_j} = e^{\beta_0 + \beta_j}, \quad j \geqslant 2$$

The ratio of the odds of disease for someone in the jth group relative to someone in the first is therefore given by

$$\psi_{j1} = \frac{p_j/(1 - p_j)}{p_1/(1 - p_1)} = \exp(\beta_2 + \beta_3 + \cdots + \beta_m + \beta_j)$$

For example, if $m = 4$ and $j = 3$, $\psi_{31} = \exp(\beta_2 + 2\beta_3 + \beta_4)$, and the variance of the corresponding estimated log odds ratio is given by

$$\text{Var}\{\log(\hat{\psi}_{31})\} = \text{Var}(\hat{\beta}_2) + 4\text{Var}(\hat{\beta}_3) + \text{Var}(\hat{\beta}_4) + 4\text{Cov}(\hat{\beta}_2, \hat{\beta}_3)$$
$$+ 4\text{Cov}(\hat{\beta}_3, \hat{\beta}_4) + 2\text{Cov}(\hat{\beta}_2, \hat{\beta}_4)$$

Each of the terms in this expression can be found from the variance–covariance matrix of the parameter estimates after fitting the linear logistic model, and an approximate confidence interval for ψ_{31} can then be obtained. Although this is reasonably straightforward, this particular coding of the indicator variables does make it rather more complicated to interpret the parameter estimates in a fitted linear logistic model.

7.5.3 Continuous exposure variable

Suppose that the value of a single continuous exposure variable is recorded for each of n individuals in a cohort study. A linear logistic model for p_i, the probability that the ith individual, $i = 1, 2, \ldots, n$, develops the disease, is then such that

$$\text{logit}(p_i) = \beta_0 + \beta_1 x_i$$

where x_i is the value of the exposure variable for the ith individual. Again, the coefficient of x_i in this model can be interpreted as a log odds ratio. Consider the ratio of the odds of disease for an individual for whom the value $x + 1$ is recorded relative to one for whom the value x is obtained. This is given by

$$\psi = \frac{\exp\{\beta_0 + \beta_1(x + 1)\}}{\exp\{\beta_0 + \beta_1 x\}} = e^{\beta_1}$$

and so $\hat{\beta}_1$ in the fitted linear logistic model is the estimated change in the logarithm of the odds ratio when x is increased by one unit.

Using a similar argument, the estimated change in the log odds when the exposure variable x is increased by r units is $r\hat{\beta}_1$, and the corresponding estimate of the odds ratio is $\exp(r\hat{\beta}_1)$. Regarding the odds ratio as an approximation to the relative risk, this would be interpreted as the approximate change in the risk of the disease for every increase of r units in the value of the variable x. The standard error of the estimated log odds ratio will be $r \times$ s.e. $(\hat{\beta}_1)$, from which confidence intervals for the true odds ratio can be derived.

The above argument shows that when a continuous variable, x, is included in a linear logistic model, the approximate relative risk of disease when x is changed by r units does not depend on the actual value of x. For example, if x were the age of a person, the relative risk of disease for an individual aged 65, relative to someone aged 60, would be the same as that for a person aged 25 relative to someone aged 20. This feature is a direct result of the assumption that there is a linear relationship between the logistic transform of the probability of disease and x. In practice, this assumption is often best avoided by using a factor whose levels correspond to different sets of values of x. However, linearity can then be checked using the procedure described in Section 3.10.

7.5.4 Models with combinations of different types of term

In the previous sub-sections, we have only considered the interpretation of parameter estimates in linear logistic models that contain a single term. Most data sets arising from aetiological studies provide information on more than just one variable, and so linear logistic models fitted to data from such studies will generally include a number of different terms. In particular, the model may include more than one exposure factor, terms that confound the relationship between the probability of disease occurrence and exposure factors, and terms that represent interactions between exposure factors. Again it turns out that with suitable coding of the indicator variables used to represent factors in the model, the parameter estimates can be interpreted as log odds ratios. When the model contains more than one term, the parameter estimates associated with a particular effect are said to be adjusted for that other term and so they correspond to adjusted log odds ratios. For example, if a model contains terms corresponding to a single exposure factor and a single confounding variable, the parameter estimates for the exposure factor will be log odds ratios, adjusted for the confounding variable. The linear logistic model therefore provides a straightforward way of obtaining estimates of odds ratios that take account of any confounding variables included in the model.

When terms that correspond to interactions between, say, an exposure factor and a confounding variable are included in a model, the estimated log odds ratios for the exposure factor will differ according to the value of the confounding variable. In this situation, the value of the confounding variable that has been used in calculating them will need to be made clear when the estimated odds ratios are presented.

As in previous sections, it is essential to know what coding of the indicator variables has been used in order to be able to interpret the parameters correctly. Instead of providing algebraic details of the derivation of odds ratios for a range of different models, the general approach will be illustrated in the following example.

Example 7.13 The Framingham study
In Example 7.9, a number of linear logistic models were fitted to the data from the
Framingham study given in Table 7.1. In this example, ratios of the odds of CHD
for the jth level of cholesterol ($j \geq 2$), relative to the first level, will be obtained for
two models:

(1) The model that includes the exposure factor chol, the confounding variables
    age, sex and the age $\times$ sex interaction,
(2) The model that contains the terms in (1), together with a term that represents
    the interaction between chol and age.

In Example 7.9, the second of these two models, model (2), was found to be the
most suitable. Nonetheless, the first, model (1), is also considered here for illustrative and comparative purposes.

Let p_{jkl} denote the probability that an individual of the lth sex in the kth age group
develops CHD when their initial cholesterol level is in the jth group, for
$j = 1, 2, 3, 4; k = 1, 2; l = 1, 2$. Model (1) can then be expressed in the form

$$\text{logit}(p_{jkl}) = \beta_0 + \text{chol}_j + \text{age}_k + \text{sex}_l + (\text{age} \times \text{sex})_{kl} \tag{7.5}$$

where β_0 is a constant, chol_j is the effect due to the jth level of chol, age_k is the
effect due to the kth age group and sex_l is the effect due to the lth sex. The ratio of
the odds of CHD for individuals exposed to the jth level of chol ($j \geq 2$) relative to
those exposed to the first level, adjusted for age, sex and age $\times$ sex, will be given by

$$\log \psi_{j1} = \text{logit}(p_{jkl}) - \text{logit}(p_{1kl}) = \text{chol}_j - \text{chol}_1$$

and the corresponding estimated odds ratio is

$$\hat{\psi}_{j1} = \exp(\hat{\text{chol}}_j - \hat{\text{chol}}_1)$$

where $\hat{\text{chol}}_j$ is the estimated effect of the jth serum cholesterol level in the model
described by equation (7.5). Since the estimates $\hat{\text{chol}}_j$ take account of age and sex
effects, this estimated odds ratio is adjusted for these confounding variables.

On fitting the model of equation (7.5) to the data given in Table 7.1, the following
parameter estimates are obtained using GLIM.

	estimate	s.e.	parameter
1	-2.976	0.1764	1
2	0.2351	0.2057	CHOL(2)
3	0.6725	0.1931	CHOL(3)
4	1.112	0.1858	CHOL(4)
5	0.9745	0.1360	AGE(2)
6	-1.384	0.1870	SEX(2)
7	0.4911	0.2424	AGE(2).SEX(2)

The parameters labelled **CHOL(2)**, **CHOL(3)**, and **CHOL(4)** in this output correspond to $\hat{chol}_2$, $\hat{chol}_3$ and $\hat{chol}_4$ in the fitted model, while in view of the coding of the indicator variables that have been used, $\hat{chol}_1 = 0$. These estimates can be interpreted as log odds ratios for level j of the factor chol, relative to level 1, adjusted for age, sex and age $\times$ sex. The corresponding estimated odds ratios, adjusted for the confounding variables are given by

$$\hat{\psi}_{21} = 1.26, \qquad \hat{\psi}_{31} = 1.96, \qquad \hat{\psi}_{41} = 3.04$$

Confidence intervals for the true odds ratios can be found from the standard errors of $\hat{chol}_j$ given in the GLIM output, using the method described in section 7.5.2.

For comparison, the estimated odds ratios obtained from fitting the model

$$\text{logit}\,(p_{jkl}) = \beta_0 + chol_j$$

to the data in Table 7.1 are

$$\hat{\psi}_{21} = 1.41, \qquad \hat{\psi}_{31} = 2.36, \qquad \hat{\psi}_{41} = 3.81$$

These are the unadjusted odds ratios, since no account has been taken of possible age and sex effects. Comparing these with the adjusted odds ratios, we see that the unadjusted odds ratios overestimate the relative risk of CHD occurring in individuals with levels 2, 3 and 4 of chol relative to those with level 1. The fact that age, sex and their interaction have an impact on estimates of the odds ratios means that age, sex and age $\times$ sex are important confounding variables.

The model that was found to be most appropriate for the data of Table 7.1 is model (2), which includes a term corresponding to the interaction between chol and age. The model then becomes

$$\text{logit}\,(p_{jkl}) = \beta_0 + chol_j + age_k + sex_l + (age \times sex)_{kl} + (chol \times age)_{jk} \qquad (7.6)$$

Under this model, for individuals in the kth age group, the ratio of the odds of disease for an individual exposed to level j of chol relative to someone who has been exposed to level 1 is given by

$$\begin{aligned} \log\,(\psi_{j1k}) &= \text{logit}\,(p_{jkl}) - \text{logit}\,(p_{1kl}) \\ &= chol_j - chol_1 + (chol \times age)_{jk} - (chol \times age)_{1k} \end{aligned}$$

This expression confirms that when an interaction between an exposure factor and a confounding variable is included in the model, the odds ratio will depend on the level of the confounding variable.

The parameter estimates shown below are obtained when the model described by equation (7.6) is fitted to the data from the Framingham study, using GLIM.

	estimate	s.e.	parameter
1	−3.195	0.2399	1
2	0.05638	0.3152	CHOL(2)
3	0.9245	0.2761	CHOL(3)
4	1.545	0.2643	CHOL(4)
5	1.422	0.3326	AGE(2)
6	−1.363	0.1880	SEX(2)
7	0.5720	0.2434	AGE(2).SEX(2)
8	0.2297	0.4221	CHOL(2).AGE(2)
9	−0.5587	0.3882	CHOL(3).AGE(2)
10	−0.8778	0.3713	CHOL(4).AGE(2)

In view of the coding used for the indicator variables corresponding to the interaction terms, $(\text{chol} \times \text{age})_{jk} = 0$ when $j = 1$ and also when $k = 1$; the estimate $\widehat{\text{chol}}_1$ is zero, as before. Consequently, the estimated log odds ratio for the occurrence of CHD in persons exposed to the jth level of chol relative to those exposed to the first level, for persons aged between 30 and 49, is given by

$$\log(\hat{\psi}_{j11}) = \widehat{\text{chol}}_j$$

while that for persons aged between 50 and 62 is

$$\log(\hat{\psi}_{j12}) = \widehat{\text{chol}}_j + (\text{chol} \ \hat{\times} \ \text{age})_{j2}$$

The corresponding odds ratios are presented in Table 7.8. The inclusion of the column for which the estimated odds ratios are 1.00 in tables such as Tables 7.8 emphasises that the odds are relative to those for the first level of serum cholesterol.

Table 7.8 Estimated odds ratios under the model in equation (7.6)

Age group	Serum cholesterol level			
	< 190	190–219	220–249	≥ 250
30–49	1.00	1.06	2.52	4.69
50–62	1.00	1.33	1.44	1.95

This table shows that the relative risk of CHD increases more rapidly with increasing initial serum cholesterol level for persons in the 30–49 age group, compared to that for persons in the 50–62 age group. The relative risk of CHD for persons with an initial cholesterol level between 190 and 219 is greater for those in the older age group, whereas if the initial cholesterol level exceeds 220 mg/100 ml, the relative risk of CHD is greater for individuals in the younger age group.

Standard errors and confidence intervals for the corresponding true odds ratios can be found using the method illustrated in Section 7.5.2. As a further illustration, a confidence interval will be obtained for the true ratio of the odds of CHD occurring in persons aged 50–62 with an initial cholesterol level between 190 and 219 mg/100 ml, relative to those whose initial cholesterol level is less than 190 mg/100 ml. This is ψ_{212} in the preceding notation. The variance of log $(\hat{\psi}_{212})$ is given by

$$\text{Var}\{\log(\hat{\psi}_{212})\} = \text{Var}(\hat{\text{chol}}_2) + \text{Var}\{(\text{chol} \hat{\times} \text{age})_{22}\}$$
$$+ 2\text{Cov}\{\hat{\text{chol}}_2, (\text{chol} \hat{\times} \text{age})_{22}\}$$

and from the variance–covariance matrix of the parameter estimates under the model in equation (7.6), $\text{Var}(\hat{\text{chol}}_2) = 0.0993$, $\text{Var}\{(\text{chol} \hat{\times} \text{age})_{22}\} = 0.178$, and $\text{Cov}\{\hat{\text{chol}}_2, (\text{chol} \hat{\times} \text{age})_{22}\} = -0.0993$. Consequently, $\text{Var}\{\log(\hat{\psi}_{212})\} = 0.079$, and so a 95% confidence interval for $\log(\psi_{212})$ ranges from -0.264 to 0.836. The corresponding interval for the true odds ratio is $(0.77, 2.31)$. The fact that this interval includes unity suggests that the risk of CHD occurring in persons aged 50–62 is not significantly different for those with initial serum cholesterol levels less than 190 mg/100 ml compared to those with cholesterol levels between 190 and 219 mg/100 ml. Confidence intervals for the other odds ratios in Table 7.8 can be obtained in a similar manner.

7.6 THE LINEAR LOGISTIC MODEL FOR DATA FROM CASE-CONTROL STUDIES

In a case-control study, the cases are sampled from a population of individuals who have the disease that is being investigated, while the controls are sampled from a corresponding population of individuals without that disease. Information on their exposure to particular factors of interest is then sought retrospectively. If the variable that describes whether or not a particular individual is a case in the study is regarded as a binary response variable, and the exposure factors are treated as explanatory variables, linear logistic modelling can again be used to analyse data from a case-control study.

Because the proportion of cases in the study will not be comparable to the proportion of diseased persons in the population, a fitted linear logistic model cannot be used to estimate the probability of disease, nor the relative risk of disease. However, it turns out that the model can still be used to examine the extent of any association between exposure factors and a disease, after allowing for confounding variables. The model also enables estimates of odds ratios to be determined, which can then be interpreted as approximate relative risks. The reason for this is presented below.

Suppose that in a case-control study, π_1 is the proportion of cases that have been sampled from the diseased population, so that π_1 is the **sampling fraction** for the cases. The value of π_1 can also be regarded as the probability that a diseased person is selected for inclusion in the study, that is,

$$\pi_1 = P[\text{individual is in the study}|\text{diseased}]$$

Similarly, let π_2 be the proportion of persons in the disease-free population who form the controls, so that π_2 is the sampling fraction for the controls, and

$$\pi_2 = P \text{ [individual is in the study|disease-free]}$$

The probability that an individual in the population develops the disease will depend on the values of k explanatory variables corresponding to relevant exposure factors and potential confounding variables. For this reason, the probability that a particular individual has the disease is denoted by $p(\mathbf{x})$, where the vector $\mathbf{x}$ consists of the values of the explanatory variables for that individual. The quantities π_1 and π_2 are assumed not to depend on $\mathbf{x}$, so that selection for inclusion in the study depends only on the disease status of an individual. This means that the probability that a diseased individual is selected for inclusion in the study as a case, and the probability that an individual who is free of the disease is selected as a control, must not depend on the values of the explanatory variables. If this assumption is violated, so that, for example, a diseased individual who had been exposed to some risk factor was more likely to be a case in the study than someone who had not been so exposed, the results of the case-control study could be very misleading.

Now consider the probability that an individual in the case-control study has the disease. This conditional probability is given by

$$p_0(\mathbf{x}) = P[\text{disease|individual is in the study}]$$

To obtain this conditional probability, we use a standard result from probability theory, according to which the probability of an event A, conditional on an event B, is given by $P(A|B) = P(A \text{ and } B)/P(B)$. Similarly, $P(B|A) = P(A \text{ and } B)/P(A)$ and combining these two formulae we get $P(A|B) = P(B|A)P(A)/P(B)$, which is a simple form of **Bayes theorem**. Using this result,

$$p_0(\mathbf{x}) = \frac{P \text{ [individual is in the study|diseased]} \times P \text{ [individual is diseased]}}{P \text{ [individual is in the study]}}$$

and so

$$p_0(\mathbf{x}) = \frac{\pi_1 p(\mathbf{x})}{P \text{ [individual is in the study]}}$$

Next, an individual is included in the study if that person is diseased and has been selected as a case, or if that person is disease-free and has been selected as a control. Consequently, P [individual is in the study] is the sum of P [individual is diseased and is a case] and P [individual is disease-free and is a control]. Using the result $P(A \text{ and } B) = P(A|B)P(B)$, P [individual is diseased and is in the study] is equal to

$$P \text{ [individual is in the study|diseased]} \times P \text{ [individual is diseased]}$$

which is $\pi_1 p(\mathbf{x})$ as before. The probability that an individual does not have the disease is $1 - p(\mathbf{x})$, and so P[individual is disease-free and is in the study] is $\pi_2\{1 - p(\mathbf{x})\}$. Hence,

$$P[\text{individual is in the study}] = \pi_1 p(\mathbf{x}) + \pi_2\{1 - p(\mathbf{x})\}$$

and so

$$p_0(\mathbf{x}) = \frac{\pi_1 p(\mathbf{x})}{\pi_1 p(\mathbf{x}) + \pi_2\{1 - p(\mathbf{x})\}}$$

From this result,

$$\frac{p_0(\mathbf{x})}{1 - p_0(\mathbf{x})} = \frac{\pi_1}{\pi_2} \frac{p(\mathbf{x})}{1 - p(\mathbf{x})}$$

and it then follows that

$$\text{logit}\{p_0(\mathbf{x})\} = \log\left(\frac{\pi_1}{\pi_2}\right) + \text{logit}\{p(\mathbf{x})\} \tag{7.7}$$

If a linear logistic model is adopted for the probability that a person in the population with explanatory variables $\mathbf{x}$ has the disease, then

$$\text{logit}\{p(\mathbf{x})\} = \beta_0 + \beta_1 x_1 + \beta_2 x_2 + \cdots + \beta_k x_k \tag{7.8}$$

and so the probability that a person with explanatory variables $\mathbf{x}$ is a case in the case-control study is given by

$$\text{logit}\{p_0(\mathbf{x})\} = \log\left(\frac{\pi_1}{\pi_2}\right) + \beta_0 + \beta_1 x_1 + \beta_2 x_2 + \cdots + \beta_k x_k$$

which, since π_1 and π_2 do not depend on $\mathbf{x}$, can be written as

$$\text{logit}\{p_0(\mathbf{x})\} = \beta_0' + \beta_1 x_1 + \beta_2 x_2 + \cdots + \beta_k x_k \tag{7.9}$$

where

$$\beta_0' = \beta_0 + \log\left(\frac{\pi_1}{\pi_2}\right) \tag{7.10}$$

The model described by equation (7.9) is a linear logistic model for the probability that an individual is a case in the case-control study. This model can be fitted using the methods described in Chapter 3, after defining a binary response variable that takes the value unity when an individual is a case, and zero when an individual is a control. On fitting this model, estimates of the parameters $\beta_0', \beta_1, \ldots, \beta_k$ are obtained, from which the estimated probability that a person in the study is a case, $\hat{p}_0(\mathbf{x})$, say, can be determined. However, this is not particularly informative. What one would really like to be able to estimate is $p(\mathbf{x})$, the probability that an individual with explanatory variables $\mathbf{x}$ has the disease. From equation (7.8), this probability depends on an estimate of β_0, which from equation (7.10) depends on estimates of β_0', π_1 and π_2. Although β_0' can be estimated, the two sampling fractions, π_1 and π_2, cannot be estimated unless information is available on the number of diseased and disease-free persons in the populations being sampled. Such information will rarely be available, and so in general it will not be possible to estimate β_0, and $p(\mathbf{x})$ itself will not be estimable.

Using a similar argument, the ratio of the estimated probability of disease for an individual with explanatory variables $\mathbf{x}_1$, relative to one whose values are $\mathbf{x}_0$, is not independent of the estimated value of β_0 and so it follows that the relative risk, ρ, is not independent of the unknown sampling fractions π_1 and π_2. Consequently, ρ cannot be estimated from the results of a case-control study. However, the ratio of the odds of disease for a person with explanatory variables $\mathbf{x}_1$ relative to someone whose values are $\mathbf{x}_0$ is given by

$$\psi = \frac{p(\mathbf{x}_1)/\{1 - p(\mathbf{x}_1)\}}{p(\mathbf{x}_0)/\{1 - p(\mathbf{x}_0)\}}$$

From equation (7.7),

$$\frac{p(\mathbf{x}_j)}{1 - p(\mathbf{x}_j)} = \frac{\pi_2}{\pi_1} \frac{p_0(\mathbf{x}_j)}{1 - p_0(\mathbf{x}_j)}$$

for $j = 0, 1$, and so

$$\psi = \frac{p_0(\mathbf{x}_1)/\{1 - p_0(\mathbf{x}_1)\}}{p_0(\mathbf{x}_0)/\{1 - p_0(\mathbf{x}_0)\}}$$

This odds ratio is independent of β_0, and so it can be estimated from data from a case-control study. Since an estimated odds ratio can be interpreted as an approximate relative risk, information about the risk of disease for an exposed relative to an unexposed person is therefore available from a case-control study. Moreover, the parameters in a linear logistic model for the probability that a person is a case in equation (7.9) are exactly those that occur in the model for the probability that a person is diseased in equation (7.8). They can therefore be interpreted as log odds ratios of disease, just as for the analysis of cohort data described in Section 7.5. The effect that each exposure factor has on the probability of disease can also be studied using an analysis of deviance, since the change in deviance on including a particular term in equation (7.8), the linear logistic model for the probability that an individual is diseased, will be the same as the change in deviance on including that term in equation (7.9), the linear logistic model for the probability that an individual is a case. This means that the effects of exposure factors on the probability that an individual has the disease, after allowing for any confounding variables, can be assessed in the manner described in Section 7.4.

Example 7.14 Age at first coitus and cervical cancer
Consider once more the data from the case-control study to investigate whether there is an association between age at first intercourse and the occurrence of cervical cancer, described in Example 7.2. Using GLIM to fit a linear logistic model to the proportion of cases in the two age groups, the following parameter estimates are obtained.

	estimate	s.e.	parameter
1	-2.156	0.3185	1
2	1.383	0.3769	AGE(2)

In the fitted model, age has been included as a factor with two levels, the first level of which corresponds to an age at first intercourse greater than fifteen years. Consequently, the parameter corresponding to the second level of age is the logarithm of the ratio of the odds of being a case for women who have first intercourse at an age of fifteen years or less, relative to those whose age is sixteen or more. This log odds ratio is also the logarithm of the ratio of the odds of cervical cancer occurring in women aged fifteen or less at first intercourse relative to those who are older than fifteen, and can be regarded as the approximate relative risk of cervical cancer. From the parameter estimates in the GLIM output, an estimate of the approximate relative risk of cervical cancer for those aged less than or equal to fifteen at first intercourse, relative to women aged more than fifteen years, is $e^{1.383} = 3.99$. This estimate is exactly the same as that found in Example 7.5. The standard error of the log odds ratio can be used to obtain a 95% confidence interval for the true odds ratio. The interval for the true log odds ratio ranges from $1.383 - (1.96 \times 0.377)$ to $1.383 + (1.96 \times 0.377)$, that is, from 0.644 to 2.122, and so the corresponding interval for the true odds ratio is (1.90, 8.35). This again is in exact agreement with that obtained in Example 7.5.

Example 7.15 Diverticular disease
A case-control study designed to investigate whether there is an association between the occurrence of colonic cancer and diverticular disease was described in Example 1.10. The data which were presented in Table 1.10 refer to the proportions of individuals with diverticular disease amongst the cases of colonic cancer and the controls. However, before a linear logistic model can be used to estimate the relative risk of colonic cancer occurring in patients with diverticular disease relative to those without, the data must be rearranged to give the proportion of cases in the study with and without diverticular disease. The resulting data are presented in Table 7.9. No cases were found amongst women aged between 50 and 54 and there

Table 7.9 Proportion of cases in the case-control study classified by age, sex and the presence of diverticular disease.

Age interval	Sex	Proportion of cases	
		Patients without DD	Patients with DD
40–49	M	3/10	0/0
	F	6/20	0/1
50–54	M	1/7	1/2
	F	0/0	0/0
55–59	M	3/15	2/5
	F	6/20	1/5
60–64	M	4/17	1/6
	F	2/8	0/2
65–69	M	3/8	1/7
	F	5/15	0/7
70–74	M	5/8	0/1
	F	10/14	3/5
75–79	M	2/2	1/1
	F	4/4	5/5
80–89	M	1/2	1/5
	F	5/5	4/4

were no cases who had diverticular disease amongst men aged between 40 and 49, which is why there are three observations of the form 0/0. For these observations, the fitted response probabilities will be zero, which means that the logistic transform of the fitted probability will be infinite. This in turn means that one or more of the parameters in the linear component of the model will become infinite. The result of this will be that the algorithm used to fit models to the data set will not converge properly. Of course, an observation of 0/0 conveys no useful information and so it can simply be omitted from the data base. Accordingly, linear logistic models are fitted to 29 observations from the data in Table 7.9. In the modelling, age interval will be taken as a factor, labelled age, with eight levels, while sex will be represented by the factor sex with two levels. The exposure factor in this example concerns whether or not the patient has diverticular disease and this will be represented by the dichotomous factor DD.

Both age and sex are likely to confound the relationship between the occurrence of diverticular disease and colonic cancer. Accordingly, the effect of the exposure factor DD has to be assessed after due allowance has been made for the age and sex of a patient. To see whether these two variables are associated with the occurrence of colonic cancer, they are included in the model that contains a constant term. Table 7.10 gives the deviances obtained on fitting a sequence of linear logistic models to the probability that a person is a case, using the data in Table 7.9. There is clearly a potential confounding effect due to age, but the sex effect is only slight. There is also some evidence that a term that represents the

Table 7.10 Deviances on fitting a sequence of models to the data given in Table 7.9

Terms fitted in model	Deviance	d.f.
Constant	73.49	28
Age	24.05	21
Sex	69.01	27
Age + sex	22.73	20
Age + sex + age × sex	11.97	14
Age + sex + age × sex + DD	9.19	13
Age + sex + age × sex + DD + DD × age	3.71	6
Age + sex + age × sex + DD + DD × sex	8.13	12

interaction between age and sex should be included in the model. For this reason, the effect of the exposure factor DD will be investigated after allowing for age, sex, and the age × sex interaction.

The change in deviance on adding DD to the model that includes age, sex and the age × sex interaction is 2.8 on one degree of freedom. This is significant at the 10% level ($P = 0.096$), and suggests that the occurrence of diverticular disease does play a part in the aetiology of bowel cancer. The reduction in deviance due to including the interaction term DD × age is 5.5 on 7 d.f., while that on including DD × sex is 1.1 on 1 d.f. Neither are significant and so there is no reason to believe that the association between diverticular disease and colonic cancer varies according to age or sex.

In the model that contains the terms age, sex, age × sex and DD, the parameter estimate corresponding to the second level of DD (DD present) is -0.722 with a standard error of 0.447. Hence the ratio of the odds of colonic cancer occurring in a patient with diverticular disease relative to one without is $e^{-0.722} = 0.49$. This can be interpreted as an estimate of the relative risk of colonic cancer occurring in patients with diverticular disease, relative to those in which this disease is absent. A 95% confidence interval for the true odds ratio is the interval (0.20, 1.17). Although this interval includes unity, these results do suggest that the occurrence of diverticular disease decreases the risk of colonic cancer; in other words, a patient with diverticular disease is less likely to be affected by colonic cancer than one without.

For comparison purposes, the estimated relative risk of colonic cancer obtained under the model that includes DD, age and sex alone is 0.48 while that under the model that contains just DD and age is 0.47. The small differences between these adjusted odds ratio estimates show that sex and sex × age have only a very slight confounding effect. However, if age is ignored, the odds ratio obtained from the model that contains DD alone increases to 0.88. Age is clearly an important confounding variable in the relationship between diverticular disease and colonic cancer.

The next step in the analysis of these data is to check that the fitted model actually fits the data. In particular, it is important to determine whether or not there are any observations to which the model is a poor fit, or observations that unduly influence the fit of the model. An index plot of the standardised deviance residuals for the model that contains DD, age, sex and age × sex is given in Fig. 7.1. The lack of pattern in this plot, and the absence of any observations with abnormally large or small residuals, suggests that the model is satisfactory and that there are no observations that are not well fitted by the model.

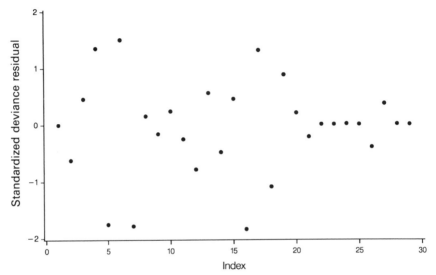

Fig. 7.1 Index plot of the standardized deviance residuals on fitting a linear logistic model to the data given in Table 7.9.

In this example, interest centres on the parameter estimate associated with the exposure factor DD, and so it will be particularly useful to determine if there are any observations that greatly change this estimate when they are deleted. Rather than investigate the effect of omitting a group of individuals, with a particular combination of levels of the exposure factors and confounding variables, on the parameter estimate associated with the second level of DD, it will be more fruitful to examine whether there are particular individuals that influence this estimate. This analysis will need to be based on the ungrouped data from the 211 individuals in the case-control study. The relevant data base will consist of the levels of the factors age, sex and DD for each individual, together with a binary response variable which takes the value zero if the individual is a control and unity if the individual is a case. When a linear logistic model is fitted to the ungrouped data, the parameter estimates will be the same as those from fitting the model to the grouped data, although the deviances obtained will differ from those given in Table 7.10. However, differences between the deviances for two alternative models will be the same as the differences between the deviances in Table 7.10. To study the influence that each individual has on the parameter estimate associated with the second level of DD, in the model that includes DD, age, sex and age × sex, the $\Delta\hat{\beta}$-statistic defined in Section 5.5.3 is used. An index plot of the values of this statistic for the parameter estimate associated with DD is given in Fig. 7.2. This plot shows that the greatest change to the parameter estimate occurs when the data from the thirty-ninth individual is omitted. This individual is a male aged between 50 and 54 with diverticular disease, who is a case in the study. If the data for this

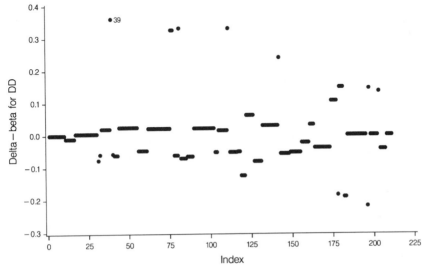

Fig. 7.2 Index plot of the $\Delta\hat{\beta}$-statistic for assessing the influence of each observation on the estimated log odds ratio of colonic cancer occurring in patients with diverticular disease, relative to those without.

individual is excluded, and the same model fitted to the remaining 210 observations, the parameter estimate for DD decreases to -0.890, with a standard error of 0.465. The corresponding ratio of the odds of colonic cancer occurring in those with diverticular disease relative to those without is $e^{-0.890} = 0.41$, and a 95% confidence interval for the true odds ratio is (0.16, 1.02). This interval barely includes unity, suggesting that the evidence that diverticular disease protects against colonic cancer is now significant at about the 5% level. Omitting the data from individual 39 from the analysis has decreased the estimated odds ratio to some extent, and so it would be prudent to check that the data for this individual, and possibly also for all other individuals for whom the $\Delta\hat{\beta}$-statistic is greater than about 0.3, have been correctly recorded.

In conclusion, these data provide some evidence that the occurrence of diverticular disease is protective against colonic cancer, although the extent to which this evidence is significant could depend on whether or not particular observations in the data set are correct.

7.7 MATCHED CASE-CONTROL STUDIES

In the previous section, we saw how data from a case-control study can be analysed using linear logistic modelling to explore the effect of particular exposure variables, after adjusting for potential confounding variables. An alternative to this method of adjusting for the effects of confounding variables is to take more account of their presence in the design of the study. A design which enables this to be achieved is the **matched case-control study**, in which each diseased person included in the study as a case is matched to one or more disease-free persons who will be included as controls. This matching is generally accomplished on the basis of potential confounding variables, such as age, ethnic group, marital status, parity, and so on. However, cases may also be matched to controls by residential area, or place of work in order to take account of confounding variables that cannot easily be quantified.

A design with M controls per case is known as a $1:M$ matched study, and the individuals that constitute the one case and the M controls to which the case has been matched are referred to as a **matched set**. The number of matched controls per case, M, will generally lie between one and five. In practice, missing data often means that the number of controls per case varies between 1 and M.

On some occasions, it may be appropriate to include more than one case, or different numbers of controls in the matched sets. However, the $1:M$ matched case-control study is the design that is most widely used and so the discussion in this section will be centred on this particular design. An example of a matched study in which there is just one control per case is given in Example 7.16.

Example 7.16 Acute herniated discs and driving
One objective of an aetiological study described by Kelsey and Hardy (1975) was to investigate whether the driving of motor vehicles is a risk factor for low back pain caused by acute herniated lumbar invertebral discs. The design used was a matched case-control study in which one control was matched to each case. The cases were selected from persons between the ages of 20 and 64 living in the area of New Haven, Connecticut, USA, who had X-rays taken of the low back between June, 1971 and May, 1973. Those diagnosed as having acute herniated lumbar invertebral discs, and who had only recently acquired symptoms of the disease, were used as the cases in the study. The controls were drawn from patients who were admitted to the same hospital as a case, or who presented at the same radiologists clinic as a case, with a condition unrelated to the spine. Individual controls were further matched to a case on the basis of sex and age. The age matching was such that the ages of the case and control in each matched pair were within ten years of one another. A total of 217 matched pairs were recruited, consisting of 89 female pairs and 128 male pairs.

After an individual had been entered into the study as either a case or a control, information on a number of potential risk factors, including place of residence and driving habits, was obtained. Those individuals who lived in New Haven itself were considered to be city residents, while others were classified as suburban residents. Data on whether or not each individual was a driver and their residential status were given by Holford, White and Kelsey (1978), and are summarized in Table 7.11. An analysis of these data will be presented later in Example 7.17.

Table 7.11 The numbers of matched sets of cases and controls according to driving and residence

Status	Driver?	Suburban resident?	Number of matched sets
Case	No	No ⎱	9
Control	No	No ⎰	
Case	No	Yes ⎱	2
Control	No	No ⎰	
Case	Yes	No ⎱	14
Control	No	No ⎰	
Case	Yes	Yes ⎱	22
Control	No	No ⎰	
Case	No	No ⎱	0
Control	No	Yes ⎰	
Case	No	Yes ⎱	2
Control	No	Yes ⎰	
Case	Yes	No ⎱	1
Control	No	Yes ⎰	
Case	Yes	Yes ⎱	4
Control	No	Yes ⎰	
Case	No	No ⎱	10
Control	Yes	No ⎰	
Case	No	Yes ⎱	1
Control	Yes	No ⎰	
Case	Yes	No ⎱	20
Control	Yes	No ⎰	
Case	Yes	Yes ⎱	32
Control	Yes	No ⎰	
Case	No	No ⎱	7
Control	Yes	Yes ⎰	
Case	No	Yes ⎱	1
Control	Yes	Yes ⎰	
Case	Yes	No ⎱	29
Control	Yes	Yes ⎰	
Case	Yes	Yes ⎱	63
Control	Yes	Yes ⎰	

7.7.1 Modelling data from matched case-control studies

As in modelling data from an unmatched case-control study, the probability that an individual is diseased will be assumed to depend on the values of k explanatory variables that have been measured on that individual. These variables may represent exposure factors and those potential confounding variables that have not been used in the matching process. In addition, the probability that a particular individual is diseased may also depend on the values of the matching variables that define the matched set in which an individual occurs. The values of these matching variables will generally differ between each of n matched sets of individuals, and so the probability that the ith person in the jth matched set is diseased will be denoted $p_j(\mathbf{x}_{ij})$, where $\mathbf{x}_{ij}$ is the vector of k explanatory variables for the ith individual, $i = 0, 1, \ldots, M$, in the jth matched set, $j = 1, 2, \ldots, n$. The vector of values of the explanatory variables for the case is denoted $\mathbf{x}_{0j}$, while the vectors $\mathbf{x}_{ij}$, for $i = 1, 2, \ldots, M$, will denote those for the M controls in the jth matched set. The disease probability $p_j(\mathbf{x}_{ij})$ will be modelled using a linear logistic model, with a different constant term, α_j, for each matched set. We therefore take

$$\text{logit}\,\{p_j(\mathbf{x}_{ij})\} = \alpha_j + \beta_1 x_{1ij} + \beta_2 x_{2ij} + \cdots + \beta_k x_{kij} \qquad (7.11)$$

where x_{hij} is the value of the hth explanatory variable x_h, $h = 1, 2, \ldots, k$, for the ith individual, $i = 0, 1, \ldots, M$, in the jth matched set. The parameter α_j in this model summarises the effect of the matching variables on the probability of disease.

By definition, the one case and M controls in a matched set will all have the same values of the variables on which they were matched. The probability that a particular individual is selected for inclusion in the study as either a case or a control will therefore depend on the values of these matching variables. However, as in an unmatched study, this probability must be independent of the values of the other explanatory variables that are being recorded.

In order to take account of the matching in the analysis of the observed data, what is termed a **conditional likelihood** is constructed. This likelihood is the product of n terms, each of which is the conditional probability that the case in a particular matched set, the jth, say, is the one with explanatory variables $\mathbf{x}_{0j}$, conditional on $\mathbf{x}_{0j}, \mathbf{x}_{1j}, \ldots, \mathbf{x}_{Mj}$ being the vectors of explanatory variables for the individuals in that matched set. The argument used to construct the relevant likelihood is quite complex and is deferred to Appendix B.2. The likelihood function is found to be

$$\prod_{j=1}^{n} \left[1 + \sum_{i=1}^{M} \exp\left\{ \sum_{h=1}^{k} \beta_h(x_{hij} - x_{h0j}) \right\} \right]^{-1} \qquad (7.12)$$

where the βs correspond to those in equation (7.11). Notice that this likelihood function does not involve the parameters $\alpha_1, \alpha_2, \ldots, \alpha_n$, and so the effects of the matching variables cannot be estimated. This in turn means that the disease probabilities $p_j(\mathbf{x}_{ij})$ are not estimable. Also, no information about the matching variables can be obtained by including them as explanatory variables in the model. This is because a matching variable will usually have the same value for the case and each of the controls in a particular matched set and so the values of $x_{hij} - x_{h0j}$ will generally be zero for that variable. However, it is possible to include a term that represents interaction between a matching variable and an exposure factor in equation (7.11) in order to investigate whether the effect of exposure is consistent across the different values of the matching variable.

The values of the βs which maximize the likelihood function given by expression (7.12) will be $\hat{\beta}_1, \hat{\beta}_2, \ldots, \hat{\beta}_k$, which are estimates of the coefficients of the explanatory variables in the model for the probability of disease given by equation (7.11). As for both cohort and unmatched case-control studies, these estimates are log odds ratios and can therefore be used to approximate the relative risk of the disease being studied, as shown in Section 7.5. In particular, for two individuals from the same matched set, the jth, say, the ratio of the odds of disease occurring in a person with explanatory variables $\mathbf{x}_{1j}$ relative to one with vector $\mathbf{x}_{2j}$ is given by

$$\log \left\{ \frac{p(\mathbf{x}_{1j})/[1 - p(\mathbf{x}_{1j})]}{p(\mathbf{x}_{2j})/[1 - p(\mathbf{x}_{2j})]} \right\} = \beta_1(x_{11j} - x_{12j}) + \cdots + \beta_k(x_{k1j} - x_{k2j})$$

This log odds ratio is independent of α_j and so the estimated ratio of the odds of disease for two matched individuals with different explanatory variables does not depend on the actual values of the matching variables.

The process of estimating the β-parameters, often referred to as **conditional logistic regression modelling**, also yields standard errors of the estimates, which can be used to construct confidence intervals for the approximate relative risk of disease using the method described in Section 7.5. For a particular model, the logarithm of the maximized likelihood function given by expression (7.12) when multiplied by -2 gives a quantity which can be treated as a deviance. The difference in deviance between two nested models has an asymptotic χ^2-distribution, and can be used to judge whether particular variables should be included in, or excluded from, the model in exactly the same way as for linear logistic modelling. The effect of a single explanatory variable can be determined by comparing the deviance for the model that contains that variable alone with the deviance for the model that has no explanatory variables. This latter model is referred to as the **null model**. For this model, $\beta_h = 0$ for $h = 1, 2, \ldots, k$ in expression (7.12), and the deviance reduces to $2n \log(1 + M)$.

In order to obtain the maximum likelihood estimates of the β-parameters in equation (7.11) by maximizing the likelihood in expression (7.12), specialist computer software is needed, such as the package EGRET or the SAS procedure **proc mcstrat**. Fuller details on EGRET are included in Chapter 9. In the particular situation where each case is matched to just one control, known as the matched pairs design, standard software for linear logistic modelling can be used. Accordingly, this special case is considered in detail in the following sub-section.

7.7.2 The 1:1 matched study

For the matched pairs design, $M = 1$, and the conditional likelihood function given by expression (7.12) reduces to

$$\prod_{j=1}^{n} \left[1 + \exp\left\{ \sum_{h=1}^{k} \beta_h(x_{h1j} - x_{h0j}) \right\} \right]^{-1}$$

Now let $z_{hj} = x_{h0j} - x_{h1j}$, so that z_{hj} is the value of the hth explanatory variable for the case minus that for the control, in the jth matched set. The likelihood function may therefore be written as

$$\prod_{j=1}^{n} \left[1 + \exp\left\{ - \sum_{h=1}^{k} \beta_h z_{hj} \right\} \right]^{-1}$$

Comparing this with the likelihood function given in equation (3.6), we see that this is the likelihood for a linear logistic model for n binary observations that are all equal to unity, where the linear part of the model contains k explanatory variables, $z_{1j}, z_{2j}, \ldots, z_{kj}, j = 1, 2, \ldots, n$, and no constant term. Consequently, standard statistical software can be used to fit the model by declaring the number of observations to be the total number of matched sets, setting the values of a putative response variable equal to unity for each observation, using as explanatory variables the differences between the original variables for the case and the control of each matched set, and specifying that a constant term is not to be included in the model. Moreover, the deviance obtained from the use of such software can be used to compare models with different explanatory variables in the usual way.

When factors are included in the model for the logistic transform of the probability of disease, the explanatory variables in equation (7.11) will correspond to indicator variables. The derived z-variables are then differences in the values of each of the indicator variables between the case and control in a matched pair. Interaction terms can be included in the model by representing them as products of the indicator variables for the main effects, as described in section 3.2.1. The differences between these products for the case and control in a matched pair are then used in the model fitting process. It is important not to use products of differences as the derived z-variables for an interaction term.

Usually, data from a matched case-control study will be presented as observations for each individual in the study. It will then be convenient to process the ungrouped data after defining a binary response variable. However, when there are a number of matched sets that all have the same values of the explanatory variables, as in Example 7.16, the data may be grouped. A binomial response can then be defined. For example, if there are s matched sets consisting of pairs of individuals with identical explanatory variables, then a binomial proportion s/s is used as the observation for that matched set.

In the previous paragraphs, we have seen how standard software for linear logistic modelling can be used to analyse data from a 1:1 matched case-control study. It is important to distinguish between this use of software for logistic modelling, in which the algorithm is simply being used to provide estimates of the β-parameters in the model described by equation (7.11) for the logistic transform of the probability of disease, and that where the intention is actually to fit a linear logistic model to observed binary response data. Indeed, the use of standard software for non-standard problems is a common procedure in advanced modelling.

Example 7.17 Acute herniated discs and driving
In the study described in Example 7.16, there are two exposure factors, namely whether or not an individual drives and whether or not an individual is a suburban resident. These two factors will be labelled as driving and residence, respectively, Each factor has two levels and so a single indicator variable is defined for each factor. Let x_1 be the indicator variable for driving and x_2 that for residence, where

$$x_1 = \begin{cases} 0 & \text{if the individual is not a driver} \\ 1 & \text{if the individual is a driver} \end{cases}$$

and

$$x_2 = \begin{cases} 0 & \text{if the individual is not a suburban resident} \\ 1 & \text{if the individual is a suburban resident} \end{cases}$$

We might also be interested in the driving $\times$ residence interaction, which corresponds to a third indicator variable x_3, obtained by multiplying together the values of x_1 and x_2 for each individual in the study. The variables z_1, z_2 and z_3 are then defined to be the difference between the values of x_1, x_2 and x_3, respectively, for the case and the corresponding matched control. For example, suppose that in a given matched set, the case is a driver and a suburban resident, while the control is not a driver, but is a suburban resident. Then, for the case $x_1 = 1, x_2 = 1$ and so $x_3 = 1$, while for the control $x_1 = 0, x_2 = 1$ and $x_3 = 0$. Then, for this particular matched pair $z_1 = 1 - 0 = 1, z_2 = 1 - 1 = 0$ and $z_3 = 1 - 0 = 1$. From Table 7.11, there are four matched pairs that have this set of explanatory variables. Consequently, if standard software for linear logistic modelling were being used to analyse these data, there would be a line of data that associates a binomial response of 4/4 with explanatory variables $z_1 = 1, z_2 = 0$ and $z_3 = 1$. This argument is adopted for the other matched sets in the study and gives the basic data in Table 7.12.

Information about the driving effect is obtained from fitting z_1, information about the residence effect is obtained from fitting z_2 and information about the driving $\times$ residence interaction from fitting z_3. The deviances obtained on fitting relevant combinations of z-variables are given in Table 7.13. Since z_1 represents the effect of driving, z_2 the effect of residence and z_3 the driving $\times$ residence interaction, the deviances in this table can be used to determine whether there is an interaction between driving and residence, and, if not, whether neither, one or both of the

Table 7.12 Input data required when using standard software for linear logistic modelling to analyse data from a 1 : 1 matched study

Observation	Response	Explanatory variables		
		z_1	z_2	z_3
1	9/9	0	0	0
2	2/2	0	1	0
3	14/14	1	0	0
4	22/22	1	1	1
5	2/2	0	0	0
6	1/1	1	−1	0
7	4/4	1	0	1
8	10/10	−1	0	0
9	1/1	−1	1	0
10	20/20	0	0	0
11	32/32	0	1	1
12	7/7	−1	−1	−1
13	1/1	−1	0	−1
14	29/29	0	−1	−1
15	63/63	0	0	0

Table 7.13 Deviances on fitting linear logistic models to the data given in Table 7.12

Terms fitted in model	Deviance	d.f.
z_1	292.57	14
z_2	296.54	14
$z_1 + z_2$	291.28	13
$z_1 + z_2 + z_3$	291.20	12

exposure factors affect the risk of a herniated disc occurring. For example, the deviance for the interaction, adjusted for driving and residence, is $291.28 - 291.20 = 0.08$. The deviances for the other relevant terms are summarised in Table 7.14. Judged against percentage points of the χ^2-distribution, the deviance due to the interaction between driving and residence is not significant, and so there is no interaction between these two risk factors for herniated disc. The deviance due to driving, adjusted for residence, is significant at the 5% level ($P = 0.022$), while the deviance due to residence, adjusted for driving, is not significant ($P = 0.256$). From the results of this analysis, we can conclude that driving is a risk factor for herniated disc, whereas area of residence is not.

To quantify the effect of driving on the risk of herniated disc, the parameter estimates can be used to give an approximate relative risk. On fitting the model

Table 7.14 Analysis of deviance based on the data given in Table 7.12

Source of variation	Deviance	d.f.
Interaction, adjusted for driving and residence	0.08	1
Driving, adjusted for residence	5.26	1
Residence, adjusted for driving	1.29	1

that contains z_1 and z_2 alone, the parameter estimates obtained from GLIM output are as follows:

	estimate	s.e.	parameter
1	0.6578	0.2934	Z1
2	0.2555	0.2257	Z2

The parameter estimate associated with z_1 is the logarithm of the ratio of the odds of herniated disc occurring in a driver ($x_1 = 1$) relative to a non-driver ($x_1 = 0$), adjusted for residence. Hence, the approximate relative risk of herniated disc in a driver relative to a non-driver is $e^{0.658} = 1.93$. A 95% confidence interval for the true log odds ratio is the interval from $0.658 - (1.96 \times 0.293)$ to $0.658 + (1.96 \times 0.293)$, that is, the interval from 0.084 to 1.232. The corresponding 95% confidence interval for the odds ratio itself is (1.09, 3.43). Similarly, the approximate relative risk of herniated disc occurring in a suburban resident ($x_2 = 1$) relative to a non-resident ($x_2 = 0$), adjusted for driving, is $e^{0.256} = 1.29$, and a 95% confidence interval for the true odds ratio is the interval (0.83, 2.01). From the results of this analysis, we may conclude that the risk of herniated disc occurring in a driver is about twice that of a non-driver, but that the risk is not affected by whether or not they are a suburban resident.

The example considered here illustrates the analysis of data from a 1 : 1 matched study. No new principles are involved in analysing data from the 1 : M matched study, although specialist software is then required to obtain the maximum likelihood estimates of the β-parameters in equation (7.11). In Section 7.8, a much larger set of data from a matched case-control study, with three controls matched to each case, is used to investigate the effect of a number of exposure factors on sudden infant death syndrome in infants.

7.7.3 Assessing goodness of fit

One of the most useful diagnostics for model checking in the analysis of data from unmatched studies is the statistic that measures the effect of omitting each observation on the parameter estimates associated with exposure factors. In this way, the influence that the data from each individual has on the parameter estimate, and hence on the approximate relative risk of disease, can be assessed. This same notion can be used in matched case-control studies.

In a matched case-control study, if an observation for a case is omitted from the data set, there is then no information in that particular matched set about the association between the exposure factors and the disease. Accordingly, the effect of omitting the record for a particular case on the β-parameters in equation (7.11) is the same as the effect of omitting the matched set which includes that case. Similarly, when there is only one control per case, as in the matched pairs design, omitting a record for a particular control has the same effect on the β-parameters as omitting the corresponding matched pair. When there is more than one control per case, it is possible to examine the influence of either individual controls, or complete matched sets, on the β-parameters. However, if a matched set is found to be influential, it will then be necessary to identify whether particular individuals in that matched set are influential.

In the particular case of the matched pairs design, when standard software for linear logistic modelling is being used to model data that are structured as in Example 7.17, the $\Delta\hat{\beta}$-statistic introduced in Section 5.5.3 can be used without modification. An index plot of the values of the $\Delta\hat{\beta}$-statistic will indicate whether there are particular matched pairs that have an undue impact on the parameter estimate of interest.

When analysing data from $1:M$ studies, where $M > 1$, a similar statistic can be used to determine the effect that particular individuals have on the β-parameters in equation (7.11). In addition, an analogue of this statistic can be used to examine the influence of an entire matched set of $M + 1$ individuals on these β-parameters. The effect of omitting each of the matched sets on the parameter estimates can be approximated by an expression whose algebraic form is a little complicated. For this reason, it will not be reproduced here, although details can be found in Pregibon (1984), for example. Fortunately, the values of the $\Delta\hat{\beta}$-statistic are often provided by standard statistical software for analysing data from $1:M$ matched studies. In particular, EGRET gives the values of the $\Delta\hat{\beta}$-statistic for individual observations, while SAS **proc mcstrat** gives the values of the corresponding statistic for individual matched sets. The use of the $\Delta\hat{\beta}$-statistic will be illustrated in an analysis of data on sudden infant death syndrome in the next section.

7.8 A MATCHED CASE-CONTROL STUDY ON SUDDEN INFANT DEATH SYNDROME

In England one hundred years ago, 150 children out of every 1000 babies born alive would die before reaching the age of one year. In the 1980s, the corresponding figure is ten deaths out of every 1000 live births. This dramatic reduction in the mortality rate has focused a great deal of attention on those children dying suddenly and unexpectedly. One hundred years ago, the vast majority of the 150 deaths per thousand live births were from malnutrition and infection, with only around two of the 150 being what would now be regarded as sudden infant deaths.

Now, out of the ten deaths per 1000 live births, two are sudden and unexpected, of which occasionally a post-mortem shows death from an unsuspected congenital deformity, an unrecongnised infection or a rare disorder. There is a large literature on possible risk factors for sudden infant death syndrome (SIDS), which has recently been reviewed by Golding, Limerick and Macfarlane (1985). Much of this literature focuses on risk factors relating to characteristics of the infant, but in the study that forms the basis of this example, interest centered on characteristics of the mother during pregnancy and of the mother and baby at birth.

The Oxford record linkage study (ORLS), which was begun in 1962, was an attempt to create a data base containing details on the births of all children born to mothers resident in the United Kingdom. Unfortunately, the study expanded only as far as Oxfordshire Regional Health Authority boundaries, but this still covers a population of some two million people. The ORLS has been used to provide data for a matched case-control study based on births to mothers resident in Oxfordshire or West Berkshire between 1976 and 1985. The cases were selected from mothers whose babies had died suddenly and unexpectedly within the first two years of life. The case notes, post-mortem results, histological and physiological data were reviewed by a consultant paediatric pathologist in order to identify those deaths that could be attributed to SIDS. This lead to 241 cases. Three controls were matched to each of these cases on the basis of six factors, namely, mother's marital status, maternal age, mother's parity, mother's social class, place of birth of the infant, and year of birth of the infant. However, for two cases, only two matched controls were identified, so that the data base contained information on 962 mothers and their offspring.

Data were recorded on over sixty different variables relating to the characteristics of the mother during pregnancy and the condition of the baby at birth. Characteristics of the mother included personal data, such as employment status, religion, smoking history, history of contraceptive use, and number of previous pregnancies. Variables relating to the pregnancy included gestation period, maximum recorded blood pressure during the pregnancy, number of ante-natal visits, weight at each ante-natal visit, numbers of X-rays and scans taken during the pregnancy, and duration of labour. Data obtained on the condition of the baby at birth included the sex of the baby, number of babies (singleton or twin), presentation and position of the baby at birth, birth weight, head circumference, whether or not the baby was transferred to a special care baby unit (SCBU) and Apgar score. Information was also collected on whether complications were experienced by the mother during the pregnancy or the delivery, the number of analgesic and anaesthetic drugs given during the birth, and the weight of the placenta. For many of the subjects in the study, the information collected was incomplete, and so many of the variables contain missing values. This feature of the data has to be borne in mind throughout the analysis.

The first step in the analysis of this data base is to eliminate those risk factors which have little or no impact on the risk of sudden infant death. The technique of conditional logistic regression modelling is used for this, and the deviance for each model fitted was obtained using the statistical package EGRET. Each of 64 potential risk factors for SIDS were fitted on their own, and the deviance compared to that for the model that contains no variables, the null model. When the values of a particular factor are missing for some of the individuals in the study, the records for these individuals have to be omitted from the data base before the deviance for the model that contains the factor can be computed. In order to assess the significance of the factor, this deviance must then be compared to the deviance on fitting the null model to the same reduced data set.

For those variables which are significant risk factors at the 10% level, the resulting change in deviance, the corresponding change in the number of degrees of freedom, and the P-value that results from comparing the change in deviance with percentage points of the χ^2-distribution, are given in Table 7.15. The risk factors

Table 7.15 Significant risk factors for sudden infant death syndrome

Risk Factor (Variate or Factor)	Abbreviated name	Change in deviance (d.f.)	P-value
Birth weight of baby in g (V)	Weight	23.806 (1)	< 0.001
Duration of baby's stay in hospital in days (V)	Duration	11.229 (1)	< 0.001
Number of anaesthetics given to mother (F)	Anaes	14.894 (3)	0.002
Pregnancy complications (F)	Pregcomp	8.527 (1)	0.003
Head circumference of the baby in mm (V)	Headcirc	8.273 (1)	0.004
Sex of baby (F)	Sex	7.242 (1)	0.007
Smoking history (F)	Smoking	7.179 (1)	0.007
Operations carried out on mother at birth (F)	Oprtns	6.172 (1)	0.013
Number of undelivered admissions (F)	Undadmns	10.485 (3)	0.015
Mother in labour on admission (F)	Labadmn	11.428 (4)	0.022
Baby premature, on time, or late (F)	Premlate	7.115 (2)	0.029
Baby transferred to SCBU (F)	Scbu	4.218 (1)	0.040
Delivery complications (F)	Delcomp	4.021 (1)	0.045
Number of scans (V)	Scans	3.776 (1)	0.052
Number of babies born (F)	Nbabies	3.771 (1)	0.054
Period of gestation (V)	Gest	3.291 (1)	0.070
Number of previous pregnancies (F)	Npreg	11.536 (6)	0.073
Weight of placenta in g (V)	Wplacenta	2.940 (1)	0.086

have been listed in increasing order of their corresponding P-values. This table also gives abbreviated names for each of the risk factors, which, for brevity, will often be used in futher discussion. In summary, the most important risk factors, when considered on their own, are the birth weight and head circumference of the baby, the period of time that the infant remains in hospital and whether or not he or she was transferred to an SCBU, whether or not the mother was operated on at birth, the number of anaesthetics given to the mother during the birth, and whether or not there were complications during pregnancy and delivery. The sex of the baby also appears to be important, as does the mothers smoking habits, the number of previous pregnancies, the condition of the mother on admission to the hospital and the number of undelivered admissions, the gestation period and whether the baby was premature or late, the number of scans given during pregnancy, the number of babies born, and the weight of the placenta.

A number of the risk factors in Table 7.15 can be divided into two distinct groups. The risk factors weight, headcirc, nbabies and gest all give some indication of the size of the baby at birth, whereas the factors duration, scbu, anaes, oprtns, pregcomp and delcomp all reflect whether or not there were any problems during the pregnancy and the birth. Consequently, the next step in the analysis of these data is to investigate which of the variables in these two groups can be used to summarise the size of the baby and the occurrence of difficulties during pregnancy and birth.

Of the four risk factors which relate to the size of the baby, weight is the dominant factor. This variable is therefore fitted and each of the other three factors are then added to the model in turn. For this part of the analysis, the data base contains 762 individuals with complete records on these four risk factors. The reduction in deviance on including headcirc in the model that contains weight is 4.670 on 1 d.f., which is significant at the 5% level ($P = 0.031$). On the other hand, the reduction in deviance brought about by adding nbabies and gest to the model that contains weight is only 0.078 and 0.406, respectively. Moreover, neither of these two variables significantly reduce the deviance when added to the model that contains both weight and headcirc, and so we conclude that weight and headcirc alone summarize the size of the baby at birth.

The data base used to identify the most relevant set of variables that relate to whether or not there were problems during the pregnancy and birth contains 906 complete records. Of duration, scbu, oprtns, anaes, pregcomp and delcomp, the single most important variable is duration. After fitting a model that contains this term, significant reductions in deviance are obtained when anaes, pregcomp and delcomp are added to the model. However, after fitting duration, pregcomp and delcomp, the reduction in deviance due to the inclusion of anaes is 6.635 on 3 d.f., which is not significant at the 5% level ($P = 0.084$). Also, the reductions in deviance when (i) pregcomp is added to the model that contains duration and delcomp, (ii) delcomp is added to duration and pregcomp, and (iii) duration is added to pregcomp and delcomp, are all highly significant. The deviance is not significantly reduced by adding scbu and oprtns to the model that contains duration, pregcomp and delcomp. This leads to the conclusion that three exposure factors, duration, pregcomp and delcomp, can be used to summarize the occurrence of difficulties during pregnancy and birth.

Having isolated weight, headcirc, duration, pregcomp and delcomp as the most important factors that relate to the size of the baby and whether a pregnancy or birth was subject to any complications, it remains to investigate whether any of the eight remaining variables in Table 7.15 are associated with the risk of an infant dying from sudden infant death syndrome. For each of these eight variables, scans, sex, smoking, npreg, wplacenta, labadmn, undadmns and premlate, a data base is constructed that has no missing values for the five risk factors, weight, headcirc, duration, pregcomp and delcomp, and the additional variable whose effect is being studied. After fitting a model that contains weight, headcirc, duration, pregcomp and delcomp, the only factor that significantly reduces the deviance is sex. When this factor is included, the reduction in deviance is 7.052 on 1 d.f., which is significant at the 1% level ($P = 0.008$). In addition, when sex is added to the model, none of the other variables can be omitted without significantly increasing the deviance.

At this point, the most appropriate model for the probability that an infant succumbs to SIDS contains the terms weight, headcirc, duration, pregcomp, delcomp and sex. Of these, weight, headcirc and duration are variates, while pregcomp, delcomp and sex are all two-level factors. For pregcomp and delcomp, the first level of the factor corresponds to the absence of complications, and the second level to the occurrence of complications. The first level of the factor sex corresponds to male and the second to female. For each of the three variates in the model, the assumption that they enter linearly was checked by grouping the values of each variable into a number of categories. The resulting categorical variables were then fitted as factors, and the procedure described in section 3.10 was used to examine the extent of any non-linearity. There was no evidence to suggest that the risk of SIDS is not linearly related to weight, headcirc and duration. Although the effect of variables used in the matching process cannot be assessed, it is possible to investigate whether there are interactions between risk factors and the matching variables. In this example, none were found. In addition, there was no evidence of any interactions between the six risk factors in the model.

The main findings from this analysis are that the birth weight, head circumference, duration of the baby's stay in hospital, whether or not there were complications during the pregnancy or at the birth, and the sex of the baby all affect the probability that an infant will succumb to SIDS. Parameter estimates associated with each of these six terms, and their standard errors, obtained using EGRET, are given in Table 7.16. The parameterisation of the terms corresponding to factors is such that the effect due to the first level of the factor is zero, the same as that used by GLIM. Estimates of the effects due to the second level of each of the three factors are therefore given in the table with a subscript of 2. The parameter estimates can be exponentiated to give odds ratios, adjusted for the other exposure factors in the model, and 95% confidence intervals for the corresponding true odds ratios can be computed from the standard errors of the parameter estimates in the manner described in Section 7.5. These adjusted odds ratios and confidence intervals are also given in Table 7.16.

Table 7.16 Parameter estimates, standard errors, odds ratios and 95% confidence intervals for weight, headcirc, duration, pregcomp, delcomp and sex

Term	Parameter estimate	s.e.	Odds ratio	95% Confidence interval
Weight	− 0.0013	0.0003	0.999	(0.998, 0.999)
Headcirc	0.0187	0.0104	1.019	(0.998, 1.040)
Duration	0.0509	0.0230	1.052	(1.006, 1.101)
Pregcomp$_2$	0.3892	0.2240	1.476	(0.951, 2.289)
Delcomp$_2$	− 0.7651	0.3326	0.465	(0.242, 0.892)
Sex$_2$	− 0.5084	0.1931	0.601	(0.412, 0.878)

The estimated odds ratio for a particular term can be interpreted as the approximate relative risk of SIDS for given values of the other variables in the model. Hence, for a baby of a particular sex with a given head circumference, and given levels of the factors pregcomp and delcomp, we see from Table 7.16 that the heavier the baby is at birth, the less the risk of SIDS for that baby, the risk decreasing multiplicatively by a factor of 0.999 for each extra gram of weight. Similarly, babies with larger heads appear to be more at risk of SIDS, the risk increasing by a factor of 1.019 for each unit increase in head circumference. Babies that remain in hospital longer are more likely to die from SIDS, the risk increasing by a factor of 1.052 for each additional day spent in hospital, and the occurrence of complications during pregnancy increases the risk of SIDS by a factor of 1.5. The risk of a SIDS baby for mothers who experience delivery complications is about half that of mothers with normal births, and the risk of SIDS occurring in female babies is only about 60% of that for males.

Finally, the influence of each of the matched sets on the parameter estimates in the fitted model, and hence on the odds ratios, will be examined using the $\Delta\hat{\beta}$-statistic. Plots of the values of this statistic against the number of the matched set, for weight, headcirc, duration, pregcomp, delcomp and sex, are given in Fig. 7.3. The plots for headcirc, pregcomp, delcomp and sex indicate that no single matched set has an undue influence on the parameter estimate associated with these variables, although the plot for delcomp is more scattered than the others. The plot of the $\Delta\hat{\beta}$-statistic for weight indicates that matched set number 63 has the greatest impact on the coefficient of this variable. In this set, the baby born to the case has a much higher birth weight than the control. The plot indicates that the effect of deleting this matched set is to decrease the coefficient of weight by about 0.3 standard errors, making the corresponding estimate of the odds ratio slightly smaller, and more significant, than that in Table 7.16. When this matched set is deleted, the coefficient of weight does in fact decrease to − 0.0014, but the effect on the odds ratio is negligible.

The plot of the $\Delta\hat{\beta}$-statistic for duration shows that matched sets numbered 22, 33 and 148 have the greatest impact on the odds ratio for this variable. Matched sets 22 and 148 both include a control whose baby spent over 30 days in hospital, while set 33 contains a case whose baby was in hospital for 37 days. When sets 22

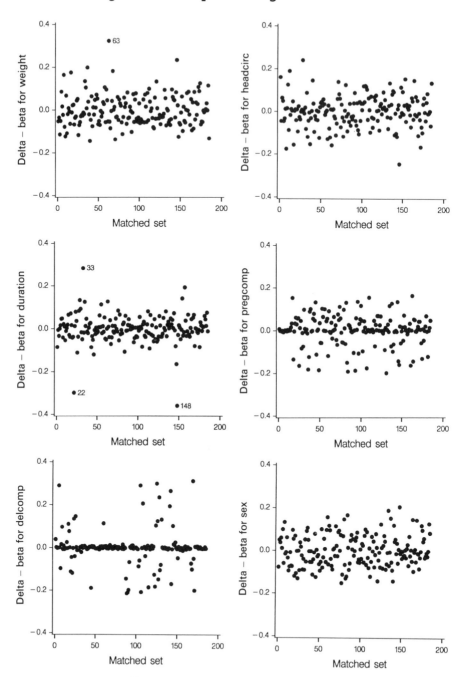

Fig. 7.3 Plots of the $\Delta\hat{\beta}$-statistic against number of matched set for weight, headcirc, duration, pregcomp, delcomp and sex.

and 148 are omitted, the coefficient of duration is increased by about 0.008, which increases the odds ratio by about 0.01. The effect of deleting matched set 33 is to reduce the estimated odds ratio by a similar amount. Again, the influence of these three matched sets on the estimated odds ratios is negligible.

In summary, there is no evidence that any particular matched set has an undue influence on the size of any of the corresponding parameter estimates, and therefore the odds ratios for the five risk factors, given in Table 7.16, provide a satisfactory summary of the magnitude of the effects of the risk factors.

FURTHER READING

General introductions to the principles of epidemiology and the different types of epidemiological study are given by MacMahon and Pugh (1970) and Lilienfeld and Lilienfeld (1980), for example. An intermediate text that covers the design, conduct and analysis of data from both unmatched and matched case-control studies is the book by Schlesselman (1982). Breslow and Day (1980, 1987) in their two-volume work provide a thorough account of the design and analysis of data from case-control and cohort studies. All of these books discuss confounding and interaction; in addition, Dales and Ury (1978) explain why it is inappropriate to use significance tests to identify confounding variables. Breslow and Day (1980) and Schlesselman (1982) include material on the more traditional methods of analysing case-control data, as well as the approach based on linear logistic modelling, although Schesselman's treatment of the modelling approach is not as detailed as that of Breslow and Day.

The Framingham study is described in fuller detail by Gordon and Kannel (1970) and Woodbury and Manton (1983). The use of linear logistic modelling in the analysis of data from the Framingham study was pioneered by Cornfield (1962) and Truett, Cornfield and Kannel (1967).

The interpretation of parameter estimates in linear logistic models fitted to data from aetiological studies is discussed in some detail by Hosmer and Lemeshow (1989). A readable summary of the derivation of the conditional likelihood function for analysing $1 : M$ matched studies is given by Breslow et al. (1978). Holford, White and Kelsey (1978) derive the conditional likelihood function for the matched pairs design and illustrate the use of this likelihood in analysing data from a number of studies, including that from Kelsey and Hardy's study of the epidemiology of acute herniated lumbar invertebral discs, given in Example 7.16. Breslow et al. (1978) also include an illustration of the methodology and compare the results of the matched analysis with those of an unmatched analysis. Hosmer and Lemeshow (1989) give a clear introduction to the analysis of data from matched studies, and illustrate the use of a number of diagnostics for model checking.

FORTRAN code for analysing data from matched case-control studies is given by Smith et al. (1981) and also in Appendix IV of Breslow and Day (1980). Adena and Wilson (1982) describe how GLIM can be used in the analysis of data from unmatched and matched case-control studies.

Techniques for assessing the goodness of fit of models for the probability of disease, fitted to data from matched case-control studies, were first proposed by Pregibon (1984). In addition to showing how the $\Delta\hat{\beta}$-statistic can be used to detect influential matched sets, he discusses the detection of outlying matched sets and methods for detecting inadequacies in the linear part of the model. Moolgavkar, Lustbader and Venzon (1985) summarize methods for detecting influential and outlying matched sets and illustrate the methods in the analysis of data from a matched pairs study. Breslow and Storer (1985) describe methods for choosing between a linear logistic model for the probability of disease and different formulations of the model.

8

Some additional topics

Although many aspects of modelling binary data have been considered in the previous seven chapters, there remains a number of topics that are either peripheral to, or beyond the scope of, this intermediate text. For completeness, some are outlined in this chapter, and references are provided to enable the interested reader to obtain further information on them. However, a full review of the literature on these additional topics will not be included, and the references will generally be restricted to recent monographs and journal articles; additional references will of course be found in these publications. Two particular monographs which include a more advanced treatment of many of the techniques described in this book are Cox and Snell (1989) and McCullagh and Nelder (1989); these texts contain much material that can be used as a springboard for further research.

8.1 ANALYSIS OF PROPORTIONS AND PERCENTAGES

This book has described methods for analysing data in which the number of successes is the response variable and the binomial denominator is known. However, the statistician can be faced with data in the form of proportions where the binomial denominator is not known, or data that are proportions derived from response variables that do not have a binomial distribution. Continuous response variables are the most common instance of this latter type, and examples include the percentage of weed cover on plots of land, the proportion of the area of a plant leaf that has been affected by fungal disease, and the percentage fat content in samples of meat. In these examples, the recorded proportion may be obtained as the ratio of the values of two measured continuous variables, but it may also be a subjective assessment of the extent of the feature of interest.

In situations where the raw data are proportions, a linear logistic model for the ith true proportion, p_i, $i = 1, 2, \ldots, n$, may still be appropriate, and the variance of the ith proportion can initially be taken to be proportional to $p_i(1 - p_i)$. The parameters in the linear component of this model can be estimated by constructing a **quasi-likelihood function**, as discussed in Chapter 9 of McCullagh and Nelder (1989). It turns out that for the model being considered here, the quasi-likelihood function behaves in a similar way to the log-likelihood function based on an underlying binomial distribution. For this reason, the model can be fitted using computer software for linear logistic modelling by defining the values of the response variable to be the observed proportions, and setting the binomial denominators equal to unity. Similarly, percentages can be modelled by setting the binomial denominators equal to 100. Because the response variable no longer has a binomial distribution, there is no reason why the scale parameter should be set equal to unity, and so appropriate standard errors of parameter estimates are usually obtained by multiplying the standard errors that assume a scale parameter of unity by the square root of the residual mean deviance. Moreover, in comparing two alternative models, the change in deviance on including a term in a model, divided by the corresponding change in the number of degrees of freedom, should now be compared to the residual mean deviance for the full model using an F-test, as in Section 6.5.

8.2 ANALYSIS OF RATES

A **rate** is an expression of the number of individuals with a particular characteristic in relation to the number of individuals that have that characteristic in the population being studied. For example, the mortality rate is the ratio of the number of deaths that occur during a particular year to the number of persons in the population being studied, and when multiplied by 1000, would be a rate per one thousand individuals per year. When rates are cross-classified according to a number of other factors, such as social class, occupation and place of residence, a modelling approach can be used to determine the extent to which the rates differ across the levels of a factor, and whether there are interactions between the effects of these factors.

Suppose that λ_i is the true rate corresponding to the ith observed rate, y_i/n_i, $i = 1, 2,..., n$. In many applications, this proportion is small, and the value of n_i is large. The Poisson approximation to the binomial distribution may then be used to construct an appropriate model. We therefore assume that y_i has a Poisson distribution with mean given by $E(y_i) = n_i\lambda_i$. Now suppose that a log-linear model of the form $\log \lambda_i = \eta_i$ is assumed for the true rate, where η_i is a linear systematic component that contains the effects due to the different factors being studied. Then, the expected value of y_i is modelled using

$$\log E(y_i) = \log n_i + \log \lambda_i = \log n_i + \eta_i$$

This log-linear model is sometimes called a **Poisson regression model**. The model includes the term $\log n_i$, which has a known coefficient of unity, a term generally known as an **offset**. Further details on modelling rates that arise in an epidemiological context are given in Chapter 4 of Breslow and Day (1987).

8.3 ANALYSIS OF BINARY DATA FROM CROSS-OVER TRIALS

In Chapters 3–6, the use of linear logistic models in modelling data from designed experiments has been illustrated in a number of examples, particularly Examples

3.3, 3.5 and 6.2. One form of experimental design which merits special consideration is the **cross-over design**, which is widely used in clinical trials and in certain types of animal experimentation. In this design, treatments are applied to individuals in distinct periods, and each individual receives each treatment during the course of the experiment. This means that the effects of the treatments can be compared within each individual.

Of particular importance is the cross-over design to compare two treatments, A and B, say, in which one group of individuals receives treatment A in the first period (period 1) and treatment B in the second (period 2), and a second group receives treatment B in the first period and treatment A in the second. This design can provide information about the main effects of treatment and period, as well as the treatment × period interaction. For example, in a trial to compare two treatments for asthma, A and B might be a standard and a new form of bronchodilator, and the binary response that is recorded might denote whether or not the patient experiences relief from symptoms. Data from a 2 × 2 cross-over design will thus consist of a pair of observations for each individual in the study, (r, s), denoting that the response in period 1 was r and that in period 2 was s, where r and s are either zero or unity. There will therefore be four possible pairs of observations, namely (0, 0), (0, 1), (1, 0) and (1, 1), for each individual, and the raw data will consist of the number of individuals in each of the two groups for whom each of these four pairs of observations has been recorded.

In Chapter 3 of the recent monograph by Jones and Kenward (1989), on the design and analysis of cross-over experiments, the authors describe a number of procedures that can be used to assess treatment effects, period effects and treatment × period interaction in the 2 × 2 cross-over experiment. They go on to develop a modelling approach for the analysis of data in which baseline information, and other explanatory variables recorded on each individual, are taken account of, and they describe how the model for the 2 × 2 cross-over design can be generalized to designs that involve more than two treatments with two or more periods. This modelling approach was also presented by Kenward and Jones (1987).

8.4 RANDOM EFFECTS MODELLING

The use of random effects in a linear logistic model is not restricted to situations where the aim is to model the occurrence of overdispersion. As in the analysis of continuous response data, a random effect can be introduced when inferences are to be drawn about some population of units, from which those used in the actual experiment have been drawn. For example, in a clinical trial to compare two treatments, the effects due to individual patients are best regarded as random effects. This is because the individuals in the trial are to be regarded as being representative of a larger population about which inferences are to be drawn. We are then only likely to be interested in the variability between the patients in this population as a source of variation in the response, and comparisons between particular subjects are of no real concern. Such models are considered by Anderson and Aitkin (1985). Stiratelli, Laird and Ware (1984) describe models for longitudinal binary response data in which one or more of the explanatory variables in the model have random coefficients for each individual. Searle (1971) gives a very readable introduction to random effects models.

In Section 6.7, the assumption was made that the random effect term had the same variance for each of a number of different groups of the data. In some situations, we might want to relax this assumption, so that, for example, in the problem discussed in Section 6.9, the variance of the random effect could be different for each of the two species. The statistical package EGRET can still be used to fit this more general random effects model.

8.5 MODELLING ERRORS IN THE MEASUREMENT OF EXPLANATORY VARIABLES

Measurement errors can occur in recording the value of a binary response variable and the values of the explanatory variables to be included in the model. We saw in section 5.7.3 that errors in recording a binary observation amount of misclassification, and so this topic need not be considered further in this section. In many circumstances, it may not be possible to measure accurately the values of explanatory variables in a model. For example, blood pressure can rarely be determined accurately, and measurements of physical quantities such as weight or temperature also tend to be subject to measurement error. When measurement error is present, it is desirable to investigate the effect of this error on model-based inferences, and, if this effect is large, to allow for the error in the model fitting process. If measurement errors are not taken into account, the estimated values of the parameters in a fitted model will be biased, and this may lead to incorrect estimates of *ED50* values, odds ratios, or other quantities based on the parameter estimates. When measurement error is taken into account, the standard error of the parameter estimates will be increased. However, the mean squared error, which measures the extent to which a possibly biased estimate deviates from its own true value, can be decreased.

The situation in which some of the explanatory variables in a linear model for the probit of a response probability are subject to measurement error was considered by Carroll *et al.* (1984). They adopted a model in which the value of an explanatory variable, x, cannot be observed directly, and it is only possible to observe $X = x + \delta$, where δ is normally distributed with zero mean and variance σ_δ^2. They further assumed that x is normally distributed with mean μ_x and variance σ_x^2, to give what is known as a model for **structural relationship**. They found that the mean squared error of the parameter estimates can be decreased when the measurement error is not too small, and the number of binary observations is large. For example, if $\sigma_x^2 > 0.08$ and $\sigma_\delta^2 = \sigma_x^2/3$, the mean squared error can be reduced if the number of binary observations exceeds 600. Stefanski and Carroll (1985) described the **functional relationship** model for linear logistic modelling, in which the variable x is assumed to be fixed, but observed with error. Burr (1988) considers a model for the situation where target values of the explanatory variables are set by the experimenter, but where these values are not realized. More general results on the effect of measurement error in explanatory variables in a generalised linear model are given by Stefanski and Carroll (1987) and Schafer (1987). A general account of regression modelling when variables are subject to error is given in Chapter 29 of Kendall and Stuart (1979) and in Chapter 13 of Wetherill (1986). See also Fuller (1987).

8.6 ANALYSIS OF BINARY TIME SERIES

In the analysis of binary data, we often make the assumption that the individual binary observations are independent. When binary observations arise successively in time, this assumption cannot be made, and methods of analysis that take account of the non-independence will be needed. For example, whether or not it rains on a particular day is a binary outcome, so that over a year, a sequence of 365 (or 366) binary observations would be recorded. Successive observations in this sequence will not be independent, since the weather on a particular day will not be independent of that on the previous day. The general approach to the analysis of such data is to make assumptions about the form of this dependence.

A model in which the occurrence of rain on day i depends only on whether or not it rained on day $i - 1$, and not on the state of the weather on day $i - 2$ and before, is known as a **first-order Markov chain**. A more complicated, but more realistic, model might assume higher-order Markov chains in which the weather on day i depends on what occurred on the two previous days, and so on. A model fitting approach to the analysis of daily rainfall data, and the use of such models in agricultural planning has been described by Stern and Coe (1984). A concise review of methods for analysing binary time series is given in Section 2.11 of Cox and Snell (1989), and the topic is treated from a mathematical viewpoint by Kedem (1980).

There is a distinction between binary time series and what is generally termed repeated measurements data. The main difference between the two types of data is that repeated measurements data consist of a relatively short sequence of observations, whereas time series data generally form a much longer series. The statistical treatment of repeated measurements data has recently been described by Crowder and Hand (1990).

8.7 MULTIVARIATE BINARY DATA

So far, we have focused exclusively on there being a single binary response variable. Frequently, several binary response variables are measured on each individual, leading to multivariate binary data. In such situations, an analysis of the separate binary response variables that form the multivariate response may be all that is needed. In other situations, interest may centre on modelling the multivariate response. For example, suppose that there are two binary responses, that is, the response is bivariate. There would then be four possible combinations of these binary responses, as in binary data that arise from the 2×2 cross-over design referred to in Section 8.3. These four combinations can be regarded as the possible values of a response variable with four outcomes, known in general as a **poly-chotomous** response. The linear logistic model may be generalized to deal with a polychotomous response, and possible models are described in Chapter 5 of McCullagh and Nelder (1989).

When it is desired to explore the association between a set of binary response variables and qualitative variables expressed as factors, the data can be considered as a **multidimensional contingency table**, and are best analysed using an approach based on the log-linear model. This approach is described in Everitt (1977), Fienberg (1980), and in Chapter 10 of Krzanowski (1988).

When the multivariate response consists of a mixture of binary response variables and other types of response, such as in clinical trial data where information is recorded on whether or not a patient has each of a number of particular symptoms, in addition to other prognostic variables, models can be built along the lines described in Cox and Snell (1989). See also the discussion in Section 10.3 of Krzanowski (1988). Other multivariate techniques, such as discriminant analysis, canonical variate analysis and multidimensional scaling, may be relevant in processing data of this type, depending on the aims of the study. These and other multivariate methods are described in general texts on multivariate analysis, such as Chatfield and Collins (1980) and Krzanowski (1988).

8.8 EXPERIMENTAL DESIGN

Prior to any data collection, an experimental study has to be planned, questions of sample size resolved, methods for allocating treatments to individuals determined, and so on. This is a very large area, but the general principles of experimental design, and sample size formulae for comparing two proportions are included in a number of texts, such as Armitage and Berry (1987) and Snedecor and Cochran (1980). See also Mead (1988).

Possible formulae for sample-size determination when comparing two or more proportions are reviewed in Abdelbasit and Plackett (1981). The choice of the number of individuals to use in a cohort study is considered in Chapter 7 of Breslow and Day (1987). This chapter also includes some material on sample size requirements for the case-control study; see also Chapter 6 of Schlesselman (1982).

Instead of fixing the sample size in advance of the data collection, a sequential approach may be adopted, in which the study continues until there are sufficient observations to be able to estimate some parameter of interest with a pre-specified precision, or until there is sufficient data to be able to distinguish between two alternative hypotheses. Sequential designs for binary data are considered by Whitehead (1983). The computer package PEST (Planning and Evaluation of Sequential Trials), documented in Whitehead and Brunier (1989), includes facilities for the design and analysis of sequential clinical trials for binary response variables.

When the aim of a study is to estimate a particular function, such as the $ED50$ value in a bioassay, or the probability of a defective component in an industrial setting, the study should be designed with this aim in mind. Determination of the dose levels to use in a bioassay are considered by Cochran (1973) and Abdelbasit and Plackett (1981). The design of biological assays that involve a binary response variable is also discussed in Chapter 19 of Finney (1978). Abdelbasit and Plackett (1983) and Minkin (1987) consider the use of certain optimality criteria for selecting the values of the explanatory variables which are to be used in linear logistic modelling. Abdelbasit and Plackett (1983) also consider sequential procedures for experimental design.

9

Computer software for modelling binary data

In this book, a systematic approach to the analysis of binary and binomial response data has been presented. In order to implement this approach, computer software for fitting linear logistic models, and models based on the probit or complementary log–log transformation of the response probability, will be essential. Accordingly, this chapter begins with a summary of the facilities for modelling binary data that are included in a number of well-known and widely used statistical packages, namely GLIM®, Genstat®, SAS®, BMDP® and SPSS®. The statistical package EGRET® is less well known, but because it incorporates a number of features that are not found in the other packages, the use of this package in modelling binary data will also be described.

The outline of the scope of each package is followed in Section 9.2 by computer-based analyses of three sets of data. The coding required when using the packages GLIM, Genstat, SAS, BMDP and EGRET to analyse these data sets is given, accompanied by notes on the input. Output from these programs is also provided, together with notes on the interpretation of the output. The use of SPSS has not been illustrated because the procedure for logistic regression in this package is restricted to the analysis of ungrouped binary data. Section 9.1.5 discusses this limitation in greater detail.

Throughout this chapter, the discussion is based on the versions of each package available to the author at the time of writing. These were as follows:

Version 3.77 of GLIM3
Release 1.3 of Genstat 5
Release 6.04 of SAS
1988 release of BMDP
Version 3.1 of SPSS/PC+
Version 0.23.25 of EGRET

Earlier or later releases of a particular package may of course require different forms of input and yield different styles of output. In addition, different packages, or different versions of a particular package, may give slightly different numerical results as a result of their using different convergence criteria in the algorithm for model fitting.

When a particular form of analysis is not a standard option in a statistical package, it is sometimes possible to perform such an analysis using programming facilities that are available. For example, in modelling binary data, the method for allowing for natural mortality described in Chapter 4 is available only as a standard option in SAS **proc probit**, but the required calculations can also be carried out within GLIM and Genstat. Similarly, the full range of diagnostics described in Chapter 5 is not automatically provided by any package, although most of the diagnostics can be computed using GLIM, Genstat, SAS, BMDP and SPSS. Section 9.3 gives algorithms that allow many of the analyses described in earlier chapters to be carried out using statistical software. GLIM macros that implement the more complex algorithms are listed in Appendix C. A summary of the relative merits of the different packages is given in Section 9.4, and the address of the distributor of each package is included in the final section of this chapter.

9.1 STATISTICAL PACKAGES FOR MODELLING BINARY DATA

In this section, the scope of GLIM, Genstat, SAS, BMDP, SPSS and EGRET is outlined, and those features of each package that are relevant to modelling binary data are summarized. All of these packages can be used in the application of many other statistical procedures, and some of them, Genstat and SAS in particular, also contain extensive facilities for data management. Further information about each of these packages can be obtained from the manuals that accompany the package, or from the distributor.

9.1.1 GLIM

GLIM (Generalized Linear Interactive Modelling) was written to enable the family of generalized linear models described by Nelder and Wedderburn (1972) to be fitted to data. The package may therefore be used to fit the general linear model to normally distributed response data, log-linear models to data in the form of counts and models to data that have a gamma distribution, as well as to fit models to binomial response data. A large number of GLIM macros distributed with the package, and additional macros described in issues of the *GLIM Newsletter*, enable GLIM to be used for a variety of non-standard statistical analyses.

In this book, we have seen that GLIM can be used to fit linear logistic models, as well as models based on the probit or complementary log–log transformations of the response probability. The different types of residual can all be computed using GLIM, and the package can be used to obtain all of the statistics for model checking described in Chapter 5. There are no standard commands that can be used to fit models that allow for natural mortality, considered in Section 4.4, although the facilities for fitting user-defined models can be used for this; see

Section 9.3.2. The method of dealing with overdispersion described in Section 6.5, which involves multiplying standard errors of parameter estimates by the square root of the mean deviance for the full model, can easily be carried out. A GLIM macro can also be used to model overdispersion using the Williams procedure; see Section 9.3.11. Linear logistic models with random effects can only be fitted using the approximate method due to Williams described in Section 6.7.2, and it is not possible to fit the non-linear logistic models to binary data mentioned at the end of Section 4.5.

When GLIM is used to fit a model that contains one or more factors, the value of the parameter that corresponds to the first level of the factor is set equal to zero, as described in Section 3.2.1. This is the parameterization that has been used in illustrative examples throughout this book.

9.1.2 Genstat

Genstat (General statistical program) is a general purpose package for the management and analysis of data. The package incorporates a wide range of data handling facilities and can provide high-resolution graphical displays. Many types of statistical analysis can be carried out using Genstat, including regression modelling, the analysis of data from designed experiments, non-linear modelling, multivariate analysis and the analysis of time series data. Genstat can also be used for fitting generalised linear models, and so can be used to fit log-linear models to contingency table data, for example, as well as in modelling binomial response variables. Genstat is a package that allows both standard and non-standard statistical tasks to be programmed using a sophisticated statistical programming language, a feature which considerably extends the scope of the package.

In regard to modelling binary data, Genstat has much the same facilities as GLIM. GLIM macros for particular types of analysis, such as the Williams procedure for modelling overdispersion, can be translated into Genstat procedures, and *vice versa*. Additionally, there is a command that yields predicted values of a response variable, along with the standard error of the predicted value. A library procedure can also be used to estimate the dose required to yield a particular response probability, and to compute confidence intervals for that dose using Fieller's theorem.

When models that include factors are fitted, Genstat uses the same parameterization as GLIM, that is, the effect due to the first level of a factor is set equal to zero.

9.1.3 SAS

SAS (Statistical Analysis System) is a computer software system for data processing and data analysis. The data management, file handling and graphics facilities are particularly extensive. SAS can be used for most types of statistical analysis, including general linear modelling, analysis of data from designed experiments, non-linear modelling, log-linear modelling of data in the form of counts, analysis of survival data, multivariate analysis, and the analysis of ordered categorical data. SAS software for use in time series analysis and forecasting, quality control and operational research is also available. The SAS package does allow macros consisting of sequences of SAS statements to be written, but non-standard analyses are often best programmed using the SAS/IML® software, which implements the programming language IML (Interactive Matrix Language).

There are three SAS procedures that can be used for linear logistic modelling, namely **proc probit**, **proc catmod** and **proc logistic**. Of these, the first is the procedure which is the most straightforward to use. In spite of its name, **proc probit** can be used for fitting models based on the logistic and complementary log–log transformations, as well as the probit transformation. The linear component of the model may include factors and explanatory variables, but not interactions. Interactions can only be included in a model after appropriate indicator variables have been defined and fitted as explanatory variables, as described in section 3.2.1. For an illustration of this, see note 3 on the input required when using SAS **proc probit** in Section 9.2.2. The procedure **proc probit** incorporates an option for specifying a value for the natural response probability, or for estimating this quantity from the observed data. An option may also be used to adjust for overdispersion using the method of Section 6.5, which entails multiplying the standard errors of the parameter estimates by a constant. The standard output from **proc probit** does not include diagnostics for model checking, but many can be derived from quantities that are made available within the program, using the methods described in Sections 9.3.3–9.3.10.

The parameterization used when a model contains a factor (a **class** variable in SAS terminology) is different from that of GLIM and Genstat. In **proc probit**, the parameter estimate corresponding to the last level of the factor is set equal to zero. It is not possible to omit a constant term from a model, and the maximized log-likelihood statistic (log $\hat{L}_c$ in the notation of Section 3.8) is given instead of the deviance. The use of this statistic is illustrated in note 2 on the SAS output in Section 9.2.1.

Linear logistic models that include interaction terms can be fitted using the procedure **proc catmod**. This procedure is primarily intended for modelling categorical data, and as a result the input required for this procedure is more cumbersome than that required for **proc probit**; see Section 9.2.2. Indeed, the procedure itself can be difficult to use, and great care should be exercised in ensuring that it has done what was intended. The procedure **proc catmod** can be used to estimate the parameters in a linear logistic model using both a weighted least squares procedure described by Grizzle, Starmer and Koch (1969), and the method of maximum likelihood. The weighted least squares procedure usually gives very similar results to the method of maximum likelihood, which has been used throughout this book, and so output obtained from the former procedure will not be discussed.

The parameterization adopted by **proc catmod** when a factor is included in the model is inconsistent with **proc probit**, in that the indicator variables set up by the procedure contrast one level of the factor with the last level. Thus, if a factor A has m levels, $m - 1$ indicator variables are generated, where the ith indicator variable has unity in row i, -1 in the last row and zero elsewhere.

The procedure **proc logistic** can be used to fit linear logistic models to binary data, but it can also be used in modelling data where the response variable has more than two categories. The procedure allows a range of model checking diagnostics to be obtained directly, and incorporates a number of sequential procedures for model selection. The procedures for automatic variable selection implemented in **proc logistic** are analogues of forward selection, backward elimination and stepwise regression used in multiple regression analysis; see Draper and Smith (1981) or Montgomery and Peck (1982) for a description of these algorithms. On the face of it, a program which allows a suitable combination of explanatory variables to be selected from a larger number might appear to be very useful. However, there are a number of drawbacks to this automatic approach to model selection, some of which are highlighted in the above references. See also the comments in Section 3.13 and the discussion on variable selection in Section A2.5 of Cox and Snell (1989).

In **proc logistic**, factors can only be fitted after first defining an appropriate set of indicator variables in the data step. This limitation, and the fact that **proc logistic** cannot be used to allow for natural mortality and overdispersion, means that **proc probit** is superior to **proc logistic** for modelling binary response data. Consequently, **proc logistic** will not be considered further in this chapter.

9.1.4 BMDP

BMDP (Biomedical Package) consists of a series of programs, each of which can be called on to process data in a prescribed manner. The programs available within the package cover a wide range of statistical procedures, and the scope of the package is similar to that of SAS. However, the user-interface is poor and the help information provided is limited.

The only program that can be used in modelling binary data is **LR**, but this program cannot be used to fit models based on the probit or complementary log–log transformations. The **LR** program is actually written for stepwise logistic regression, so that it is possible to use this program for the automatic selection of variables. The program **LR** can also be used to fit models one by one, in the manner of GLIM, Genstat and SAS **proc probit**. A number of goodness of fit statistics are provided by **LR**, including the maximized log-likelihood and the deviance. An extensive set of model checking diagnostics is available to the user, and graphical displays are easily obtained. Those diagnostics not provided in the standard output can be computed using commands for variable transformations. **LR** does not allow any adjustments to be made for natural mortality and overdispersion.

The parameterization used when factors are included in a model is such that each level of the factor is contrasted to the first, a parameterization similar to that used in SAS **proc catmod**. In BMDP, if a factor A had m levels, $m - 1$ indicator variables are generated, where the first row of each is -1. The ith indicator variable then has unity in row $i + 1$ and zero elsewhere. However, it is possible to change the default parameterization to that used by GLIM and Genstat. Interactions can be included in a model, and the constant term can be omitted. However, the coding of the indicator variables used in fitting a factor are dependent on a constant term being in the model, and so the constant cannot be omitted when the model contains a factor.

9.1.5 SPSS

SPSS (Statistical Package for Social Scientists) is a comprehensive data analysis system. The scope of the package is broadly similar to SAS and BMDP, in that the package incorporates facilities for a wide range of statistical procedures.

The advanced statistics module includes the procedure **LOGISTIC REGRESSION** for linear logistic modelling, which can be used for automatic variable selection, as in SAS **proc logistic** and BMDP. The default coding for indicator variables corresponding to factors contrasts each factor level with the last level, the same as that used by SAS **proc catmod**, and it is possible to incorporate interaction terms in a model. A wide range of diagnostics are made available, but the procedure cannot be used to fit probit or complementary log–log models, no adjustment can be made for natural mortality, and there are no facilities for modelling overdispersion. However, the main limitation of this procedure is that it is designed only for use with ungrouped binary response data, such as that in Example 1.9 concerning nodal involvement in patients with prostatic cancer. To use this procedure in the analysis of grouped binary data, such as that encountered in bioassay and other areas of application, the raw data would first need to be restructured. For example, suppose that SPSS is to be used to fit a linear logistic model to the data from Example 1.6 on the extent to which mice are protected against pneumonia by a particular anti-pneumococcus serum. If the value of the binary response variable for a mouse which does not develop pneumonia is coded as zero, and unity otherwise, the data from this experiment, given in Table 1.6, would have to be presented to the **LOGISTIC REGRESSION** procedure in the following form:

$$
\left.\begin{array}{ll} 0.0028 & 0 \\ \quad . \ . \ . & \\ 0.0028 & 0 \end{array}\right\} \; 5 \text{ rows}
$$

$$
\left.\begin{array}{ll} 0.0028 & 1 \\ \quad . \ . \ . & \\ 0.0028 & 1 \end{array}\right\} \; 35 \text{ rows}
$$

$$
. \ . \ .
$$

$$
\left.\begin{array}{ll} 0.0450 & 0 \\ \quad . \ . \ . & \\ 0.0450 & 0 \end{array}\right\} \; 39 \text{ rows}
$$

$$
0.0450 \quad 1
$$

It is not straightforward to carry out this re-structuring within the package. After the data have been so transformed, considerable care has to be taken in interpreting the resulting output; for example, the deviance for a particular model cannot be used as a measure of lack of fit (see Section 3.8.1). In view of this, the SPSS procedure **LOGISTIC REGRESSION** will not be described in greater detail in this book.

9.1.6 EGRET

EGRET (Epidemiological, Graphics, Estimation and Testing program) is a package developed for the analysis of data from epidemiological studies. Apart from its use in linear logistic modelling, it can be used to analyse data from matched studies using the technique of conditional logistic regression, described in Section 7.7.1. EGRET may also be used to fit Cox's proportional hazards model to survival data (Cox and Oakes, 1984), and the Poisson regression model to cohort data (Breslow and Day, 1987). Exact analyses of $2 \times k$ contingency tables and k sets of 2×2 tables of counts can also be carried out using the methods described in Breslow and Day (1980). The package consists of two modules: DEF for data definition and PECAN (Parameter Estimation through Conditional probability Analysis) for data analysis. EGRET is not designed for data management, and so the facilities for data manipulation within the package are extremely limited.

EGRET can be used for linear logistic modelling, and the models may also include random effects to allow overdispersion to be modelled using the method described in Section 6.7. The beta-binomial model described in section 6.6 can also be fitted. The output from this package is quite concise, but does include parameter estimates, the deviance, and optionally, fitted probabilities. The package does not compute any type of residual, and the only model checking diagnostic that is provided is the delta–beta ($\Delta\hat{\beta}$) statistic described in Section 5.5.3. There are no facilities for constructing additional diagnostics after a model has been fitted.

When a factor is included in the model, the same parameterisation as used by GLIM and Genstat is adopted, in that the effect due to the first level of the factor is set equal to zero. As with BMDP, the coding of the indicator variables assumes that the model contains a constant term, so that a constant term should not be excluded from a model that contains one or more factors. Moreover, if an interaction is included, both main effects must be explicitly included in the model.

9.2 COMPUTER-BASED ANALYSES OF EXAMPLE DATA SETS

Three examples will now be used to illustrate how GLIM, Genstat, SAS, BMDP, and EGRET can be used in linear logistic modelling. The examples chosen are as follows:

1. Data on the proportion of mice who develop pneumonia after being inoculated with different doses of a protective serum. These data were described in Example 1.6, and subsequently analysed in Examples 4.1, 4.2, and 5.1.
2. Data on the proportion of cuttings from the roots of plum trees that survive, when the cuttings are either long or short and are planted at once or in spring.

These data were given in Example 1.2, and subsequently analysed in Examples 2.5 and 3.5.
3. Data on the proportion of rotifers of each of two species that remain in suspension in solutions of different relative densities. These data were given in Table 6.10, and analysed in Section 6.9.

For each of these three data sets, GLIM, Genstat, SAS and BMDP programs used in modelling the data are given, together with comments on the form of the input. The basic instructions that are required by each of these packages will be given in full, although in most cases they can be considerably shortened. Relevant output from these packages is also provided, complemented by notes on the interpretation of the output. The input code given for these examples can usually be enhanced to make the resulting output more comprehensible, for example, by including labels for factor levels. The manuals that accompany the individual packages contain information on how the basic output can be refined.

With EGRET, the DEF module is first used to read the data from a file, and to provide information about the type of model to be fitted. The model fitting process is then carried out using PECAN. Instructions for the EGRET package are supplied through a menu system, and so the actual commands will not be given. However, summary information provided by EGRET, which indicates the form of the required input, is included with the output that is reproduced.

9.2.1 Analysis of data set (1): anti-pneumococcus serum

For this data set, the manner in which the computer packages are used to fit a logistic regression line to the data will be illustrated. The way that each of the packages is used to obtain parameter estimates under the fitted model, the variance–covariance matrix of the estimates and the fitted values will also be shown. The explanatory variable which represents the dose of the serum is transformed to log (dose) within the programs prior to the model fitting.

GLIM input:

```
$units 5
$data dose y n
$read
0.0028 35 40
0.0056 21 40
0.0112  9 40
0.0225  6 40
0.0450  1 40
$calculate logdose = %log(dose)
$error b n
$yvariate y
$fit logdose $
$display e v r
```

Notes on the GLIM input:

1. After the data have been input, the **$calculate** directive is used to define the identifier **logdose** which will contain the natural logarithm of the doses. The directive **$calculate n=40** could have been used to set the values of the binomial denominators, instead of reading them in as data.
2. The directive **$error b n** specifies that the data are binomial and that the identifier **n** contains the binomial denominators. By default, a logistic model is fitted to binomial data, but models based on other link functions, such as the probit and complementary log–log links, can be fitted by including the directive **$link**.
3. The **$yvariate** directive declares the identifier **y** to be the binomial response variable.
4. The **$fit** directive is used to fit a model that contains a constant term and the explanatory variable **logdose**. A constant term alone would be fitted using **$fit** on its own. A constant can be omitted from the fit by incorporating −1 in the **$fit** directive, a feature which is illustrated in the example to be considered in section 9.2.3.
5. The arguments **e**, **v** and **r** used in the **$display** directive produce parameter estimates, their variance–covariance matrix, and the fitted values. The only output produced by the **$fit** statement is the deviance, the number of degrees of freedom of this deviance, and the number of cycles of the iterative process used in fitting the model.

GLIM output:

```
scaled deviance = 2.8089 at cycle  4
            d.f. = 3

        estimate        s.e.       parameter
    1    -9.189        1.255       1
    2    -1.830        0.2545      LOGD
    scale parameter taken as  1.000

(Co)variances of parameter estimates
    1        1.575
    2        0.3158      0.06479
                1           2
    scale parameter taken as  1.000

unit   observed    out of    fitted    residual
    1       35        40      33.085      0.801
    2       21        40      22.950     -0.623
    3        9        40      10.987     -0.704
    4        6        40       3.823      1.171
    5        1        40       1.155     -0.146
```

Notes on the GLIM output:

1. The deviance on fitting a model is labelled as **scaled deviance** by GLIM. This is the same as the quantity referred to as deviance throughout this book. In

general, if the scale parameter is denoted by σ^2, which in modelling binary data is such that Var $(y_i) = n_i p_i (1 - p_i)\sigma^2$, the scaled deviance is the deviance divided by σ^2. Unless overdispersion is being modelled, $\sigma^2 = 1$, and the scaled deviance is identical to the deviance. The value of the scale parameter is given after the parameter estimates and their variance–covariance matrix.

2. GLIM recognises only the first four letters of an identifier, and so the labels for the terms fitted in the model, which are given alongside the parameter estimates and their standard errors, are so abbreviated. The parameter labelled **1** is the constant term.

3. The rows and columns of the variance–covariance matrix are labelled according to the order in which the parameter estimates appear in the output from the **$display** directive. If a scale parameter different from unity has been set, the variances and covariances in this matrix will be multiplied by the scale parameter.

4. The fitted values of the binomial response variable, $\hat{y}_i$, are given, together with the (unstandardized) Pearson residuals defined by equation (5.1).

Genstat input:

```
variate [nvalues=5] dose,y,n
read dose,y,n
0.0028 35 40
0.0056 21 40
0.0112  9 40
0.0225  6 40
0.0450  1 40 :
calculate logdose=log(dose)
model [distribution=binomial] y;nbinomial=n
terms logdose
fit [print=m,s,e,f] logdose
rkeep vcov=v
print v
```

Notes on the Genstat input:

1. The colon after the last row of data signifies the end of the data. The variable **dose** is then transformed to the natural logarithm of **dose** using the **calculate** statement. The binomial denominators of 40 could also have been set using a **calculate** statement.

2. The option **distribution = binomial** in the **model** statement declares the data to be binomial and a logistic model is fitted by default. Other transformations of the response probability can be adopted through the use of the **link** option of the **model** statement.

3. In the **model** statement, the identifier **y** is the binomial response variable, and the parameter **nbinomial** declares **n** to be the binomial denominator.

4. The **terms** statement can be omitted. When it is included, subsequent model fitting is computationally more efficient, particularly when a range of different models are being fitted. If the terms statement is omitted, less information is given in the analysis of deviance table; see note 2 on the Genstat output.
5. The **print** option in the **fit** statement yields a description of the fitted model, an analysis of deviance, parameter estimates and a table of fitted values of the binomial response variable. If this option is omitted, the fitted values are not included in the output.
6. When a **fit** statement is used, a constant term is included in the model unless the option **constant = omit** is incorporated in the statement. If it is desired to fit a model with a constant term alone, the statement **fit** with no arguments should be included in the program.
7. The **rkeep** command stores the variance–covariance matrix of the parameter estimates in a matrix **v**, which can be printed and used in subsequent calculations.
8. The default convergence criterion used in the iterative algorithm for model fitting is that iteration stops when the absolute change in the deviance in two successive cycles of the iterative process is less than $0.0004 \times (1 + v)$, where v is the number of residual degrees of freedom. This is not as strict as the criterion used in some other packages, and as a result, small differences between Genstat output and output from other packages can be expected. The criterion can be made more strict by reducing the factor of 0.0004 using the statement **rcycle [tolerance = 0.000001]** after the **model** statement.

Genstat output:

```
***** Regression Analysis *****

Response variate: y
Binomial totals: n
   Distribution: Binomial
  Link function: Logit
   Fitted terms: Constant, logdose


*** Summary of analysis ***
    Dispersion parameter is 1

               d.f.      deviance    mean deviance
Regression        1        87.243         87.2427
Residual          3         2.809          0.9363
Total             4        90.052         22.5129

Change           -1       -87.243         87.2427


*** Estimates of regression coefficients ***

               estimate        s.e.              t
Constant          -9.19        1.25          -7.33
logdose          -1.830        0.254         -7.20


* MESSAGE: s.e.s are based on dispersion parameter with value 1
```

*** Fitted values and residuals ***

	Binomial			Standardized	
Unit	total	Response	Fitted value	residual	Leverage
1	40	35	33.08	1.28	0.58
2	40	21	22.95	-0.81	0.41
3	40	9	10.99	-0.90	0.36
4	40	6	3.82	1.39	0.38
5	40	1	1.15	-0.17	0.27
Mean				0.16	0.40

v

1	1.5700	
2	0.3148	0.0646
	1	2

Notes on the Genstat output:

1. In two parts of the output, a message is given about the dispersion parameter having the value unity. The dispersion parameter in Genstat is the same as the scale parameter in GLIM. A dispersion parameter of unity is the default, and simply means that the variance of the binomial response variable is taken to be that under binomial sampling. It is possible to switch off this warning message using the **nomessage** option of the **fit** statement.

2. The deviance for the fitted model is contained in the line labelled **Residual** in the summary of analysis, and here it is 2.809 on 3 d.f. The **Total deviance** of 90.052 on 4 d.f. is that for a model that contains only a constant, and so the **Regression deviance** is the deviance due to the explanatory variables in the model. The **Change** line at the foot of the analysis of deviance table gives the change in deviance on including in the model the variables specified in the **fit** statement, after fitting a constant. If the **add** command were used to modify the model fitted in a previous **fit** statement, the **Change** line would give the deviance due to including the additional variables specified in the **add** statement. If the **terms** statement were omitted, the deviances, and other quantities in the **Regression** line and the **Total** line of the analysis of deviance, would be replaced by asterisks.

3. In addition to the parameter estimates and their standard errors, the Genstat output gives the values of the ratio of the estimate to its standard error in the column headed **t**. These statistics can be used to test the null hypothesis that a particular coefficient is zero. Because no dispersion parameter is being estimated, these statistics have a standard normal distribution, rather than a t-distribution. Of the two t-values (or more correctly z-values) given, that for the constant can be used to test the hypothesis that logit $(p) = 0$, or $p = 0.5$, when the value of **logdose** is zero. This hypothesis is not usually of interest. The t-value for **logdose** is used in testing the hypothesis that the coefficient of **logdose** is zero,

that is that there is no relationship between the probability that a mouse is affected by pneumonia and **logdose**.

4. The display of residuals and fitted values gives the fitted values of the binomial response variable, $\hat{y}_i$, standardized deviance residuals, r_{Di}, as defined by equation (5.3), and the leverage, h_i, defined in Section 5.1. The means of the standardised deviance residuals and the leverage values are not particularly informative.

5. The rows and columns of the variance–covariance matrix correspond to the order in which the parameter estimates are given in the output, so that the first row and column correspond to the constant term and the second row and column to the coefficient of **logdose**. The variance–covariance matrix produced by Genstat is slightly different from that given by GLIM, SAS, BMDP and EGRET. If the **rcycle** statement is used, as described in note 8 on the Genstat input, Genstat gives the same matrix as given by GLIM and other packages.

SAS input:

```
data serum1;
input dose y n;
logdose = log(dose);
cards;
0.0028 35 40
0.0056 21 40
0.0112  9 40
0.0225  6 40
0.0450  1 40
proc probit;
model y/n = logdose / d = logistic;
output out=serum2 prob=phat;
proc print;
```

Notes on the SAS input:

1. The input data set is named **serum1**.

2. The transformation of dose is carried out within the data step. SAS **proc probit** includes an option that allows the first named variable that appears after the = sign in the **model** statement to be transformed to its natural logarithm. The binomial denominators could have been set equal to 40 by using the statement **n=40;** in the data step.

3. The procedure being used for linear logistic modelling is **proc probit**. The binomial response variable is specified to be **y**, and the corresponding denominator to be **n**, through **y/n** in the **model** statement. Explanatory variables to be included in the model follow the = sign in this statement. If no variables appear after the = sign, a model that contains only a constant term is fitted.

4. The option **d=logistic** in the **model** statement specifies that a logistic model is to be fitted. If this option is omitted, a probit model is fitted. More than one **model** statement can be used in this procedure.
5. The standard output from **proc probit** does not include fitted values. However, the **output** statement allows certain quantities from the fit to be saved, including the fitted probabilities. The option **prob=phat** saves the fitted probabilities in the identifier **phat**, and the option **out=serum2** names the data set in which the values of the identifier **phat** are to be stored.
6. The use of **proc print** gives a listing of all variables in the most recently created data set, which here is named **serum2**. SAS carries over all variables in the previous data set, named **serum1**, to this new data set, and so the values of **dose**, **y**, **n**, **logdose** and **phat** will all be printed.
7. If fitted values of the binomial response variable **y** are required, they can be obtained by appending the following statements to the preceding program:

> **data serum3;**
> **set serum2;**
> **yhat=n*phat;**

The values of **yhat** can be printed by following these statements by **proc print**.

SAS output:

```
                              SAS
                        Probit Procedure

   Data Set          =WORK.SERUM1
   Dependent Variable=Y
   Dependent Variable=N
   Number of Observations=    5
   Number of Events     =    72    Number of Trials =      200

   Log Likelihood for LOGISTIC -87.06231266


      Variable  DF   Estimate  Std Err  ChiSquare  Pr>Chi Label/Value

      INTERCPT   1 -9.1893917  1.25511   53.6056   0.0001 Intercept
      LOGDOSE    1 -1.8296213 0.254547  51.66384   0.0001


                     Estimated Covariance Matrix

                             INTERCPT           LOGDOSE

               INTERCPT      1.575300           0.315849
               LOGDOSE       0.315849           0.064794

        OBS    DOSE     Y    N    LOGDOSE      PHAT

         1    0.0028   35   40   -5.87814    0.82712
         2    0.0056   21   40   -5.18499    0.57375
         3    0.0112    9   40   -4.49184    0.27468
         4    0.0225    6   40   -3.79424    0.09558
         5    0.0450    1   40   -3.10109    0.02887
```

Notes on the SAS output:

1. The number of events and number of trials in the first block of the output give the total number of mice who develop pneumonia accumulated over the five groups, out of the 200 mice used in the experiment.
2. The line **Log Likelihood for LOGISTIC** gives the maximized log-likelihood after fitting a logistic regression on **logdose** to the data, omitting the logarithm of the combinatoric term. It is not the same as the deviance, which is given by $D = -2[\log \hat{L}_c - \log \hat{L}_f]$, where $\hat{L}_f$ is the maximized likelihood under the full model, the model that is a perfect fit to the data. In order to obtain the deviance, $\log \hat{L}_f$ is required. Using SAS **proc probit**, this is most easily obtained by fitting a model that contains a single factor that is defined to have the same number of levels as there are observations, so that level i of this factor occurs only for the ith observation. On fitting this model, the value of $\log \hat{L}_f$ will then appear alongside the statement **Log Likelihood for LOGISTIC** in the resulting output. For the example being considered here, a factor A with 5 levels is fitted, one level for each observation, and the value of $\log \hat{L}_f$ is found from the output to be -85.6579. The deviance on fitting **logdose** is therefore $-2(-87.0623 + 85.6579) = 2.809$, in agreement with that given by GLIM and Genstat. Very often, the actual deviance will not be required, and it will only be important to be able to compare two alternative models. The difference in deviance between two models does not involve $\hat{L}_f$, and is $-2[\log \hat{L}_{c1} - \log \hat{L}_{c2}]$, where $\hat{L}_{c1}$ and $\hat{L}_{c2}$ are the maximized likelihoods under the two models; see Section 3.9. The output from fitting the two models using SAS **proc probit** can therefore be used to calculate this difference in deviance, which can then be compared with percentage points of the χ^2-distribution in the usual way. For example, the model statement

$$\textbf{model y/n} = \text{ / } \textbf{d} = \textbf{logistic;}$$

would be used to fit a linear logistic model that includes only a constant term. From the resulting output (not shown here), the maximized log-likelihood is -130.6836. Consequently, the change in deviance on adding **logdose** into the model is $-2(-130.6836 + 87.0623) = 87.243$, in agreement with that given in the Genstat output.
3. The **ChiSquare** value given alongside the parameter estimates and their standard errors is $(\textbf{Estimate/Std Err})^2$, and is the square of the statistic labelled **t** in the Genstat output. In note 3 on the Genstat output, it was pointed out that **t** = **Estimate/Std Err** has a standard normal distribution, and so the square of **t** has a χ^2-distribution on one degree of freedom. The column headed **Pr > Chi** gives the P-value for these chi-squared statistics. Again, the P-value for the variable labelled **INTERCPT** relates to the hypotheses that $p = 0.5$ when **logdose** = 0, which is not usually relevant.
4. The variance–covariance matrix of the parameter estimates is self-explanatory, and is the same as that produced by GLIM, BMDP and EGRET.

5. The output below the variance–covariance matrix is that obtained from the **proc print** statement at the end of the program, and gives the fitted probabilities in the column headed **PHAT**. If these are multiplied by 40, the fitted values of **y** are formed, which agree with those given in the GLIM and Genstat output.

BMDP input:

/problem	title='serum'.
/input	variables=3.
	format=free.
/variable	names=dose,y,n.
/transformation	logdose=ln(dose).
/regress	count=n.
	scount=y.
	interval=logdose.
	model=logdose.
/print	cells=model.
	cova.
	no corr.
/end	
0.0028 35 40	
0.0056 21 40	
0.0112 9 40	
0.0225 6 40	
0.0450 1 40	
/end	

Notes on the BMDP input:

1. After entitling the program **serum**, specifying that three variables are to be read in using free format, and assigning names to the variables, the variable **dose** is transformed to log (**dose**) using the **transformation** paragraph. The **transformation** paragraph could also have been used to set the binomial denominators equal to 40, instead of reading them in as data.
2. In the **regress** paragraph, **n** is declared to be the binomial denominator and **y** the variable associated with the number of successes, **scount**. An **interval** statement is used to specify that the variable **logdose** is to be treated as a continuous explanatory variable. Variables not declared to be interval variables are assumed to be factors.
3. The terms to be fitted in the model are specified in the **regress** paragraph using the **model** statement. A constant term is automatically included, unless the statement **constant = out** is included in the **regress** paragraph.
4. The **print** paragraph is needed to request that the fitted values and other statistics be printed, that the variance–covariance matrix of the parameter

estimates be given, and that printing of the correlation matrix of the parameter estimates be suppressed.

5. The data follows the **end** statement. If additional models are to be fitted within the same program, the **problem** and **regress** paragraphs have to be repeated, together with the data. For example, the following additional statements would be needed to fit a linear logistic model that includes the untransformed variable **dose**:

> /**problem**
> /**regress**   count=n.
> scount=y.
> interval=dose.
> model=dose.
> /**end**
> ⟨*data*⟩
> /**end**

To avoid having to repeat the data for each model to be fitted, the data can be incorporated in a data file, or a BMDP file can be created.

BMDP output:

```
***** TRAN     PARAGRAPH IS USED *****

     NEW VARIABLES DEFINED IN TRAN     PARAGRAPH
          logdose

     PROBLEM TITLE IS
     serum

     NUMBER OF VARIABLES TO READ IN. . . . . . . .      3
     NUMBER OF VARIABLES ADDED BY TRANSFORMATIONS. .     1
     TOTAL NUMBER OF VARIABLES . . . . . . . . . .      4
     CASE FREQUENCY VARIABLE . . . . . . . . . . .
     CASE LABELING VARIABLES . . . . . . . . . . .
     NUMBER OF CASES TO READ IN. . . . . . . . . . TO END
     MISSING VALUES CHECKED BEFORE OR AFTER TRANS. . NEITHER
     BLANKS ARE. . . . . . . . . . . . . . . . . . MISSING
     NUMBER OF WORDS OF DYNAMIC STORAGE. . . . . . .  16298

     VARIABLES TO BE USED
          1 dose        2 y          3 n          4 logdose

     INPUT FORMAT IS
     FREE

     MAXIMUM LENGTH DATA RECORD IS    80 CHARACTERS.

     DEPENDENT VARIABLE. . . . . . . . . . . . . . .  0
          COUNT VARIABLE. . . . . . . . . . . . . .  3 n
          SCOUNT VARIABLE. . . . . . . . . . . . . .  2 y
          FCOUNT VARIABLE. . . . . . . . . . . . . .  0
     METHOD TO SELECT NEXT TERM TO REMOVE OR ENTER .  ACE
     HIERARCHICAL TERM INCLUSION RULE USED . . . . .  SING
     REMOVE LIMIT (P-VALUE MUST BE GREATER). . . . .  0.1500  0.1500
     ENTER LIMIT (P-VALUE MUST BE LESS). . . . . . .  0.1000  0.1000
     TOLERANCE . . . . . . . . . . . . . . . . . . .  0.0001000
```

```
CONVERGENCE CRITERION . . . . . . . . . . . .   0.0000010
MAXIMUM NUMBER OF ITERATIONS. . . . . . . . .        10
              STEP HALVINGS . . . . . . . .         5

NUMBER OF CASES TO BE PRINTED . . . . .     10

CASE    1        2        3        4
NO.  dose    y        n      logdose
-----  --------  -------- -------- --------
  1   .002800    35       40       -5.88
  2   .005600    21       40       -5.18
  3    0.01       9       40       -4.49
  4    0.02       6       40       -3.79
  5    0.05       1       40       -3.10

NUMBER OF CASES READ. . . . . . . . . . . .        5

  TOTAL NUMBER OF RESPONSES USED IN THE ANALYSIS     200.
                          y      . . . . . .          72.
                       FAILURE   . . . . . .         128.

  NUMBER OF DISTINCT COVARIATE PATTERNS . . . . .      5
```

DESCRIPTIVE STATISTICS OF INDEPENDENT VARIABLES

```
VARIABLE                                STANDARD
NO. N A M E   MINIMUM  MAXIMUM   MEAN   DEVIATION  SKEWNESS  KURTOSIS

 4 logdose   -5.8781  -3.1011  -4.4901   0.9846    0.0013   -1.3185
```

STEP NUMBER   0

```
                      LOG LIKELIHOOD = -87.062
GOODNESS OF FIT CHI-SQ  (2*O*LN(O/E))  = 2.809  D.F.=  3  P-VALUE= 0.422
GOODNESS OF FIT CHI-SQ (HOSMER-LEMESHOW)= 2.917  D.F.=  3  P-VALUE= 0.405
GOODNESS OF FIT CHI-SQ  ( C.C.BROWN )  = 1.871  D.F.=  2  P-VALUE= 0.392

                            STANDARD
   TERM        COEFFICIENT   ERROR   COEFF/S.E. EXP(COEFFICIENT)

logdose         -1.8296     0.2545   -7.188      0.1605
CONSTANT        -9.1894     1.255    -7.322      0.1021E-03
```

COVARIANCE MATRIX OF COEFFICIENTS

```
             logdose     CONSTANT

logdose      0.06479
CONSTANT     0.31585     1.57530
```

STATISTICS TO ENTER OR REMOVE TERMS

```
               APPROX.              APPROX.
   TERM         F TO   D.F. D.F.     F TO   D.F. D.F.
               ENTER                REMOVE              P-VALUE

LogDose                              50.37   1   197    0.0000
LogDose                              IS IN        MAY NOT BE REMOVED.
CONSTANT                             52.26   1   197    0.0000
CONSTANT                             IS IN        MAY NOT BE REMOVED.
```

  NO TERM PASSES THE REMOVE AND ENTER LIMITS (0.1500 0.1000) .

```
SUMMARY DESCRIPTION OF CELLS.
CELLS ARE FORMED BY ALL COMBINATIONS OF VALUES OF VARIABLES IN THE MODEL
-----------------------------------------------------------------------
```

NUMBER y	NUMBER FAILURE	OBSERVED PROPORTION y	PREDICTED PROB.OF y	S.E. OF PREDICTED PROB.	OBS-PRED -------- S.E.RES.	PRED. LOG ODDS
1	39	0.0250	0.0289	0.0137	-0.1710	-3.5156
6	34	0.1500	0.0956	0.0288	1.4926	-2.2474
9	31	0.2250	0.2747	0.0423	-0.8797	-0.9710
21	19	0.5250	0.5738	0.0501	-0.8116	0.2972
35	5	0.8750	0.8271	0.0454	1.2314	1.5654

CHI	DEVIANCE	HAT MATRIX DIAGONAL	INFLUENCE	logdose
-0.1463	-0.1496	0.2686	0.011	-3.10
1.1707	1.0901	0.3848	1.393	-3.79
-0.7039	-0.7186	0.3598	0.435	-4.49
-0.6235	-0.6210	0.4098	0.457	-5.18
0.8008	0.8344	0.5771	2.069	-5.88

```
MINIMUM EXPECTED CELL FREQUENCY           =      1.15
NUMBER OF EXPECTED VALUES LESS THAN 5.0   =      2
```

Notes on the BMDP output:

1. The output from BMDP includes much information about the variables that have been read in and transformations used. Criteria to be used by stepwise algorithms for model fitting are also given, even when this facility is not being used. No detailed comments on this part of the output will be included here, and in future illustrations, this part of the BMDP output will not be reproduced.

2. After the line **STEP NUMBER 0**, a number of statistics concerned with the fit of the model are given. The first of these is the maximized log-likelihood under the model, $\log \hat{L}_c$ in the notation of Section 3.8. See the comments on this in note 2 on the SAS output.

3. The line labelled **GOODNESS OF FIT CHI-SQ (2*O*LN(O/E))** gives the deviance for the fitted model, 2.809 on 3 d.f., which agrees with that given by GLIM and Genstat. The *P*-value given compares the deviance with a χ^2-distribution on 3 d.f.

4. The other two lines which begin **GOODNESS OF FIT CHI-SQ** are measures of the goodness of fit of a model, proposed by Hosmer and Lemeshow (1980) and Brown (1982), respectively. The test due to Hosmer and Lemeshow is based on the statistic labelled $\hat{H}_g$ in their paper. To construct $\hat{H}_g$, a $2 \times g$ contingency table is formed, the two rows of which correspond to success ($y = 1$) and failure ($y = 0$), and the columns are g intervals for the estimated success probabilities. The numbers of binary observations which are classified according to each of the $2g$ cells of the table are then compared to those which would have been expected if the fitted linear logistic model were correct. The test due to Brown compares the logistic transformation with a more general transformation of the

response probability introduced by Prentice (1976). This test is therefore a test of the adequacy of the logistic transformation of the response probability, and is an alternative to the method described in Section 5.3. Both of these test statistics have approximate χ^2-distributions, but the original papers should be consulted for fuller details on their use.

5. Following the results of the goodness of fit tests, the parameter estimates and their standard errors are given. The ratio **COEFF/S.E.** is the same as the statistic labelled **t** in Genstat output, while **EXP(COEFFICIENT)** can be interpreted as an odds ratio. In particular, the value of this statistic for the term **logdose** is the ratio of the odds of developing pneumonia for a mouse exposed to a log (dose) of $x + 1$ units of serum, relative to a mouse exposed to a log (dose) of x units. The corresponding value of the statistic for the constant term cannot be meaningfully interpreted, and should be ignored. A full discussion on the interpretation of odds ratios is given in Section 7.5.

6. The variance–covariance matrix of the parameter estimates is the same as that of GLIM, SAS and EGRET, but note that the constant term appears in the final row and column of the matrix.

7. That part of the output which relates to the addition of terms to, or the removal of terms from, a model is not relevant when the BMDP procedure **LR** is not being used for automatic model selection and can be ignored.

8. The statement **cells = model** in the **print** paragraph gives a number of statistics that can be used in assessing the adequacy of the fitted model. The column headed **NUMBER FAILURE** are the values of $n_i - y_i$, and the observed proportions are simply the ratios y_i/n_i. The column labelled **PREDICTED PROB. OF y** gives the values of the fitted probabilities, $\hat{p}_i$, and can be compared with the corresponding quantities obtained using SAS. The values in the column headed **S.E. OF PREDICTED PROB.** are the standard errors of the fitted probabilities under the logistic model. The next column gives the values of the standardized Pearson residuals, r_{Pi}, defined by equation (5.2), and this column is followed by that headed **PRED. LOG ODDS**, which gives the values of $\log\{\hat{p}_i/(1 - \hat{p}_i)\}$, that is, the values of the linear predictor, denoted by $\hat{\eta}_i$ in section 3.7. The column labelled **CHI** gives the unstandardized Pearson residuals, X_i, defined by equation (5.1), and this is followed by the (unstandardized) deviance residuals, d_i, which are the signed square roots of the contributions of each observation to the overall deviance. The column **HAT MATRIX DIAGONAL** gives the values of the leverage, h_i, and that headed **INFLUENCE** gives the values of $h_i r_{pi}^2/(1 - h_i)$, which apart from a factor of p^{-1} is the D-statistic, defined by equation (5.13), and used in identifying influential observations. The values of the transformed variable, **logdose**, are given in the final column.

9. The final two lines of the output give information about the smallest model-based estimate of the number of successes and failures, $\hat{y}_i$ and $n_i - \hat{y}_i$, and the number of these estimates which are smaller than five. Since the deviance is not well-approximated by a χ^2-distribution when some of the fitted numbers of successes or failures are small, this information may be used in judging the adequacy of the chi-squared approximation to the distribution of the deviance.

EGRET input:

Because the instructions required by this package are input using a menu system, details of these instructions are not included.

EGRET output:

```
                        3 variables and 5 observations

-------------------------------VARIABLES------------------------------
   1. dose          2. y         3. n
----------------------------------------------------------------------

            ANALYSIS MODEL:    Logistic regression
               RISK TYPE:      Relative risk (multiplicative)
              GROUP SIZE:      Group Size Given by Variable:  n
     OUTCOME SPECIFICATION:    Outcome Variable Name:  y
          REPETITION COUNT:    -none-

----------------------------REGRESSION TERMS--------------------[LR]--
   a. %GM                   b. %log.dose


   Cycle   Deviance
     1    90.05154
     2     9.519132
     3     3.160813
     4     2.810561
     5     2.808889
     6     2.808889
-------------------------------RESULTS--------------------------[LR]--
   TERM                COEFFICIENT    STD ERROR    P-VALUE    ODDS RATIO

   %GM                    -9.189       (1.26)      <.001      .1021E-03
   dose.%LOG              -1.830       (.255)      <.001      .1605

               DEVIANCE ON     3 DF =       2.809
     LIKELIHOOD RATIO STATISTIC ON  2 DF =    103.134, p < .001


   COVARIANCE MATRIX:

                 {  1}        {  2}

      {  1}      1.575
      {  2}       .3158       .6479E-01

-------------------------------DIAGNOSTICS----------------------------
              PLOT FITTED VALUES:      [maximum fitted-value = 1.3336]
              SELECT OBSERVATIONS:     all observations
     SELECT FITTED-VALUE THRESHOLD:    .0
                  SELECT OUTPUT:       print to .LST file

----------------------------------------------------------------------
   obs  num denom   fitted values
    1.   35   40       .827
    2.   21   40       .574
    3.    9   40       .275
    4.    6   40       .096
    5.    1   40       .029
```

Notes on the EGRET output:

1. The deviances at each cycle of the iterative procedure used in fitting the model are followed by a summary of the parameter estimates, their standard errors, and *P*-values for tests of the hypotheses that the value of each parameter is zero, based on the ratio of the estimate to its standard error. The constant term is denoted by **%GM**, while **dose.%LOG** signifies that **dose** has been transformed to its natural logarithm. The column headed **ODDS RATIO** gives the values of $\exp(\hat{\beta})$, where $\hat{\beta}$ is the parameter estimate. See the discussion on this in note 5 on the BMDP output.
2. The deviance on fitting the model follows the parameter estimates. The likelihood ratio statistic is the decrease in the deviance that results from extending a previously fitted model. Here a single model has been fitted, and the comparision is between this fitted model and the deviance for a model with no terms in it, under which $\hat{p} = 0.5$ for each observation. The deviance for this latter model can be found by putting $\hat{y}_i = n_i\hat{p}_i = n_i/2$ in the expression for the deviance in equation (3.5), which in this case gives 105.943. The deviance for the model that contains a constant and log (dose) is 2.809, and the difference $105.943 - 2.809$ is the likelihood ratio statistic given by EGRET. This statistic is not informative in this context, and can be ignored.
3. The rows of the variance–covariance matrix are presented in the same order as the parameter estimates are given, and can be output to the file PECAN.LST after requesting this using the configuration screen in PECAN.
4. The fitted values are obtained by requesting that they be printed in the file PECAN.LST by using the postfit menu in PECAN.

9.2.2 Analysis of data set (2): propagation of plum root stocks

This data set involves two factors, length of cutting and time of planting, and is used to illustrate how a binomial response from a 2 × 2 factorial experiment can be modelled using GLIM, Genstat, SAS, BMDP and EGRET. Many of the notes on the input required for these packages that were given in Section 9.2.1 apply to this example, and will not be repeated. Similarly, a number of notes on the output from these packages, given in the previous section, apply equally to the output that is described in this section. In discussing the input required when using the various packages to analyse the data on the propagation of plum root stocks, some remarks will be made on how these packages can be used to fit different models to the data. However, for illustrative purposes, only the input required to fit a model that contains the main effects of length of cutting and time of planting will be shown. In the case of SAS, the model that includes both main effects and the interaction between them will also be fitted, so that the use of indicator variables in fitting an interaction term using SAS **proc probit** can be illustrated. Additionally, input and output from using SAS **proc catmod** to fit the model with the two main effects will be described.

GLIM input:

```
$units 4
$factor length 2 time 2
$data length time y n
$read
1 1 107 240
1 2  31 240
2 1 156 240
2 2  84 240
$error b n
$yvariate y
$fit length + time $
$display e
```

Notes on the GLIM input:

1. The **$factor** directive is used to declare **length** and **time** to be factors, each with two levels.
2. Additional **$fit** statements can be used to fit alternate models. In GLIM, the interaction is denoted by **length.time**. By beginning a **$fit** statement with a + or − operator, the previous **$fit** statement is modified, so that a model that contains a constant and terms that represent length of cutting, time of planting and the length × time interaction could be fitted using **$fit length + time + length.time**, or by including **$fit + length.time** after the statement **$fit length + time**. In GLIM, the notation **length*time** stands for the combination of three terms, **length, time** and **length.time**, and so this model could also be fitted using **$fit length*time**.

GLIM output:

```
        scaled deviance = 2.2938 at cycle  3
                   d.f. = 1

             estimate        s.e.      parameter
        1     -0.3039       0.1172     1
        2      1.018        0.1455     LENG(2)
        3     -1.428        0.1465     TIME(2)
        scale parameter taken as  1.000
```

Note on the GLIM output:

When a factor with m levels is included in a fitted model, $m − 1$ indicator variables are set up, the ith of which has a zero in the first row, unity in row $i + 1$, and zero elsewhere. In this example, there is one indicator variable for each of the two factors, which has the value zero at the first level of the factor and unity at the second. Denoting the two indicator variables for **length** and **time** by z_1 and z_2, respectively, their values are as in the following two tables. If p_{jk} is the survival

Length of cutting	z_1		Time of planting	z_2
1. Short	0		1. At once	0
2. Long	1		2. In spring	1

probability for a cutting of the jth length planted at the kth time, $j = 1, 2; k = 1, 2$, the model that includes the main effects of length and time can be expressed as

$$\text{logit}\,(p_{jk}) = \beta_0 + \beta_1 z_{1j} + \beta_2 z_{2k}$$

The GLIM output gives estimates $\hat{\beta}_0$, $\hat{\beta}_1$ and $\hat{\beta}_2$ of β_0, β_1 and β_2, respectively, which are labelled **1**, **LENG(2)** and **TIME(2)** in the output. This notation emphasises the fact that these estimates will only affect the estimated survival probabilities when the corresponding factor occurs at the second level. Fitted values of the probabilities under this model are obtained from

$$\text{logit}\,(\hat{p}_{jk}) = \hat{\beta}_0 + \hat{\beta}_1 z_{1j} + \hat{\beta}_2 z_{2k}$$

so that, for example, the fitted survival probability for short cuttings planted in spring, $\hat{p}_{12}$, is given by

$$\text{logit}\,(\hat{p}_{12}) = -0.304 - 1.428 = -1.732$$

as in Example 3.5.

Genstat input:

```
variate [nvalues=4] y, n
factor [nvalues=4;levels=2] length, time
read length,time,y,n
1 1 107 240
1 2  31 240
2 1 156 240
2 2  84 240 :
model [distribution=binomial] y;nbinomial=n
terms length*time
fit length+time
```

Notes on the Genstat input:

1. The terms **length** and **time** are declared to be factors with two levels using the **factor** statement.
2. Possible terms for inclusion in linear logistic models to be fitted to these data are the main effects of **length** and **time** and their interaction. As in GLIM, the

interaction is denoted by **length.time**, and the notation **length*time** in the **terms** statement stands for the combination of three terms, **length**, **time** and **length.time**.

3. The fitted values are not required here and so the **print** option need not be included in the **fit** statement.

4. Other models can be fitted using additional **fit** statements, or by modifying the current model using the **add** and **drop** commands. For example, the model that contains a constant term, both main effects and their interaction can be fitted using either **fit length + time + length.time**, **fit length*time**, or by including **add length.time** after the statement **fit length + time**.

Genstat output:

```
***** Regression Analysis *****

Response variate: y
   Binomial totals: n
      Distribution: Binomial
     Link function: Logit
      Fitted terms: Constant + length + time


*** Summary of analysis ***
    Dispersion parameter is 1

                   d.f.     deviance    mean deviance
Regression           2      148.725        74.363
Residual             1        2.294         2.294
Total                3      151.019        50.340

Change              -2     -148.725        74.363


*** Estimates of regression coefficients ***

                   estimate         s.e.            t
Constant             -0.304        0.117        -2.59
length 2              1.018        0.145         7.00
time 2               -1.428        0.146        -9.75
* MESSAGE: s.e.s are based on dispersion parameter with value 1
```

Notes on the Genstat output:

1. The coding used for the indicator variables corresponding to the two factors is the same as that used by GLIM, and described in the note on the GLIM output. The resulting estimates and their standard errors are therefore the same as those given by GLIM.

2. The *t*-value given for **length 2** can be used to test the hypothesis that there are no differences between the two lengths of cutting, after adjusting for time of planting. Similarly, that for **time 2** can be used to test the hypothesis that there are no differences between the two times of planting, after adjusting for length of cutting. However, hypotheses such as these are best examined using an analysis of deviance, as shown in Section 3.9.

SAS input (using **proc probit***):*

```
data plums;
input length time y n lt;
cards;
1 1 107 240 1
1 2  31 240 0
2 1 156 240 0
2 2  84 240 0
proc probit;
class length time;
model y/n=length time / d=logistic;
model y/n=length time lt / d=logistic;
```

Notes on the SAS input:

1. The variables **length** and **time** are declared to be factors using the **class** statement in the procedure **proc probit**.
2. Different models can be fitted simply by including additional **model** statements in the procedure **proc probit**.
3. When a factor with m levels is included in a model, SAS defines $m - 1$ indicator variables, the ith of which has unity in the ith row and zero elsewhere. The indicator variables used by SAS to represent the effect of each of the factors, z_1 and z_2, say, are therefore defined to be as in the following two tables. The

Length of cutting	z_1
1. Short	1
2. Long	0

Time of planting	z_2
1. At once	1
2. In spring	0

interaction between length of cutting and time of planting is then represented by the product of the values of these two indicator variables, and so if the identifier **lt** is defined to be the product $z_1 \times z_2$, the values taken by **lt** are as in the following table. This variable is assigned values in the data step of the program, and is included in the fitted model through the second of the two model statements in **proc probit**. Alternatively, the values of **lt** could have been calculated by including the statement **lt = (2−length)*(2−time);** in the data step.

Length of cutting	Time of planting	**lt**
Short	At once	1
Short	In spring	0
Long	At once	0
Long	In spring	0

SAS output:

```
                                    SAS
                            Probit Procedure

                        Class Level Information

                        Class     Levels     Values

                        LENGTH       2        1 2
                        TIME         2        1 2

                    Number of observations used = 4
```

```
Data Set            =WORK.PLUMS
Dependent Variable=Y
Dependent Variable=N
Number of Observations=   4
Number of Events      =    378    Number of Trials =       960

Log Likelihood for LOGISTIC -569.2174083
```

Variable	DF	Estimate	Std Err	ChiSquare	Pr>Chi	Label/Value
INTERCPT	1	-0.7137712	0.121669	34.41576	0.0001	Intercept
LENGTH	1			48.93569	0.0001	
	1	-1.0176915	0.14548	48.93569	0.0001	1
	0	0	0	.	.	2
TIME	1			95.00046	0.0001	
	1	1.42754237	0.146462	95.00046	0.0001	1
	0	0	0	.	.	2

```
Data Set            =WORK.PLUMS
Dependent Variable=Y
Dependent Variable=N
Number of Observations=   4
Number of Events      =    378    Number of Trials =       960

Log Likelihood for LOGISTIC -568.0704886
```

Variable	DF	Estimate	Std Err	ChiSquare	Pr>Chi	Label/Value
INTERCPT	1	-0.6190392	0.135333	20.92324	0.0001	Intercept
LENGTH	1			30.02857	0.0001	
	1	-1.2893078	0.235282	30.02857	0.0001	1
	0	0	0	.	.	2
TIME	1			41.84648	0.0001	
	1	1.23807842	0.19139	41.84648	0.0001	1
	0	0	0	.	.	2
LT		1	0.45274834	0.300894	2.264049	0.1324

Notes on the SAS output:

1. For each of the two factors, **length** and **time**, there are three lines of output in the section that relates to the parameter estimates. The first line of this output

gives the result of an overall test of the hypothesis that there are no differences between the effects of the different levels of the factor. The remaining lines give the coefficients of indicator variables corresponding to the factors. The final line, which begins with three zeros, is a result of the value of the indicator variable corresponding to the last level of the factor being set equal to zero. Also from this part of the output, the coefficient of z_1, the indicator variable associated with length of cutting, is -1.018, and the coefficient of z_2, the indicator variable associated with time of planting, is 1.428. The logistic transform of the fitted survival probability for short cuttings planted in spring is therefore given by

$$\text{logit}(\hat{p}_{12}) = -0.7138 - 1.0177 = -1.7315$$

which is marginally more accurate than that obtained using the Genstat output.

2. The model that contains the main effects of length of cutting and time of planting, together with their interaction, is the full model, and so it is a perfect fit to the data. Consequently, the log-likelihood given in the second part of the output is $\log \hat{L}_f$, the maximized log-likelihood for the full model. Using $\log \hat{L}_f = -568.0705$, the deviance for the model that does not include the interaction, is found from $-2(\log \hat{L}_c - \log \hat{L}_f)$, where from the first part of the output, $\log \hat{L}_c = -569.2174$. The resulting deviance is 2.294, in agreement with the GLIM and Genstat output.

3. From the second part of the output, the logistic transform of the fitted survival probability for a short cutting planted in spring is now given by

$$\text{logit}(\hat{p}_{12}) = -0.6190 - 1.2893 + (0.4527 \times 0) = -1.9083$$

from which $\hat{p}_{12} = 0.129$. This is the same as the observed proportion of 31/240, since the model fitted is a perfect fit to the data.

SAS input (using **proc catmod***):*

```
data plums1;
input survival length time count;
cards;
1 1 1 107
2 1 1 133
1 1 2 31
2 1 2 209
1 2 1 156
2 2 1 84
1 2 2 84
2 2 2 156
proc catmod;
model survival=length time / nogls ml;
weight count;
```

Notes on the SAS input:

1. When analysing binomial data, the procedure **proc catmod** requires the data to be in a different format to that used by **proc probit**. In particular, each of the n binomial observations of the form y_i/n_i has to be split up into a pair of observations, the number of successes, y_i, and the number of failures, $n_i - y_i$. A factor is then defined which labels each observation in a pair according to whether it is the number of successes or the number of failures. In the coding given, this factor is labelled **survival** and takes the value unity for the number of cuttings that survive, and two for the number that fail at particular combinations of the levels of the factors **length** and **time**. When processing ungrouped binary data, the raw data can be presented to **proc catmod** without doubling the number of observations, but then a linear logistic model is fitted to the probability of a response of $y = 0$. Consequently, if $y = 1$ actually denotes a success, and it is desired to model the success probability, the response variable will first need to be transformed to $1 - y$.
2. In **proc catmod**, the **model** statement has to specify that **survival** is the dependent variable, while as in **proc probit**, the terms in the model are the factors **length** and **time**. The **proc catmod** procedure assumes that each term in the **model** statement is a factor unless it is declared to be a continuous variable using the **direct** statement. A **class** statement is not required. When analysing ungrouped binary data, the identifier that contains the binary observations is specified as the dependent variable.
3. The option **nogls** in the **model** statement specifies that output from using the method of generalised least squares to fit the model is not to be given. The option **ml** specifies that output from using the method of maximum likelihood is to be provided.
4. The **weight** statement is used to inform the program of the variable that contains the counts of the numbers of successes and failures. In this case the variable is named **count**. This statement is not required when analysing ungrouped binary data.
5. In order to fit additional models, the series of statements from and including **proc catmod;** are repeated, so that in order to fit the model that contains the main effects of length of cutting and time of planting, and their interaction, the following additional statements are needed.

```
proc catmod;
model survival = length time length*time / nogls ml;
weight count;
```

Note that in SAS, an interaction is denoted using the * operator, and the notation **length|time** stands for the terms **length**, **time**, and **length*time**.

SAS output:

```
                              SAS
                        CATMOD PROCEDURE
```

Response: SURVIVAL		Response Levels (R)=	2
Weight Variable: COUNT		Populations (S)=	4
Data Set: PLUMS1		Total Frequency (N)=	960
		Observations (Obs)=	8

```
                       POPULATION PROFILES
                                            Sample
              Sample  LENGTH  TIME           Size
              -------------------------------------
                 1       1      1            240
                 2       1      2            240
                 3       2      1            240
                 4       2      2            240
```

```
                        RESPONSE PROFILES

                      Response   SURVIVAL
                      ------------------
                         1          1
                         2          2
```

```
                   MAXIMUM LIKELIHOOD ANALYSIS
```

| | Sub | -2 Log | Convergence | Parameter Estimates | | |
Iteration	Iteration	Likelihood	Criterion	1	2	3
0	0	1330.8426	1.0000	0	0	0
1	0	1142.0232	0.1419	-0.4250	-0.4250	0.6167
2	0	1138.4464	0.003132	-0.5040	-0.5040	0.7082
3	0	1138.4348	0.0000102	-0.5088	-0.5088	0.7138
4	0	1138.4348	1.349E-10	-0.5088	-0.5088	0.7138

```
          MAXIMUM LIKELIHOOD ANALYSIS OF VARIANCE TABLE
```

Source	DF	Chi-Square	Prob
INTERCEPT	1	48.94	0.0000
LENGTH	1	48.94	0.0000
TIME	1	95.00	0.0000
LIKELIHOOD RATIO	1	2.29	0.1299

```
          ANALYSIS OF MAXIMUM LIKELIHOOD ESTIMATES
```

Effect	Parameter	Estimate	Standard Error	Chi-Square	Prob
INTERCEPT	1	-0.5088	0.0727	48.94	0.0000
LENGTH	2	-0.5088	0.0727	48.94	0.0000
TIME	3	0.7138	0.0732	95.00	0.0000

Notes on the SAS output:

1. The first part of the output gives summary information about the structure of the data set.
2. Below the heading **MAXIMUM LIKELIHOOD ANALYSIS** is the history of the iterative process used to fit the model. Note that $-2 \times$ log-likelihood for the model is recorded here, that is $-2\hat{L}_c$, rather than the log-likelihood itself. Consequently, the value at the final iteration, 1138.4348, is equal to $-2(-569.2174)$, where -569.2174 was the maximized log-likelihood on fitting the same model using **proc probit**.
3. Following the heading **MAXIMUM LIKELIHOOD ANALYSIS OF VARI-ANCE TABLE**, the values in the column headed **Chi-Square** are the values of **(Estimate/Standard Error)**^{2}, which are also given in the output that follows the heading **ANALYSIS OF MAXIMUM LIKELIHOOD ESTIMATES**, and their corresponding P-values relative to a χ^2-distribution. Note that these chi-squared values are the same as those found using **proc probit**.
4. The value labelled **LIKELIHOOD RATIO** is actually the deviance for the fitted model.
5. The coding of the indicator variables generated by SAS **proc catmod** differs from that of **proc probit**. In **proc catmod**, $m - 1$ indicator variables are generated when a factor with m levels is included in a model, where the ith indicator variable has unity in the ith row, -1 in the final row and zero elsewhere. For the factors **length** and **time**, the two indicator variables used in **proc catmod** are defined to be as shown in the following tables. From the output, the coefficients of z_1 and

Length of cutting	z_1
1. Short	1
2. Long	-1

Time of planting	z_2
1. At once	1
2. In spring	-1

z_2 are -0.5088 and 0.7138, respectively. The logistic transform of the survival probability for short cuttings planted in spring will then be given by

$$\text{logit}(\hat{p}_{12}) = -0.5088 + (-0.5088 \times 1) + (0.7138 \times -1) = -1.7315$$

which agrees with that obtained using SAS **proc probit** and other packages.
6. Finally, the earlier warning about **proc catmod** will be repeated. Here, the procedure has been used for a relatively simple problem. Interpretation of the output can be more difficult in more complicated problems. However, since SAS **proc catmod** cannot be generally recommended for linear logistic modelling, fuller details will not be given here.

BMDP input:

/problem	title='plums'.
/input	variables=4.
	format=free.
/variable	names=length,time,y,n.
/group	codes(time)=1,2.
	codes(length) = 1,2.
/regress	scount=y.
	count=n.
	model=length,time.
/print	no corr.
/end	
1 1 107 240	
1 2 31 240	
2 1 156 240	
2 2 84 240	
/end	

Notes on the BMDP input:

1. The **group** paragraph is used to define the categories of the variables **length** and **time** which are to be used as factors in the model.
2. Additional models can be fitted by repeating the **problem** and **regress** paragraphs, and also the data, as explained in note 5 on the BMDP input in Section 9.2.1.
3. The statement **dvar** = **partial**. could be used in the **regress** paragraph to force BMDP to use the same coding for the indicator variables corresponding to a factor as used by GLIM and Genstat.
4. The **print** paragraph is used to suppress the printing of the correlation matrix of the parameter estimates.

BMDP output:

```
DESCRIPTIVE STATISTICS OF INDEPENDENT VARIABLES
-------------------------------------------------
```

VARIABLE NO. N A M E	GROUP INDEX	FREQ	DESIGN VARIABLES (1)
1 length	1	480	-1
	2	480	1
2 time	1	480	-1
	2	480	1

```
STEP NUMBER    0
---------------
```

```
                        LOG LIKELIHOOD =  -569.217
GOODNESS OF FIT CHI-SQ    (2*O*LN(O/E))  =  2.294  D.F.= 1  P-VALUE= 0.130
GOODNESS OF FIT CHI-SQ (HOSMER-LEMESHOW)=  2.270  D.F.= 2  P-VALUE= 0.321
GOODNESS OF FIT CHI-SQ   ( C.C.BROWN )   =  0.000  D.F.= 0  P-VALUE= 1.000
```

TERM	COEFFICIENT	STANDARD ERROR	COEFF/S.E.	EXP(COEFFICIENT)
length	0.50885	0.7274E-01	6.995	1.663
time	-0.71377	0.7323E-01	-9.747	0.4898
CONSTANT	-0.50885	0.7274E-01	-6.995	0.6012

Notes on the BMDP output:

1. Much of the irrelevant output from **BMDP** has been omitted here.
2. The first part of the output that has been reproduced concerns the coding of dummy variables that have been used by **BMDP**. The default coding for the $m - 1$ indicator variables corresponding to an m-level factor is to set the first row to be -1, the $(i + 1)$th row to be unity, and zero elsewhere. Thus the two indicator variables (or design variables as **BMDP** calls them), are as shown in the two following tables.

Length of cutting	z_1
1. Short	-1
2. Long	1

Time of planting	z_2
1. At once	-1
2. In spring	1

3. The maximized log-likelihood for the fitted model is given by $\log \hat{L}_c = -569.217$, the same as that obtained using SAS. The deviance of 2.294 on 1 d.f. is also in agreement with that given by other packages.
4. The coefficients of z_1 and z_2 are, respectively -0.7138 and -0.5088, and so the logistic transform of the fitted survival probability for short cuttings planted in spring is

$$\text{logit}\,(\hat{p}_{12}) = -0.5088 + (0.5088 \times -1) + (-0.7138 \times 1) = -1.7315$$

as before.

5. In view of the coding used for indicator variables that correspond to factors included in a model, the values in the column labelled **EXP(COEFFICIENT)** cannot be directly interpreted as odds ratios. In fact, the ratio of the odds of a cutting surviving when treated at the second level of a factor, relative to the first level, is $\exp(2\hat{\beta})$, where $\hat{\beta}$ is the coefficient given in the BMDP output. If the statement **dvar = partial.** is included in the **regress** paragraph, the same parameter estimates as given by GLIM and Genstat are obtained, and the values in the column headed **EXP(COEFFICIENT)** are the ratios of the odds of a

cutting surviving when treated at the second level of the factor, relative to the first.

Notes on the EGRET input:

As before, full details of the input will not be given. However, note that factors are declared by the process of factoring variables, which is done through the menu system within PECAN. Interactions are denoted using the dot operator to combine the names of two factors, as in GLIM and Genstat.

EGRET output:

<pre>
                        4 variables and 4 observations

─────────────────────────────VARIABLES───────────────────────────────
   1. length   2. time   3. y   4. n
──

              ANALYSIS MODEL: Logistic regression
                  RISK TYPE: Relative risk (multiplicative)
                 GROUP SIZE: Group Size Given by Variable: n
     OUTCOME SPECIFICATION: Outcome Variable Name: y
          REPETITION COUNT: −none−

     FACTORED VARIABLES: ┌─────────────────────────┐
                             │ VARIABLE #LEVELS BASE   │
                             │   │
                             │ length   2   1   │
                             │ time   2   1   │
                             └─────────────────────────┘

─────────────────────────REGRESSION TERMS───────────────────────[LR]──
   a. %GM   b. length   c. time
──

   Cycle   Deviance
     1   151.0193
     2   4.525810
     3   2.298428
     4   2.293825
     5   2.293825

──────────────────────────────RESULTS──────────────────────────────[LR]──
   TERM   COEFFICIENT   STD ERROR   P-VALUE   ODDS RATIO

   %GM   −.3039   (.117)   .009   .7379
   length='2'   1.018   (.145)   <.001   2.767
   time='2'   −1.428   (.146)   <.001   .2399

                  DEVIANCE ON   1 DF =   2.294
   LIKELIHOOD RATIO STATISTIC ON   3 DF =   192.408, p < .001
</pre>

Note on the EGRET output:

The coding of the indicator variables used by EGRET is the same as that used by GLIM and Genstat, and so the parameter estimates and their standard errors are the same as those produced by these packages. The odds ratio for a factor is the ratio of the odds of a cutting surviving when treated at the second level of the factor relative to the first.

9.2.3 Analysis of data set (3): relative density of rotifers

In modelling the data from this example, separate logistic regression lines are first fitted to the data from each of the two species of rotifer. This model is then compared with a model under which the two logistic regression lines are parallel, and with a common logistic regression relationship for the two species. In the illustrations of the use of GLIM and other packages, only the output from fitting parallel logistic regression lines is given. Notes on the input required for each package include a description of how the other models can be fitted. In section 6.9, it was found that these data exhibit overdispersion, and so some comments are included on the way in which GLIM, Genstat and SAS can be used to allow for overdispersion. In this example, the data are read from a file labelled **rotifers.dat**. The instructions for this are dependent upon the computer hardware being used, and some changes to the code will be necessary if using these packages on a mainframe rather than a personal computer, for example.

GLIM input:

> **$units 40**
> **$factor species 2**
> **$data species density y n**
> **$dinput 8 $**
> *File name?* **rotifers.dat**
> **$error b n**
> **$yvariate y**
> **$fit species+density−1 $**
> **$display e**

Notes on the GLIM input:

1. The **$dinput** directive is used to specify that channel 8 is to be used as the input channel for the data file. The filename is then given when prompted by GLIM.
2. In order to fit a pair of parallel logistic regression lines to these data, a factor **species** is used to index the data from the two species. By including **species** in a model along with a constant, there will be three parameters which are associated with the intercepts of the two regression lines, and the model is over-parameterized. In GLIM, the parameter associated with the first species would then be set equal to zero, and the two estimated intercepts could be calculated from the resulting parameter estimates. If the constant term is omitted from the model, by adding −1 to the **$fit** statement, the output gives the two estimated intercepts directly.

3. Separate logistic regression lines can be fitted using the directive **$fit species + species.density − 1**, and a common line can be fitted using **$fit density**.

GLIM output:

```
        scaled deviance = 434.25 at cycle  4
                   d.f. =  37

            estimate         s.e.      parameter
   1         -111.8          3.176      SPEC(1)
   2         -113.2          3.192      SPEC(2)
   3          107.6          3.052      DENS
   scale parameter taken as  1.000
```

Notes on the GLIM output:

1. If $\hat{p}_{jk}$ is the fitted probability that a rotifer of the jth species remains in the kth density of the supernatant, the equations of the fitted parallel logistic regression lines are as follows:

    Species 1: $\text{logit}(\hat{p}_{1k}) = -111.8 + 107.6 \text{ density}_k$

    Species 2: $\text{logit}(\hat{p}_{2k}) = -113.2 + 107.6 \text{ density}_k$

   The forthcoming discussion on the EGRET output for this example indicates how the output from GLIM could be interpreted if a constant term had not been omitted from the model.

2. One way of allowing for the overdispersion in the data is to multiply the standard errors of the parameter estimates by the square root of the mean deviance, or by the square root of the value of the X^2-statistic divided by its degrees of freedom, as discussed in Section 6.5. This is achieved by using the **$scale** directive in GLIM. To multiply the standard errors by the square root of the mean deviance for the full model, which in this example is 11.74, the directive **$scale 11.74** would be used before fitting alternative models.

Genstat input:

```
units [nvalues=40]
factor [levels=2] species
variate density,y,n
open 'rotifers.dat';channel=8
read [channel=8] species,density,y,n
model [distribution=binomial] y;nbinomial=n
fit [constant=omit] species+density
```

Notes on the Genstat input:

1. The **open** statement enables the data file to be read on channel 8 by using the option **channel = 8** in the **read** statement.

2. As in GLIM, **species** is declared as a factor and a constant omitted when fitting the model. See note 2 on the GLIM input.
3. Two separate linear logistic regression lines can be fitted to these data using the statement **fit [constant = omit] species + species.density** and a common line can be fitted using **fit density**.

Genstat output:

```
***** Regression Analysis *****


Response variate: y
  Binomial totals: n
     Distribution: Binomial
    Link function: Logit
     Fitted terms: species + density


*** Summary of analysis ***
    Dispersion parameter is 1

                  d.f.     deviance     mean deviance
Regression         3           *             *
Residual          37         434.3         11.74
Total             40           *             *


* MESSAGE: The following units have large residuals:
                              2            3.30
                              6            6.18
                              8           -3.81
                              9           -2.33
                             12            5.10
                             14            2.42
                             15            3.29
                             16           -5.31
                             17           -2.53
                             20           -3.31
                             22           -3.38
                             23            3.28
                             26           -4.10
                             28           -6.88
                             30            5.76
                             31           -5.68
                             32           -2.61
                             34            4.51
                             35           -6.44
                             37            4.10
                             38            2.94
                             40            6.81


* MESSAGE: The following units have high leverage:
                             37            0.226
                             38            0.172


*** Estimates of regression coefficients ***

                 estimate       s.e.           t
species 1        -111.76        3.19        -35.07
species 2        -113.18        3.20        -35.33
density           107.62        3.06         35.14
* MESSAGE: s.e.s are based on dispersion parameter with value 1
```

Notes on the Genstat output:

1. Because a **terms** statement has not been included, the total deviance and the deviance due to the regression are not provided in the analysis of deviance table. However, if the **terms** statement is included, it is not possible to omit the constant term! Since the deviance for the fitted model is really all that is n of interest in the analysis of deviance table, it is better to omit the terms statement in this type of problem.
2. The warning message about the large residuals is due to the fact that the data exhibit overdispersion, as discussed in section 6.9. The warning about observations with relatively high leverage should not be ignored, and it is important to verify that these two observations, which have extreme values of the variable **density**, are not influential.
3. The parameter estimates given by Genstat are the same as those given by GLIM; see note 1 on the GLIM output.
4. If the overdispersion in the data is to be allowed for by multiplying the standard errors of the parameter estimates by the square root of the residual mean deviance for the full model, 11.74, as in note 2 on the GLIM output, the option **dispersion = 11.74** is included in the **model** statement. The option **dispersion = *** has the effect of multiplying the standard errors by the square root of the residual mean deviance for the current model.

SAS input:

```
data rotifers;
filename d 'rotifers.dat';
infile d;
input species density y n;
proc probit;
class species;
model y/n = species density / d = logistic;
```

Notes on the SAS input:

1. The species effect is represented by a **class** variable, or factor. It is not possible to omit the constant term using SAS **proc probit**.
2. In order to allow for the overdispersion in the data, the option **lackfit** could be used in the **proc probit** statement, or in the **model** statement. The effect of this is to multiply the standard errors by $\sqrt{(X^2/v)}$, where X^2 is the value of Pearson's X^2-statistic for the fitted model and v is the number of degrees of freedom of this statistic. The value of the maximized log-likelihood statistic is unaffected.
3. A common line can be fitted to the data using the statement **model y/n = density / d = logistic;** in **proc probit**. It is not possible to include an interaction term such as **species*density** in the **model** statement. One way of fitting the

separate lines model is to define an indicator variable, **s**, say, in the data step, which takes the value zero for the first species and unity for the second. Next, a new variable is defined, **sd**, say, where **sd** = **s** × **density**. Then the terms **species**, **density** and **sd** are specified in the **model** statement, with **species** declared to be a class variable. If $\hat{\beta}_1$ and $\hat{\beta}_2$ are the coefficients of the explanatory variables **density** and **sd** in the resulting output, the slopes of the fitted lines will be $\hat{\beta}_1$ for the first species and $\hat{\beta}_1 + \hat{\beta}_2$ for the second. Alternatively, the procedure **proc catmod** could be used.

SAS output:

```
                                    SAS
                             Probit Procedure

                        Class Level Information

                     Class    Levels    Values

                     SPECIES     2       1 2

                     Number of observations used = 40


   Data Set         =WORK.ROTIFERS
   Dependent Variable=Y
   Dependent Variable=N
   Number of Observations=  40
   Number of Events     =    1881    Number of Trials =      4365

   Log Likelihood for LOGISTIC -1610.435519


       Variable  DF   Estimate  Std Err ChiSquare  Pr>Chi Label/Value

       INTERCPT   1 -113.17779 3.203053 1248.516  0.0001 Intercept

       SPECIES    1                     209.7655  0.0001
                  1 1.42168205  0.09816 209.7655  0.0001              1
                  0          0        0       .       .              2

       DENSITY    1 107.622744 3.062551 1234.928  0.0001
```

Note on the SAS output:

If $\hat{p}_{jk}$ is the fitted probability for the jth species of rotifer at the kth density, the fitted model can be expressed as

$$\text{logit}(\hat{p}_{jk}) = \hat{\beta}_0 + \widehat{\text{species}}_j + \hat{\beta}_1 \, \text{density}_k$$

where $\hat{\beta}_0$ is the estimated constant term. Since this model is overparameterized, a constraint needs to be placed on the effects due to species. As a result of the way in which indicator variables are used to represent the effects of different levels of a factor, SAS **proc probit** will set $\widehat{\text{species}}_2$ to be zero. Then, the two estimated intercepts are $\hat{\beta}_0 + \widehat{\text{species}}_1$ for the first species, and simply $\hat{\beta}_0$ for the second species. From the SAS output, $\hat{\beta}_0 = -113.178$, and $\widehat{species}_1 = 1.422$, so that the equations of the fitted logistic regression lines are

Species 1: $\quad \text{logit}(\hat{p}_{1k}) = -113.178 + 1.422 + 107.623 \, \text{density}_k$

Species 2: $\quad \text{logit}(\hat{p}_{2k}) = -113.178 + 107.623 \, \text{density}_k$

which agree with those obtained from GLIM and Genstat output.

BMDP input:

/problem	title='rotifers'.
/input	variables=4.
	format=free.
	file='rotifers.dat'.
/variable	names=species,density,y,n.
/group	codes(species)=1,2.
/regress	scount=y.
	count=n.
	interval=density.
	model=species,density.
/print	no corr.
/end	

Notes on the BMDP input:

1. The **file** statement in the **input** paragraph is used to specify the source of the data.
2. The variable **species** is declared to be a factor using the **group** statement, and only **density** is declared to be an **interval** variable in the **regress** paragraph.
3. Two separate logistic regression lines can be fitted using **model = species,species∗density.** in the **regress** paragraph, and a common line can be fitted using **model density.** It is not possible to obtain meaningful output after omitting a constant, and so the constant must be retained.

BMDP output:

```
STEP NUMBER    0
---------------

                         LOG LIKELIHOOD = -1610.436
GOODNESS OF FIT CHI-SQ    (2*0*LN(O/E))   = 300.691 D.F.= 25  P-VALUE=0.000
GOODNESS OF FIT CHI-SQ (HOSMER-LEMESHOW)= 194.560 D.F.= 8   P-VALUE=0.000
GOODNESS OF FIT CHI-SQ   ( C.C.BROWN )   = 151.035 D.F.= 2   P-VALUE=0.000

                                   STANDARD
       TERM          COEFFICIENT     ERROR    COEFF/S.E.  EXP(COEFFICIENT)

    species          -0.71084    0.4908E-01   -14.48         0.4912
    density          107.62       3.063        35.14         0.0000
    CONSTANT         -112.47      3.195        -35.21         0.0000
```

Notes on the BMDP output:

1. The maximized log-likelihood is the same as that given by SAS. However, the deviance of 300.691 on 25 d.f. is not the same as that given by GLIM and Genstat, which is 434.25 on 37 d.f. The reason for this difference is that BMDP has aggregated the data over those individual experimental units for which the explanatory variables included in the model are the same. For example, there are three binomial observations for species 1 (*Polyarthra major*) at a density of 1.030, namely, 19/83, 9/56 and 21/73. These three observations have been combined to give a binomial observation of $(19 + 9 + 21)/(83 + 56 + 73)$, that is, 49/212. In fact, there are only 14 different values of density, and so there are only 28 distinct combinations of species and density. The fitted model has three unknown parameters, and so after aggregation, there are 25 degrees of freedom for the deviance, as reported by BMDP. Because this aggregation is not desirable, it is recommended that this part of the output be ignored. Instead, models are compared using differences in $-2 \times$ maximized log-likelihood, in the manner in which SAS **proc probit** can be used. See note 2 on the SAS output in Section 9.2.1

2. The indicator variable used when the two-level factor **species** is included in the model takes the value -1 for the first species and $+1$ for the second. The coefficient of this indicator variable is -0.711, and so the fitted logistic regression lines are

$$\text{Species 1:} \quad \text{logit}(\hat{p}_{1k}) = -112.47 + (-0.711 \times -1) + 107.62 \text{ density}_k$$

$$\text{Species 2:} \quad \text{logit}(\hat{p}_{2k}) = -112.47 + (-0.711 \times 1) + 107.62 \text{ density}_k$$

which agree with the fitted equations found using the other packages.

EGRET input:

When using this package to fit models to the data on the relative density of rotifers, there are two points to note. First, after factoring the term **species**, meaningful output is not obtained when a constant is omitted. Second, individual logistic regression lines are obtained by fitting a model with terms **species**, **density**, and **species.density**. Since the indicator variable that is formed when a factor with two levels is included in a model takes the value zero at the first level, the term **density** must be included in the model in addition to the interaction **species.density**. If **density** is omitted, the logistic regression line fitted to the first species of rotifer will have a slope of zero! A common line can be fitted by just including the term **density** in the model, in addition to the constant.

EGRET output:

4 variables and 40 observations

```
-----------------------------VARIABLES-----------------------------
1. species      2. density      3. y              4. n
```

```
          ANALYSIS MODEL:    Logistic regression
              RISK TYPE:    Relative risk (multiplicative)
             GROUP SIZE:    Group Size Given by Variable:  n
   OUTCOME SPECIFICATION:    Outcome Variable Name:  y
        REPETITION COUNT:    -none-
```

```
        FACTORED VARIABLES:  | VARIABLE  #LEVELS BASE
                             |
                             | species     2     1
```

```
---------------------------REGRESSION TERMS---------------------------[LR]---
   a. %GM                    b. species              c. density
```

Cycle	Deviance
1	3180.989
2	664.8591
3	451.2112
4	434.4450
5	434.2516
6	434.2516
7	434.2516

```
-----------------------------RESULTS-----------------------------[LR]---
```

TERM	COEFFICIENT	STD ERROR	P-VALUE	ODDS RATIO
%GM	-111.8	(3.19)	<.001	.0000
species='2'	-1.422	(.982E-01)	<.001	.2413
density	107.6	(3.06)	<.001	>.170E+39

```
              DEVIANCE ON      37 DF =       434.252
LIKELIHOOD RATIO STATISTIC ON   3 DF =    2830.305, p < .001
```

Notes on the EGRET output:

1. The deviance agrees with that given by GLIM and Genstat.
2. The indicator variable that is defined when the factor **species** is included in the model takes the value zero for the first species and unity for the second, and so the effect of the first species is set equal to zero, as in GLIM and Genstat, when a constant term is not omitted. From the EGRET output, the equations of the parallel logistic regression lines are therefore

$$\text{Species 1:} \quad \text{logit}\,(\hat{p}_{1k}) = -111.8 + 107.6\ \text{density}_k$$

$$\text{Species 2:} \quad \text{logit}\,(\hat{p}_{2k}) = -111.8 - 1.422 + 107.6\ \text{density}_k$$

which, apart from rounding error resulting from the fact that only four significant figures are given in the output, are the same as those found using the other four packages.

9.3 USING PACKAGES TO PERFORM SOME NON-STANDARD ANALYSES

Many packages provide a wide range of statistical information after fitting a model to binary or binomial response data. Some of the statistics that are ouput will be useful, particularly, the deviance and parameter estimates, while other information will not be so useful. Frequently, the user of any particular package will want to use that software to provide results that are not part of the standard output of the package. For example, the values of the leverage, h_i, may be required when using GLIM, or the Williams procedure for modelling overdispersion may be needed when using Genstat. When such information is not provided as a standard option, it can often be obtained from quantities that are given, using the facilities of a package for calculation.

In this section, computational methods are presented for computing non-standard statistics and for performing a number of non-standard analyses that have been described in this book. An outline of the coding required when using GLIM, Genstat, SAS and BMDP is given for the calculation of some of these statistics, but for the more complicated algorithms, GLIM macros that implement them are given in Appendix C. These algorithms could be encoded for use in other packages, but details will not be given here.

9.3.1 Prediction of a response probability

In Section 3.14, we saw how to obtain the predicted value of a response probability for an individual with known values of the explanatory variables used in a linear logistic model for this probability. We also saw how a confidence interval for the corresponding true response probability could be derived from the expression for the variance of the linear predictor, the predicted value of the linear part of the model being given by equation (3.8). In some packages, there are standard commands that can be used to provide the values of a predictor and its standard error. For example, Genstat includes the command **predict**, which when used with the option **print = d,p,s**, will give the required information.

In packages with no such facility, but which do enable observations to be omitted from an analysis by giving them a weight of zero, the value of the linear predictor and its variance can be obtained in the following manner. The input data set is first extended by the addition of a row that contains the values of the explanatory variables for each of the individuals for whom the predicted response probability is required. A value of the binomial response variable, and a value for the corresponding binomial denominator will also need to be provided, but the actual numbers given for these quantities are arbitrary. A new variable is then defined which takes the value unity for the original observations in the data base, and zero for those that have been appended. This variable is then declared to be a **weight variable**, and the additional observations which have been assigned zero weight will not be taken account of in the fitting process. However, standard output available from the package will contain the desired information. This procedure can be adopted using GLIM and SAS.

When using GLIM, the variable that defines the weights, say, w, is declared to be a weight vector using **$weight w**, prior to fitting the model. After the model has been fitted, predicted values for all of the observations in the data set, are found using the directives

$$\text{\$calculate phat} = \%\text{fv}/\%\text{bd}$$
$$\text{\$look phat}$$

where **%fv** and **%bd** are GLIM system vectors that contain the fitted values of the response variable and the values of the binomial denominators. The variance of the linear predictor for each observation can then be found using the following sequence of GLIM directives:

$$\text{\$extract }\%\text{vl}$$
$$\text{\$look }\%\text{vl}$$

where **%vl** is a GLIM system vector that contains the variance of the linear predictor. Note that this vector needs to be extracted before it can be used in subsequent GLIM directives. In SAS, the variable to be used as weights, w, is declared as such using the statement **weight w;** in **proc probit**. The parameters **prob = phat** and **std = selp** are then included in an **output** statement. After the model has been fitted, the identifiers **phat** and **selp** will contain the predicted values of the response probabilities and the standard errors of the linear predictor, for each observation in the data set, including those assigned zero weight.

In packages which have no direct facilities for obtaining predicted values, and do not allow weights to be used, such as BMDP and EGRET, predicted values have to be obtained directly from the parameter estimates in the linear component of the model. The variance of the linear predictor has then to be obtained from equation (3.8), which involves the elements of the variance–covariance matrix of the parameter estimates.

9.3.2 Adjusting for a natural response

In Section 4.4, a model that allowed for a natural response was proposed. If the probability of a natural response is π, and p_i^* is the observable response probability, the dependence of p_i^* on the ith value of a single explanatory variable, d, was given in equation (4.11) as

$$\log\left\{\frac{p_i^* - \pi}{1 - p_i^*}\right\} = \beta_0 + \beta_1 d_i$$

More generally, the model can be expressed as

$$\log \left\{ \frac{p_i^* - \pi}{1 - p_i^*} \right\} = \eta_i \tag{9.1}$$

where η_i is the linear systematic component of the model. When using SAS to fit this model, the option $c = \pi_0$ in the **proc probit** statement can be used in the case where π is being estimated by π_0. When π has to be estimated from the data, simultaneously with the parameters in the linear component of the model, the option **optc** is used.

When π has the value π_0, the model described by equation (9.1) can also be fitted using the facilities in GLIM and Genstat for fitting user-defined models. With GLIM, this model can be fitted by providing expressions for four quantities. These are (i) an expression that shows how the fitted values, $\hat{y}_i$, of the binomial response variable, y_i, can be obtained from the linear predictor, $\hat{\eta}_i$; (ii) an expression for the derivative of η_i with respect to $E(y_i) = n_i p_i$, where n_i is the ith binomial denominator, in terms of $\hat{y}_i$; (iii) an expression for the variance of y_i in terms of $\hat{y}_i$; (iv) the value of the deviance for an individual experimental unit. Denoting the linear predictor by $\hat{\eta}_i$, for the model described by equation (9.1) these four quantities are, respectively, as follows

$$n_i \{ \pi_0 + (1 - \pi_0)/(1 + \exp(-\hat{\eta}_i)) \}$$
$$n_i (1 - \pi_0)/\{ (n_i - \hat{y}_i)(\hat{y}_i - n_i \pi_0) \}$$
$$\hat{y}_i (n_i - \hat{y}_i)/n_i$$
$$2[y_i \log(y_i/\hat{y}_i) + (n_i - y_i) \log \{ (n_i - y_i)/(n_i - \hat{y}_i) \}]$$

After the value of π_0 has been specified, the model can be fitted using the GLIM directive **$own**. Similar expressions are needed when using the Genstat library procedure **GLM** to fit user-defined models. The value $\hat{\pi}$ which minimizes the deviance for the model described by equation (9.1) can be found by fitting the model for a range of values of π. Alternatively, a GLIM macro or a Genstat procedure written for function minimization, could be used to obtain the value $\hat{\pi}$ for which the deviance is minimized. A GLIM macro for function minimization has been given by Nelder (1985).

9.3.3 Computation of the leverage

A quantity that is needed to obtain many of the diagnostics described in Chapter 5 is the leverage h_i. Because of the importance of this quantity in computing diagnostics for model checking, full details are given on how the values of h_i can be obtained using GLIM, Genstat, SAS and BMDP. In Chapter 5, the values of the leverage were defined to be the diagonal elements of the matrix $H = W^{1/2} X (X'WX)^{-1} X' W^{1/2}$, where W is the $n \times n$ diagonal matrix of iterative weights used in fitting the model, and X is the design matrix. The diagonal elements of H

are the same ·as those of $X(X'WX)^{-1}X'W = \text{Var}(X\hat{\beta})\ W$, where $\text{Var}(X\hat{\beta})$ is the variance–covariance matrix of the linear predictor, and so the ith value of the leverage is $h_i = \text{Var}(\hat{\eta}_i)\ w_i$, the variance of the linear predictor multiplied by the iterative weight.

In GLIM, the iterative weights are stored in the system vector **%wt**, and so the values of the leverage can be obtained in an identifier **h** using the following directives:

$extract %vl
$calculate h=%vl*%wt

When using Genstat, the values of the leverage can be obtained in an identifier **h** using the command **rkeep leverage**=**h** after a model has been fitted.

The iterative weights are actually given by $w_i = n_i \hat{p}_i (1 - \hat{p}_i)$, where $\hat{p}_i$ is the fitted value of the response probability for the ith observation. If the numerator and denominator of the binomial observations are contained in identifiers **y** and **n**, **h** can be obtained using the following code when using SAS **proc probit**:

```
proc probit;
model y/n= < model terms >/d=logistic;
output out=data1 prob=phat std=selp;
data data2;
set data1;
wt=n*phat*(1-phat);
h=wt*selp*selp;
```

The values of the leverage are provided as standard output when using the **LR** procedure in BMDP. For use in calculations, values of the leverage and other model checking diagnostics can be saved in a BMDP file using a **save** paragraph which contains

```
/save               file=' filename'.
                    code='codename'.
                    new.
                    content=cell.
```

The BMDP program **1D** is then used to read the saved file and carry out transformations to produce the desired statistics, using the following instructions:

```
/problem
/input              file=' filename'.
                    code='codename'.
                    type=data.
/transformation     transformations.
/print              data.
/end
```

It is not possible to obtain the values of the leverage within EGRET.

9.3.4 Computation of Anscombe residuals

In expression (5.5), Anscombe residuals were defined by

$$r_{Ai} = \frac{A(y_i) - A(\hat{y}_i)}{\text{s.e.} \{A(y_i) - A(\hat{y}_i)\}}$$

where $A(y_i)$ is a function of the ith binomial observation, y_i, which is approximately normally distributed. The appropriate function A for binomial data is given by

$$A(u) = \int_0^{u/n_i} t^{-1/3}(1 - t)^{-1/3} \, dt, \quad 0 \leqslant u \leqslant n_i \qquad (9.2)$$

and the standard error in the denominator of r_{Ai} is approximately equal to $[\hat{p}_i(1 - \hat{p}_i)]^{1/6} \sqrt{\{(1 - h_i)/n_i\}}$. The function $A(u)$ can be computed from the **incomplete beta function**, defined by

$$I_z(\alpha, \beta) = \frac{1}{B(\alpha, \beta)} \int_0^z t^{\alpha - 1}(1 - t)^{\beta - 1} \, dt$$

and so $A(u) = B(\frac{2}{3}, \frac{2}{3}) I_{u/n_i}(\frac{2}{3}, \frac{2}{3})$. The value of the beta function with parameters $\frac{2}{3}, \frac{2}{3}$ is 2.05339. Some software packages, including SAS, provide the incomplete beta function as a standard function, and so the values of $A(y_i)$, for example, are calculated in SAS using the following statement in a data step:

p = y/n;
a = 2.05339∗probbeta(p, 0.66667, 0.66667);

where **y** and **n** are identifiers that contains the values of y_i and n_i, and **a** is destined to contain the values of $A(y_i)$.

When the incomplete beta function is not available as a standard function the value of the integral in equation (9.2) can be obtained using the series expansion

$$A(u) = \frac{3}{2}\left(\frac{u}{n_i}\right)^{2/3} + \sum_{r=1}^{\infty} a_r \left(\frac{u}{n_i}\right)^{(3r+2)/3}$$

where

$$a_r = [3^{r-1}(3r + 2)]^{-1} \prod_{j=1}^{r} (3j - 2)/j$$

for $0 \leqslant u \leqslant n_i/2$, with $A(u) = 2.05339 - A(n_i - u)$, for $n_i/2 < u \leqslant n_i$. This series converges rapidly, and eight terms at the most are needed to give $A(u)$ correct to four decimal places. A GLIM macro, named **ansc**, which gives the Anscombe residuals after a fit, is listed in Appendix C.1.

9.3.5 Constructing simulated envelopes for half-normal plots

The procedure described in Section 5.2.2 for obtaining simulated envelopes to aid in the interpretation of a half-normal plot of residual involves generating 19 sets of n observations from a binomial distribution, where n is the number of binomial observations in the original data base. In each of these sets, the ith observation is simulated from a binomial distribution with parameters n_i and $\hat{p}_i$, where n_i is the binomial denominator and $\hat{p}_i$ is the estimated response probability for the ith observation, $i = 1, 2, ..., n$. The same linear logistic model that was fitted to the original data set is then fitted to each of the 19 simulated data sets, and the standardized deviance (or likelihood) residuals computed. These are then ordered, and the maximum and minimum absolute values of the standardized deviance (or likelihood) residuals corresponding to each observation, across the 19 simulated sets of data, are calculated. These values then form the boundaries of the envelopes to the plot.

Pseudo-random numbers from a binomial distribution with parameters n and p can be obtained using the following algorithm:

1. Set $y = 0$ and $r = 0$.
2. Simulate a pseudo-random number, u, from a uniform distribution over the range $(0, 1)$. Most software packages can supply uniform random variates as a standard function.
3. If $u \leqslant p$, set $y = y + 1$.
4. Set $r = r + 1$. If $r < n$, go to step 2, otherwise return y as the required random number.

This algorithm has been incorporated in a GLIM macro, named **hnp** and listed in Appendix C.2, which can be used to obtain the simulated envelope.

9.3.6 Smoothing residual plots for binary data

In Section 5.7.1, we saw that added variable, constructed variable and partial residual plots for binary observations are more easily interpreted after fitting a smoothed line. When packages that do not include smoothing algorithms as a standard facility are being used for linear logistic modelling, an algorithm for locally weighted regression due to Cleveland (1979) can be used to construct the smoothed line. This algorithm is readily coded as a GLIM macro, and can easily be implemented in other packages.

Suppose that the coordinates of the points to be plotted in either an added variable plot, a constructed variable plot or a partial residual plot are denoted (u_i, v_i), $i = 1, 2, ..., n$, where the us are arranged in increasing order so that u_1 is the smallest and u_n the largest. For example, in an added variable plot, u_i would be the *i*th smallest added variable residual and v_i the corresponding value of the Pearson residual. For a given value of u, u_i, say, weights are then defined which are such that the largest weight attaches to u_i, and decrease as values of u become distant from u_i until the weights are zero at and beyond the *r*th nearest neighbour of u_i. Using these weights, a polynomial of degree d in u_i is fitted to the values of v_i using the method of weighted least squares, and the fitted value at u_i is used as the smoothed value of v_i. This procedure is repeated for each value of u_i, $i = 1, 2, ..., n$.

The values of d and r determine the degree of smoothness in the line. The values $d = 1$ and $r = [nf]$, where $[nf]$ denotes the value of nf rounded off to the nearest integer, $f = 0.5$, and n is the number of observations, are generally recommended. Indeed, these values were used in the production of Figs. 5.19 and 5.20. The line can be made less smooth by decreasing f, and more smooth by increasing f.

The algorithm is presented more formally below.

1. Set $i = 1$ and fix values of r and d.

2. Determine s_i, the distance from u_i to the *r*th nearest neighbour of u_i, using the fact that s_i is the *r*th smallest value of $|u_i - u_j|$, for $j = 1, 2, ..., n$.

3. For $k = 1, 2, ..., n$, define weights

$$w_{ki} = \begin{cases} (1 - |a_i|^3)^3, & |a_i| < 1 \\ 0, & \text{otherwise} \end{cases}$$

where $a_i = (u_k - u_i)/s_i$.

4. Fit a polynomial of degree d in u_k, using weights w_{ki}, to the values v_k, $k = 1, 2, ..., n$. Thus, parameter estimates $\hat{\beta}_{0i}, \hat{\beta}_{1i}, ..., \hat{\beta}_{di}$ are found, which are such that

$$\sum_{k=1}^{n} w_{ki}(v_k - \beta_{0i} - \beta_{1i}u_k - \cdots - \beta_{di}u_k^d)^2$$

is minimized.

5. Obtain the fitted value of v at u_i from $\hat{v}_i = \hat{\beta}_{0i} + \hat{\beta}_{1i}u_i + \cdots + \hat{\beta}_{di}u_i^d$.

6. If $i < n$, set $i = i + 1$ and repeat from step 1.

On completion of the algorithm, the smoothed values will be contained in $\hat{v}_1, \hat{v}_2, ..., \hat{v}_n$.

A GLIM macro named **smooth** that can be used to produce the smoothed values for use in added variable, constructed variable and partial residual plots for binary observations is provided in Appendix C.3.

9.3.7 Computation of the $\Delta\hat{\beta}$-statistic

The $\Delta\hat{\beta}$-statistic, introduced in Section 5.5.3, measures the influence of the ith observation on the jth parameter estimate, $\hat{\beta}_j$. This statistic was defined by equation (5.14) to be given by

$$\Delta_i\hat{\beta}_j = \frac{(X'WX)_{j+1}^{-1}\,x_i\,(y_i - \hat{y}_i)}{(1 - h_i)\,\text{s.e.}\,(\hat{\beta}_j)}$$

where $(X'WX)_{j+1}^{-1}$ is the $(j+1)$th row of the variance–covariance matrix of the parameter estimates, x_i is the vector of explanatory variables for the ith observation, $\hat{y}_i$ is the fitted value corresponding to the ith observation y_i, and h_i is the ith value of the leverage. In order to calculate the values of the $\Delta\hat{\beta}$-statistic, note that the values of $(X'WX)_{j+1}^{-1}\,x_i$ are the elements of the $(j+1)$th row of the matrix $(X'WX)^{-1}X'$, $j = 0, 1, \ldots, k$. If a weighted least squares regression of a variate z on the explanatory variables that make up X is carried out, where the weights are the diagonal elements of W, the resulting vector of parameter estimates is $(X'WX)^{-1}X'Wz$. Now suppose that z is taken to be the vector whose ith component is w_i^{-1}, and zero elsewhere, where w_i is the ith diagonal element of W. Then the set of parameter estimates from the weighted least squares regression analysis are the $(k+1)$ values of $(X'WX)_{j+1}^{-1}\,x_i$, for $j = 0, 1, \ldots, k$. By repeating this linear regression analysis for $i = 1, 2, \ldots, n$ in turn, all the required values of $(X'WX)^{-1}\,x_i$ can be found.

A GLIM macro named **dbeta** that computes the values of $\Delta_i\hat{\beta}_j$ using this algorithm is given in Appendix C.4. This algorithm may also be implemented in other packages. An alternative way of computing the values of this statistic is to use a package that has facilities for matrix algebra. The vector $(X'WX)^{-1}\,x_i$ can then be obtained directly.

9.3.8 Computation of the ΔD-statistic

The statistic $\Delta_i D_j$ is used to assess the influence of the jth observation on the fit of the model to the ith observation. This statistic was introduced in section 5.5.7, and is defined by

$$\Delta_i D_j = \frac{2X_j h_{ij} X_j}{1 - h_j} + \frac{X_j^2 h_{ij}^2}{(1 - h_j)^2}$$

where X_i, X_j are the Pearson residuals for the ith and jth observations and h_{ij} is the (i,j)th element of the hat matrix H. The only quantity which is not straightforwardly obtained from any of the major software packages is h_{ij}. The values of h_{ij} form the jth column of the matrix H, and to compute them, suppose that the dependent variable z is regressed on the explanatory variables that make up X, using weights that are the diagonal elements of W. The resulting linear predictor is $\hat{\eta}_0 = X(X'WX)^{-1}X'Wz$. If z is the vector that has unity in the jth position and zero elsewhere, the ith element of $\hat{\eta}_0$ is $w_i^{-1/2}h_{ij}w_j^{1/2}$, and so the values of h_{ij}, $i = 1, 2, \ldots, n$, are found from $\hat{\eta}_{0i}\sqrt{(w_i/w_j)}$. This procedure is repeated for $j = 1, 2, \ldots, n$.

A GLIM macro named **dstat** in Appendix C.5 gives the values of the ΔD-statistic after a model has been fitted. As with the $\Delta \hat{\beta}$-statistic, it may be easier to obtain the values h_{ij} using a package that has facilities for matrix algebra.

9.3.9 Modelling misclassification in binary data

A model that allows a misclassification of a binary response variable to be modelled was given in Section 5.7.3. If γ is the misclassification probability, the observable success probability, p_i^* can be modelled using equation (5.18), according to which

$$\log\left\{\frac{p_i^* - \gamma}{1 - p_i^* - \gamma}\right\} = \eta_i \tag{9.3}$$

where η_i is the linear systematic component of the model. This model is rather similar to that considered in Section 9.3.2, and can be fitted using either GLIM or Genstat in the same way as the model described by equation (9.1). The four expressions that are required in the specification of a user-defined model in these packages, are, for a given value of γ, γ_0, say,

$$[(1 - \gamma_0)\exp(\hat{\eta}_i) + \gamma_0]/\{1 + \exp(\hat{\eta}_i)\}$$

$$(1 - 2\gamma_0)/\{\hat{y}_i(1 - \hat{y}_i) - \gamma_0(1 - \gamma_0^2)\}$$

$$\hat{y}_i(1 - \hat{y}_i)$$

$$-2\{y_i \log(\hat{y}_i) + (1 - y_i)\log(1 - \hat{y}_i)\}$$

As in Section 9.3.2, a search procedure, or a function minimisation routine could be used to identify that value $\hat{\gamma}$ which minimizes the deviance on fitting the model described by equation (9.3) to binary data.

9.3.10 Computation of the C-statistic

The C-statistic defined in Section 5.7.5 measures the influence of each observation in a data set on the classification of a future individual whose vector of explanatory variables is x_0. The value of this statistic for the ith observation is

$$C_i = \frac{x_0'(X'WX)^{-1}x_i(y_i - \hat{p}_i)}{(1 - h_i)}$$

where $(X'WX)^{-1}$ is the variance–covariance matrix of the parameter estimates in the original data set, $\mathbf{x}_i$ is the vector of explanatory variables, h_i is the leverage and $\hat{p}_i$ is the fitted response probability corresponding to the ith observation, y_i, in that data set.

To calculate $\mathbf{x}_0'(\mathbf{X}'\mathbf{W}\mathbf{X})^{-1}\mathbf{x}_i$, simply obtain the vector $(\mathbf{X}'\mathbf{W}\mathbf{X})_{j+1}^{-1}\mathbf{x}_i$, for $j = 0, 1, \ldots, k$, using the method described in Section 9.3.7, multiply each element of the vector by the corresponding element of $\mathbf{x}_0$, and sum the resulting $(k + 1)$ values. The GLIM macro **cstat** listed in Appendix C.6 uses this procedure to compute the values of the C-statistic. As for the calculation of the $\Delta\hat{\beta}$-statistic, an alternative way of obtaining the values of $\mathbf{x}_0'(\mathbf{X}'\mathbf{W}\mathbf{X})^{-1}\mathbf{x}_i$ is to use matrix algebra.

9.3.11 The Williams procedure for modelling overdispersion

In Section 6.2, a general model for overdispersion was described which involved the estimation of a quantity denoted by ϕ. The procedure for fitting this model, outlined in Section 6.4, requires ϕ to be estimated. An algorithm for this estimation procedure, described by Williams (1982a), is given in the sequel.

Suppose that the n binomial observations have associated weights w_i. The X^2-statistic is then given by

$$X^2 = \sum_{i=1}^{n} \frac{w_i(y_i - n_i\hat{p}_i)^2}{n_i\hat{p}_i(1 - \hat{p}_i)} \tag{9.4}$$

and the approximate expected value of this statistic is

$$\sum_i w_i(1 - w_i v_i d_i)\{1 + \phi(n_i - 1)\} \tag{9.5}$$

where $v_i = n_i p_i(1 - p_i)$ and d_i is the ith diagonal element of the variance–covariance matrix of the linear predictor, $\hat{\eta}_i = \sum \hat{\beta}_j x_{ji}$. The fact that expression (9.5) involves ϕ suggests the following iterative procedure for estimating this parameter. First, the full linear logistic model, the most complicated model that one is prepared to consider, is fitted to the data using $w_i = 1$ for $i = 1, 2, \ldots, n$. The value of the Pearson X^2-statistic is then calculated. When the initial weights are unity, expression (9.5) becomes

$$n - p + \phi \sum_i \{(n_i - 1)(1 - v_i d_i)\} \tag{9.6}$$

where p is the number of parameters in the linear predictor. An initial estimate of ϕ from expression (9.6) is therefore

$$\hat{\phi}_0 = \{X^2 - (n - p)\}/\sum_i \{(n_i - 1)(1 - v_i d_i)\}$$

from which initial estimates of the weights are $w_{i0} = [1 + (n_i - 1)\hat{\phi}_0]^{-1}$. After a weighted fit of the model, $\hat{\beta}$ is recalculated, and so is X^2 using equation (9.4). Then expression (9.5) is used to get a revised estimate of ϕ, $\hat{\phi}_1$, from

$$\hat{\phi}_1 = [X^2 - \sum_i \{w_i(1 - w_i v_i d_i)\}]/\sum_i \{w_i(n_i - 1)(1 - w_i v_i d_i)\}$$

If X^2 remains large relative to its number of degrees of freedom, an additional cycle of this iterative procedure is carried out. The procedure is terminated when X^2 becomes very close to its number of degrees of freedom. As a result, the final deviance will also be approximately equal to its degrees of freedom, and it can then no longer be used to assess the goodness of fit of the model, as noted in Section 6.4.

Once ϕ has been estimated by $\hat{\phi}$, weights of $[1 + (n_i - 1)\hat{\phi}]^{-1}$ are used in fitting models that have fewer terms than the full model. An analysis of deviance can then be used to compare alternative models, after allowing for overdispersion, as described in Section 6.4.

A GLIM macro that implements this algorithm, named **disp**, is given in Appendix C.7. After the algorithm has been used, the standard error of the parameter estimates given in the GLIM output will have been scaled appropriately, and provide a correct reflection of the precision of the parameter estimates. A similar procedure can be used to fit a linear logistic model that contains a random effect, using the approximate method described in Section 6.7.2. The required modifications to the macro **disp** are indicated in the GLIM code in Appendix C.7.

9.4 SUMMARY OF THE RELATIVE MERITS OF PACKAGES FOR MODELLING BINARY DATA

Much can be deduced about the relative merits of the different packages from their scope described in Section 9.1, the discussion on the input and output in Section 9.2, and the details of how some non-standard computations can be carried out within the packages given in Section 9.3. However, some of the main points are summarized in this section.

Of the packages considered in this chapter, GLIM and Genstat are the most versatile and can be used to carry out all the analyses described in the first seven chapters of this book with four exceptions. These exceptions are non-linear logistic modelling, modelling overdispersion using the beta-binomial model or by including random effects in a model, and the analysis of data from matched case-control studies where there is more than one control per case. These last three analyses can all be carried out using EGRET. In addition, matched case-control studies can also be analysed using the SAS procedure **proc mcstrat**, which is part of the SAS supplemental library distributed with release 5 of the package.

Genstat and SAS have excellent facilities for data management, while the graphics facilities in Genstat, SAS, and BMDP are superior to those of the other packages. Apart from EGRET, all packages incorporate extensive facilities for data transformation, and for the computation of statistics such as model checking diagnostics after a model has been fitted. The corresponding facilities in EGRET are extremely limited, and so this package would generally need to be used in conjunction with other software.

Although each package can be used in fitting a linear logistic model, BMDP, SPSS and EGRET cannot be used in modelling other transformations of the response probability. For GLIM, Genstat, BMDP and EGRET, the linear component of a model may include continuous variables, factors and interactions. In BMDP, only hierarchic models are permitted, and so it is not possible to include an interaction term in the model unless the corresponding main effects are present. Similarly, a constant term cannot be omitted when the model contains a factor. EGRET does not take account of the other terms in a model when indicator variables for factors are constructed, and so particular care must be taken when using this package to fit models that contain interactions. Interactions cannot be fitted directly using SAS **proc probit**, and so if such terms are to be fitted, appropriate indicator variables have to be defined explicitly in the data step before using **proc probit**.

SAS **proc probit** is the only program of those being considered in this chapter that includes a standard option for allowing for a natural response. In the presence of overdispersion, the standard errors of the parameter estimates can easily be adjusted by multiplication by a constant scale parameter in GLIM, Genstat and SAS. It is also straightforward to implement the Williams procedure for modelling overdispersion in both GLIM and Genstat.

The volume of output provided by the packages varies widely. At one extreme, the standard output provided by GLIM after fitting a model is exceptionally brief, in that only the deviance, the number of degrees of freedom, and the number of iterations used in fitting the model are given. However, in the model selection phase of an analysis, only the deviance, and its corresponding number of degrees of freedom, are required. After a model has been fitted, the parameter estimates and futher information about the fit of the model can be obtained using additional GLIM directives. The standard output from EGRET is also rather concise, and contains similar information to that in the default output produced by Genstat. Output from SAS **proc probit** is more verbose, although it does not include the deviance, and BMDP includes much irrelevant information. Most of the output provided by GLIM and Genstat is statistically meaningful and reliable. Other packages can produce meaningless output, such as odds ratios for a constant term given by EGRET, or perform undesirable analyses, such as the pooling of binomial data over experimental units with common values of the explanatory variables in BMDP.

Of the packages considered in this chapter, BMDP has the poorest user-interface, and the help information provided by this package is more limited than that given by the others. The manuals that accompany each package are generally satisfactory, although the sheer size of the SAS documentation can make it difficult to locate any particular piece of information. Each package can be used interactively, and all apart from EGRET can also be run in batch mode. Versions of all packages, are available for both a mainframe and a personal computer, again apart from EGRET which is only available for a personal computer.

FURTHER READING

Healy (1988) gives a general introduction to the use of GLIM and a chapter on modelling binary data using GLIM is included in Aitkin, *et al.* (1989). Collett and Roger (1988) provide a description of how GLIM can be used to obtain a number of diagnostics after fitting a generalised linear model, including the $\Delta\hat{\beta}$- and ΔD-statistics, and give appropriate GLIM macros. GLIM code to enable overdispersion to be modelled using the Williams procedure was given in Williams (1982a).

The principal references for the material presented in this chapter are the manuals that accompany GLIM, Genstat, SAS, BMDP, SPSS and EGRET. These manuals are normally multi-authored, and in some cases the authors' names are not even listed. This means that it is not possible to give a reference to each manual in standard form, and so instead the name and address of the distributor of each package is listed below. When there is a distributor based in both the UK and the USA, both have been given.

GLIM and Genstat: NAG Ltd.
Wilkinson House
Jordan Hill Road
Oxford OX2 8DR, UK

NAG Inc.
1400 Opus Place, Suite 200, Downers Grove
Illinois 60515–5702, USA

SAS: SAS Institute Inc.
Cary
North Carolina 27512–8000, USA

SAS Software Ltd.
Wittington House
Henley Road, Medmenham
Marlow SL7 2EB, Bucks, UK

BMDP: BMDP Statistical Software Inc.
1440 Sepulveda Blvd.
Los Angeles, California 90025, USA

Statistical Software Ltd.
Cork Technology Park
Cork, Ireland

SPSS: SPSS Inc.
444 North Michigan Avenue
Chicago, Illinois 60611, USA

SPSS UK Ltd.
SPSS House, 5 London Street
Chertsey KT16 8AP, Surrey, UK

EGRET: Statistics & Epidemiology Research Corporation
            909 Northeast 43rd Street, Suite 310
            Seattle, Washington 98105, USA

Appendix A

Table of the values of logit (p) and probit (p) for p = 0.01(0.01)1.00

p	logit (p)	probit (p)	p	logit (p)	probit (p)
0.01	−4.5951	−2.3263	0.51	0.0400	0.0251
0.02	−3.8918	−2.0537	0.52	0.0800	0.0502
0.03	−3.4761	−1.8808	0.53	0.1201	0.0753
0.04	−3.1781	−1.7507	0.54	0.1603	0.1004
0.05	−2.9444	−1.6449	0.55	0.2007	0.1257
0.06	−2.7515	−1.5548	0.56	0.2412	0.1510
0.07	−2.5867	−1.4758	0.57	0.2819	0.1764
0.08	−2.4423	−1.4051	0.58	0.3228	0.2019
0.09	−2.3136	−1.3408	0.59	0.3640	0.2275
0.10	−2.1972	−1.2816	0.60	0.4055	0.2533
0.11	−2.0907	−1.2265	0.61	0.4473	0.2793
0.12	−1.9924	−1.1750	0.62	0.4895	0.3055
0.13	−1.9010	−1.1264	0.63	0.5322	0.3319
0.14	−1.8153	−1.0803	0.64	0.5754	0.3585
0.15	−1.7346	−1.0364	0.65	0.6190	0.3853
0.16	−1.6582	−0.9945	0.66	0.6633	0.4125
0.17	−1.5856	−0.9542	0.67	0.7082	0.4399
0.18	−1.5163	−0.9154	0.68	0.7538	0.4677
0.19	−1.4500	−0.8779	0.69	0.8001	0.4959
0.20	−1.3863	−0.8416	0.70	0.8473	0.5244
0.21	−1.3249	−0.8064	0.71	0.8954	0.5534
0.22	−1.2657	−0.7722	0.72	0.9445	0.5828
0.23	−1.2083	−0.7388	0.73	0.9946	0.6128
0.24	−1.1527	−0.7063	0.74	1.0460	0.6433
0.25	−1.0986	−0.6745	0.75	1.0986	0.6745
0.26	−1.0460	−0.6433	0.76	1.1527	0.7063
0.27	−0.9946	−0.6128	0.77	1.2083	0.7388
0.28	−0.9445	−0.5828	0.78	1.2657	0.7722
0.29	−0.8954	−0.5534	0.79	1.3249	0.8064
0.30	−0.8473	−0.5244	0.80	1.3863	0.8416

Table—*contd*

p	logit (p)	probit (p)	p	logit (p)	probit (p)
0.31	-0.8001	-0.4959	0.81	1.4500	0.8779
0.32	-0.7538	-0.4677	0.82	1.5163	0.9154
0.33	-0.7082	-0.4399	0.83	1.5856	0.9542
0.34	-0.6633	-0.4125	0.84	1.6582	0.9945
0.35	-0.6190	-0.3853	0.85	1.7346	1.0364
0.36	-0.5754	-0.3585	0.86	1.8153	1.0803
0.37	-0.5322	-0.3319	0.87	1.9010	1.1264
0.38	-0.4895	-0.3055	0.88	1.9924	1.1750
0.39	-0.4473	-0.2793	0.89	2.0907	1.2265
0.40	-0.4055	-0.2533	0.90	2.1972	1.2816
0.41	-0.3640	-0.2275	0.91	2.3136	1.3408
0.42	-0.3228	-0.2019	0.92	2.4423	1.4051
0.43	-0.2819	-0.1764	0.93	2.5867	1.4758
0.44	-0.2412	-0.1510	0.94	2.7515	1.5548
0.45	-0.2007	-0.1257	0.95	2.9444	1.6449
0.46	-0.1603	-0.1004	0.96	3.1781	1.7507
0.47	-0.1201	-0.0753	0.97	3.4761	1.8808
0.48	-0.0800	-0.0502	0.98	3.8918	2.0537
0.49	-0.0400	-0.0251	0.99	4.5951	2.3263
0.50	0.0000	0.0000	1.00	∞	∞

Appendix B

B.1 AN ALGORITHM FOR FITTING A GENERALIZED LINEAR MODEL TO BINOMIAL DATA

In this appendix, an algorithm for fitting a linear model to the transformed value of a binomial response probability is described. As explained in Section 3.7, the process is iterative, and so we begin with a general description of iterative maximum likelihood estimation.

B.1.1 Iterative maximum likelihood estimation

Suppose that n observations are to be used to estimate the value of q parameters, $\beta_1, \beta_2, \ldots, \beta_q$, and denote the likelihood function by $L(\boldsymbol{\beta})$. The q derivatives of the log-likelihood function with respect to $\beta_1, \beta_2, \ldots, \beta_q$, are called the **efficient scores**, and can be assembled to give a $q \times 1$ vector of efficient scores, whose jth component is $\partial \log L(\boldsymbol{\beta})/\partial \beta_j$, for $j = 1, 2, \ldots, q$. Denote this vector by $\mathbf{u}(\boldsymbol{\beta})$. Now let $H(\boldsymbol{\beta})$ be the $q \times q$ matrix of second partial derivatives of $\log L(\boldsymbol{\beta})$, where the (j, k)th element of $H(\boldsymbol{\beta})$ is

$$\frac{\partial^2 \log L(\boldsymbol{\beta})}{\partial \beta_j \partial \beta_k}$$

for $j = 1, 2, \ldots, q;\ k = 1, 2, \ldots, q$. The matrix $H(\boldsymbol{\beta})$ is sometimes called the **Hessian matrix**.

Now consider $\mathbf{u}(\hat{\boldsymbol{\beta}})$, the vector of efficient scores evaluated at the maximum likelihood estimate of $\boldsymbol{\beta}$, $\hat{\boldsymbol{\beta}}$. Using a Taylor series to expand $\mathbf{u}(\hat{\boldsymbol{\beta}})$ about $\boldsymbol{\beta}^*$, where $\boldsymbol{\beta}^*$ is near to $\hat{\boldsymbol{\beta}}$, we get

$$\mathbf{u}(\hat{\boldsymbol{\beta}}) \approx \mathbf{u}(\boldsymbol{\beta}^*) + H(\boldsymbol{\beta}^*)(\hat{\boldsymbol{\beta}} - \boldsymbol{\beta}^*) \tag{B.1}$$

By definition, the maximum likelihood estimates of the βs must satisfy the equations

$$\left. \frac{\partial \log L(\boldsymbol{\beta})}{\partial \beta_j} \right|_{\hat{\boldsymbol{\beta}}} = 0$$

for $j = 1, 2, \ldots, q$, and so $\mathbf{u}(\hat{\boldsymbol{\beta}}) = \mathbf{0}$. From equation (B.1), it then follows that

$$\hat{\boldsymbol{\beta}} \approx \boldsymbol{\beta}^* - \boldsymbol{H}^{-1}(\boldsymbol{\beta}^*)\mathbf{u}(\boldsymbol{\beta}^*)$$

which suggests an iterative scheme for estimating $\hat{\boldsymbol{\beta}}$, in which the estimate of $\boldsymbol{\beta}$ at the $(r + 1)$th cycle of the iteration is given by

$$\hat{\boldsymbol{\beta}}_{r+1} = \hat{\boldsymbol{\beta}}_r - \boldsymbol{H}^{-1}(\hat{\boldsymbol{\beta}}_r)\mathbf{u}(\hat{\boldsymbol{\beta}}_r) \tag{B.2}$$

for $r = 0, 1, 2, \ldots$, where $\hat{\boldsymbol{\beta}}_0$ is a vector of initial estimates of $\boldsymbol{\beta}$. This is the **Newton-Raphson procedure** for obtaining the maximum likelihood estimator of $\boldsymbol{\beta}$.

The algorithm used by GLIM, and other packages that allow generalized linear models to be fitted, is a modification of this scheme, in which $\boldsymbol{H}(\boldsymbol{\beta})$ is replaced by the matrix of the expected values of the second partial derivatives of the log-likelihood function. When this matrix is multiplied by -1, we get the **information matrix**, which is the matrix whose (j, k)th element is

$$-E\left\{\frac{\partial^2 \log L(\boldsymbol{\beta})}{\partial \beta_j \partial \beta_k}\right\}$$

for $j = 1, 2, \ldots, q; k = 1, 2, \ldots, q$. This matrix is denoted by $\boldsymbol{I}(\boldsymbol{\beta})$, and plays a particularly important role in maximum likelihood estimation, since the inverse of $\boldsymbol{I}(\boldsymbol{\beta})$ is the **asymptotic variance–covariance matrix** of the maximum likelihood estimates of the parameters. Using the information matrix in the iterative scheme defined by equation (B.2) gives

$$\hat{\boldsymbol{\beta}}_{r+1} = \hat{\boldsymbol{\beta}}_r + \boldsymbol{I}^{-1}(\hat{\boldsymbol{\beta}}_r)\mathbf{u}(\hat{\boldsymbol{\beta}}_r) \tag{B.3}$$

The iterative procedure based on equation (B.3) is known as **Fisher's method of scoring**. Note that as a by-product of this iterative scheme, an estimate of the asymptotic variance–covariance matrix of the parameter estimates, $\boldsymbol{I}^{-1}(\hat{\boldsymbol{\beta}})$, is obtained.

Both the Newton-Raphson and Fisher-scoring procedures converge to the maximum likelihood estimate of $\boldsymbol{\beta}$. However, the standard errors of the parameter estimates, obtained as the square roots of the diagonal elements of $-\boldsymbol{H}^{-1}(\hat{\boldsymbol{\beta}})$ or $\boldsymbol{I}^{-1}(\hat{\boldsymbol{\beta}})$, can be different in some circumstances. When these algorithms are used to fit a linear logistic model to binomial data, $\boldsymbol{I}^{-1}(\hat{\boldsymbol{\beta}}) = -\boldsymbol{H}^{-1}(\hat{\boldsymbol{\beta}})$, and so the two algorithms will give the same standard errors of the parameter estimates. For other transformations of the response probability, the variance–covariance matrices obtained using the two procedures will differ. In large samples, this difference will usually be small, but the difference can be quite marked when the number of observations in the data set is small.

B.1.2 Fitting a generalized linear model to binomial data

Suppose that n independent binomial observations $y_1, y_2, \ldots, y_n$ are such that the ith observation, $i = 1, 2, \ldots, n$, has a binomial distribution with parameters n_i and p_i. Also suppose that the transformed value of the response probability for the ith observation is related to a linear combination of q explanatory variables, that is,

$$g(p_i) = \beta_1 x_{1i} + \cdots + \beta_q x_{qi}$$

where $g(p)$ may be the logistic transformation of p, the probit of p or the complementary log–log transformation of p. The linear component of this model will be denoted by η_i, so that $g(p_i) = \eta_i$. The function g is similar to the link function of a generalized linear model, except that it relates p_i rather than the expected value of y_i, $n_i p_i$, to η_i.

To fit this model using the iterative scheme based on equation (B.3), we need expressions for $\mathbf{u}(\boldsymbol{\beta})$ and $I(\boldsymbol{\beta})$. The log-likelihood function for n binomial observations is given by

$$\log L(\boldsymbol{\beta}) = \sum_{i=1}^{n} \left\{ \log \binom{n_i}{y_i} + y_i \log p_i + (n_i - y_i)\log(1 - p_i) \right\}$$

and

$$\frac{\partial \log L(\boldsymbol{\beta})}{\partial \beta_j} = \sum_{i=1}^{n} \left(\frac{\partial \log L}{\partial p_i} \frac{\partial p_i}{\partial \eta_i} \frac{\partial \eta_i}{\partial \beta_j} \right)$$

Now,

$$\frac{\partial \log L(\boldsymbol{\beta})}{\partial p_i} = \sum_i \left\{ \frac{y_i}{p_i} - \frac{n_i - y_i}{1 - p_i} \right\} = \sum_i \frac{y_i - n_i p_i}{p_i(1 - p_i)}$$

Also, $\partial \eta_i / \partial \beta_j = x_{ji}$, and writing $g'(p_i)$ for $\partial \eta_i / \partial p_i$, we get

$$\frac{\partial \log L(\boldsymbol{\beta})}{\partial \beta_j} = \sum_{i=1}^{n} \frac{y_i - n_i p_i}{p_i(1 - p_i)} \frac{1}{g'(p_i)} x_{ji}$$

Now, let $y_i^* = (y_i - n_i p_i)g'(p_i)/n_i$, and $w_i = n_i / \{ p_i(1 - p_i)[g'(p_i)]^2 \}$, so that

$$\frac{\partial \log L(\boldsymbol{\beta})}{\partial \beta_j} = \sum_i w_i y_i^* x_{ji}$$

If X is the $n \times q$ matrix of values of the q explanatory variables, W is the $n \times n$ diagonal matrix whose ith diagonal element is w_i, and $\mathbf{y}^*$ is the $n \times 1$ vector whose

ith component is y_i^*, the vector of efficient scores, $\mathbf{u}(\boldsymbol{\beta})$, can be written as

$$\mathbf{u}(\boldsymbol{\beta}) = X'W\mathbf{y}^*$$

Next, to obtain the (j, k)th element of the information matrix, we use the standard result that

$$-E\left\{\frac{\partial^2 \log L(\boldsymbol{\beta})}{\partial \beta_j \partial \beta_k}\right\} = E\left(\frac{\partial \log L(\boldsymbol{\beta})}{\partial \beta_j}\frac{\partial \log L(\boldsymbol{\beta})}{\partial \beta_k}\right)$$

Then, the (j, k)th element of $I(\boldsymbol{\beta})$ is

$$E\left\{\sum_{i=1}^{n}\frac{y_i - n_i p_i}{p_i(1 - p_i)}\frac{1}{g'(p_i)}x_{ji}\sum_{i'=1}^{n}\frac{y_{i'} - n_{i'}p_{i'}}{p_{i'}(1 - p_{i'})}\frac{1}{g'(p_{i'})}x_{ki'}\right\}$$

Now,

$$E\left[(y_i - n_i p_i)(y_{i'} - n_{i'}p_{i'})\right] = \mathrm{Cov}(y_i, y_{i'}) = 0, \quad \text{for } i \neq i'$$

since the observations are assumed to be independent, while if $i = i'$, this expectation becomes

$$E\left[(y_i - n_i p_i)^2\right] = \mathrm{Var}(y_i) = n_i p_i(1 - p_i)$$

It then follows that

$$-E\left\{\frac{\partial^2 \log L(\boldsymbol{\beta})}{\partial \beta_j \partial \beta_k}\right\} = \sum_{i=1}^{n}\frac{n_i}{p_i(1 - p_i)[g'(p_i)]^2}x_{ji}x_{ki}$$

$$= \sum_{i}w_i x_{ji}x_{ki}$$

and, in matrix form, the information matrix is $X'WX$. Substituting for $\mathbf{u}(\hat{\boldsymbol{\beta}})$ and $I(\hat{\boldsymbol{\beta}})$ in equation (B.3), we find that the estimate of $\boldsymbol{\beta}$ at the $(r + 1)$th iteration is

$$\hat{\boldsymbol{\beta}}_{r+1} = \hat{\boldsymbol{\beta}}_r + (X'W_r X)^{-1}X'W_r \mathbf{y}_r^*$$

where W_r is the diagonal matrix of the values of w_i at the rth cycle and $\mathbf{y}_r^*$ is the vector of values of y_i^* at the rth cycle. Then,

$$\hat{\boldsymbol{\beta}}_{r+1} = (X'W_r X)^{-1}[X'W_r(X\hat{\boldsymbol{\beta}}_r + \mathbf{y}_r^*)]$$

$$= (X'W_r X)^{-1}X'W_r \mathbf{z}_r$$

where $\mathbf{z}_r = X\hat{\boldsymbol{\beta}}_r + \mathbf{y}_r^* = \hat{\boldsymbol{\eta}}_r + \mathbf{y}_r^*$, and $\hat{\boldsymbol{\eta}}_r$ is the vector of values of the linear predictor at the rth cycle of the iteration. Hence, $\hat{\beta}_{r+1}$ is obtained by regressing values of the adjusted dependent variable z_r, whose ith element is

$$\hat{\eta}_{ir} + (y_i - n_i \hat{p}_{ir})g'(\hat{p}_{ir})/n_i$$

on the q explanatory variables $x_1, x_2, ..., x_q$, using weights w_{ir}, where $w_{ir} = n_i/\{\hat{p}_{ir}(1 - \hat{p}_{ir})[g'(\hat{p}_{ir})]^2\}$, and $\hat{p}_{ir}, \hat{\eta}_{ir}$ are estimated from the values of $\hat{\beta}$ after the rth cycle of the iteration, $\hat{\beta}_r$.

To start the iterative process, initial estimates of the p_i are taken to be $\hat{p}_{i0} = (y_i + 0.5)/(n_i + 1)$. The initial weights are then $w_{i0} = n_i/\{\hat{p}_{i0}(1 - \hat{p}_{i0})[g'(\hat{p}_{i0})]^2\}$, and the initial values of the adjusted dependent variable can be taken to be $z_{i0} = \hat{\eta}_{i0} = g(\hat{p}_{i0})$. The values of z_{i0} are then regressed on the q explanatory variables $x_1, x_2, ..., x_q$ using the method of weighted least squares, with weights w_{i0}. The resulting estimates of the coefficients of $x_1, x_2, ..., x_q$ give $\hat{\beta}_1$, and lead to revised estimates of the linear predictor $\hat{\eta}_{i1} = \Sigma \hat{\beta}_{j1} x_{ji}$, revised fitted probabilities $\hat{p}_{i1} = g^{-1}(\hat{\eta}_{i1})$, revised weights $w_{i1} = n_i/\{\hat{p}_{i1}(1 - \hat{p}_{i1})[g'(\hat{p}_{i1})]^2\}$, and revised values of the adjusted dependent variable $z_{i1} = \hat{\eta}_{i1} + (y_i - n_i \hat{p}_{i1})g'(\hat{p}_{i1})/n_i$. Revised values of the parameter estimates, $\hat{\beta}_2$, are then obtained by regressing z_{i1} on the explanatory variables using weights w_{i1}, and so on. At each stage in the process, the deviance can be computed and the iteration is continued until the change in deviance in successive cycles is sufficiently small.

B.1.3 The special case of the linear logisitic model

In the special case where $g(p) = \log\{p/(1 - p)\}$, some simplification is possible. Here, $g'(p) = 1/\{p(1 - p)\}$, and the values of the adjusted dependent variable are $z_i = \hat{\eta}_i + (y_i - n_i \hat{p}_i)/\{n_i \hat{p}_i(1 - \hat{p}_i)\}$. The iterative weights, w_i, become $n_i p_i(1 - p_i)$, which is the variance of y_i.

B.2 THE LIKELIHOOD FUNCTION FOR A MATCHED CASE-CONTROL STUDY

Consider a $1:M$ matched case-control study, where the jth matched set contains one case and M controls, for $j = 1, 2, ..., n$. Let the vector of explanatory variables measured on each of these individuals be $x_{0j}, x_{1j}, ..., x_{Mj}$, where x_{0j} is the vector of values of the explanatory variables $x_1, x_2, ..., x_k$ for the case, and $x_{ij}, i \geq 1$, is the vector of the values of $x_1, x_2, ..., x_k$ for the ith control, $i = 1, 2, ..., M$, in the jth matched set. The probability that an individual in the study whose vector of explanatory variables is x_{0j} is actually the case, conditional on the values $x_{ij}, i = 0, 1, ..., M$, being observed for the individuals in the jth matched set, will now be derived.

Let $P(x_{ij}|d)$ be the probability that a diseased person in the jth matched set has explanatory variables x_{ij}, for $i = 0, 1, ..., M$, and let $P(x_{ij}|\bar{d})$ be the probability that a disease-free individual has explanatory variables x_{ij}. If x_{0j} is the vector of explanatory variables for the case in the study, that individual must have the disease that is being studied. Hence, the joint probability that $\mathbf{x}_{0j}$ corresponds to the case, and $\mathbf{x}_{ij}$ to the controls is

$$P(\mathbf{x}_{0j}|d) \prod_{i=1}^{M} P(\mathbf{x}_{ij}|\bar{d}) \tag{B.4}$$

Next, the probability that one of the $M + 1$ individuals in the jth matched set is the case, and the remainder are controls, is the union of the probability that the person with explanatory variables $\mathbf{x}_{i0}$ is diseased and the rest disease-free, the probability that the individual with explanatory variables $\mathbf{x}_{i1}$ is diseased and the rest are disease-free, and so on. This is,

$$P(\mathbf{x}_{0j}|d) \prod_{i=1}^{M} P(\mathbf{x}_{ij}|\bar{d}) + P(\mathbf{x}_{1j}|d) \prod_{i \neq 1} P(\mathbf{x}_{ij}|\bar{d}) + \cdots + P(\mathbf{x}_{Mj}|d) \prod_{i \neq M} P(\mathbf{x}_{ij}|\bar{d})$$

which is

$$\sum_{i=0}^{M} P(\mathbf{x}_{ij}|d) \prod_{r \neq i} P(\mathbf{x}_{rj}|\bar{d}) \tag{B.5}$$

It then follows that the required conditional probability is the ratio of the probabilities given by expressions (B.4) and (B.5), namely,

$$\frac{P(\mathbf{x}_{0j}|d) \prod_{i=1}^{M} P(\mathbf{x}_{ij}|\bar{d})}{\sum_{i=0}^{M} P(\mathbf{x}_{ij}|d) \prod_{r \neq i} P(\mathbf{x}_{rj}|\bar{d})} \tag{B.6}$$

We now use Bayes theorem, according to which the probability of an event A, conditional on an event B, can be expressed as $P(A|B) = P(B|A)P(A)/P(B)$. Applying this result to each of the conditional probabilities given by expression (B.6), the terms $P(d)$, $P(\bar{d})$ and $P(\mathbf{x}_{ij})$ cancel out, and we are left with

$$\frac{P(d|\mathbf{x}_{0j}) \prod_{i=1}^{M} P(\bar{d}|\mathbf{x}_{ij})}{\sum_{i=0}^{M} P(d|\mathbf{x}_{ij}) \prod_{r \neq i} P(\bar{d}|\mathbf{x}_{rj})}$$

$$= \left[1 + \frac{\sum_{i=1}^{M} P(d|\mathbf{x}_{ij}) \prod_{r \neq i} P(\bar{d}|\mathbf{x}_{rj})}{P(d|\mathbf{x}_{0j}) \prod_{i=1}^{M} P(\bar{d}|\mathbf{x}_{ij})} \right]^{-1}$$

which reduces to

$$\left[1 + \sum_{i=1}^{M} \frac{P(d|\mathbf{x}_{ij}) P(\bar{d}|\mathbf{x}_{0j})}{P(d|\mathbf{x}_{0j}) P(\bar{d}|\mathbf{x}_{ij})}\right]^{-1} \tag{B.7}$$

Now suppose that the probability that a person in the jth matched set with explanatory variables $\mathbf{x}_{ij}$ is diseased, $P(d|\mathbf{x}_{ij})$, is described by a linear logistic model, where

$$\text{logit}\,\{P(d|\mathbf{x}_{ij})\} = \alpha_j + \beta_1 x_{1ij} + \cdots + \beta_k x_{kij}$$

and where α_j is an effect due to the jth matched set. Then,

$$\frac{P(d|\mathbf{x}_{ij})}{P(\bar{d}|\mathbf{x}_{ij})} = \exp\,[\alpha_j + \beta_1 x_{1ij} + \cdots + \beta_k x_{kij}]$$

for $i = 1, 2, \ldots, M$, and

$$\frac{P(d|\mathbf{x}_{0j})}{P(\bar{d}|\mathbf{x}_{0j})} = \exp\,[\alpha_j + \beta_1 x_{10j} + \cdots + \beta_k x_{k0j}]$$

and so the conditional probability given by expression (B.7) becomes

$$\left[1 + \sum_{i=1}^{M} \exp\,\{\beta_1(x_{1ij} - x_{10j}) + \cdots + \beta_k(x_{kij} - x_{k0j})\}\right]^{-1} \tag{B.8}$$

Finally, to obtain the conditional probability of the observed data over all the n matched sets, we take the product of the terms given by expression (B.8) to give

$$\prod_{j=1}^{n} \left[1 + \sum_{i=1}^{M} \exp\,\{\beta_1(x_{1ij} - x_{10j}) + \cdots + \beta_k(x_{kij} - x_{k0j})\}\right]^{-1}$$

This conditional probability is then interpreted as the conditional likelihood of the parameters $\beta_1, \beta_2, \ldots, \beta_k$.

Appendix C

This appendix gives a listing of the macros that allow many of the analyses described in this book to be carried out using GLIM. The macros are based on the algorithms described in Section 9.3. Each section of this appendix gives the main macro, which includes a short description of its function, followed by the subsidiary macros called by the main macro. The macros are available from the author in machine readable form.

C.1 GLIM MACRO FOR COMPUTING ANSCOMBE RESIDUALS

```
$macro ansc $
!
!This macro computes the Anscombe residuals after fitting a linear logistic
!model.  On exit, the residuals are contained in the vector ar_.  This macro
!can be used when prior weights of 0 or 1 have been defined to omit
!particular observations.
!
!macro arguments: none
!
!output: ar_ (bd_ pw_)
!
!scalars used: %z1 %z2
!
!vectors used: a1_ a2_ hi_ se_ d_ s0_ s1_ u_ ar_
!
!macros called: an1_ sum_ pwt_
!
$warn $cal pw_=1 $switch %pwf pwt_ $use an1_ %yv a1_ $use an1_ %fv a2_
$ext %vl  $cal hi_=%vl*%wt/%sc  : se_=%fv/%bd  : se_=(se_*(1-se_))**(1/6)
: se_=se_*%sqrt((1-hi_)/%bd)  : ar_=pw_*(a1_-a2_)/se_
$del a1_ a2_ hi_ se_ d_ s0_ s1_ u_  $warn
$$endmac

$macro an1_
$cal u_ =%if(%le(%1,%bd/2),%1,%bd-%1)  : %z1=%z2=1  : s0_=0  : ar_=1/5
$while %z1 sum_  $cal %2=1.5*((u_/%bd)**(2/3))+s0_
: %2=%if(%le(%1,%bd/2),%2, 2.05339-%2)
$$endmac

$macro sum_
$cal s1_=s0_+ ar_*((u_/%bd)**((3*%z2+2)/3))
: ar_=ar_*(3*%z2+2)*(3*%z2+1)/(3*(3*%z2+5)*(%z2+1))  : %z2=%z2+1
: d_=s0_-s1_  : d_=%sqrt(d_*d_)  : %z1=%cu(d_>0.00005)  : s0_=s1_
$$endmac

$macro pwt_  $cal pw_=%pw  $$endmac
```

C.2 GLIM MACRO FOR CONSTRUCTING A HALF-NORMAL PLOT OF THE STANDARDIZED DEVIANCE RESIDUALS WITH SIMULATED ENVELOPES

```
$macro hnp
!
!This macro gives a half normal plot of the standardised deviance residuals,
!with simulated envelopes, after fitting a linear logistic model.  On exit,
!the minimum, mean and maximum simulated residuals are contained in the
!vectors min_, mid_, and max_, the residuals are in the vector rd_, and the
!expected values of the half normal order statistics are in nd_.  By changing
!the macro res_, half normal plots of other types of residual can be
!produced.
!
!macro arguments: none
!
!output: min_ mid_ max_ rd_ nd_
!
!scalars used: %z1 - %z3
!
!vectors used: i_ yv_ rd_ hi_ p_ min_ mid_ max_ nd_ rb_ rd1_ n_ r_ ran_ u_
!
!macros called: env_ res_ bin_ gen_
!
$cal %z1=%coc  $out  $warn  $cal yv_=%yv  : %z2=%z3=1  : i_=%cu(1)
$use res_ rd_  $sort rd_  $cal p_=%fv/%bd  : min_=max_=mid_=0
$while %z2 env_  $cal nd_=%nd((i_+%nu-0.125)/(2*%nu+0.5))
: %yv=yv_  $f +  $out %z1  $plot min_ mid_ max_ rd_ nd_  '-.-*' $
$del yv_ i_ u_ hi_ p_ rb_ rd1_ n_ r_ ran_  $warn
$$endmac

$macro env_
$use bin_ %bd p_ rb_  $cal %yv=rb_  $f +  $use res_ rd1_  $sort rd1_
$cal min_=%if(%eq(%z3,1), rd1_,min_)
: mid_=%if(%eq(%z3,1),rd1_/19,mid_+(rd1_/19))
: max_=%if(%eq(%z3,1),rd1_,max_)
: min_=%if(%lt(min_,rd1_), min_,rd1_)
: max_=%if(%gt(max_,rd1_), max_,rd1_)
: %z3=%z3+1  : %z2=%z3<20
$$endmac

$macro res_
$ext %vl  $cal hi_=%vl*%wt/%sc
: %1=%sqrt(2**%yv*%log(%yv/%fv)+2*(%bd-%yv)*%log((%bd-%yv)/(%bd-%fv)))
: %1= %1/%sqrt(1-hi_)
$$endmac

$macro bin_
$cal n_=%1 : p_=%2  : ran_=r_=0  : %z2=1 $while %z2 gen_  $cal %3=ran_
$$endmac

$macro gen_
$cal u_=%sr(0)  : ran_=ran_+((r_<n_)*(u_<p_))  : r_=r_+1  : %z2=%cu(r_<n_)
$$endmac
```

C.3 GLIM MACRO FOR CONSTRUCTING A SMOOTHED LINE

```
$macro smooth
!
!This macro constructs and plots a smoothed line using Cleveland's algorithm.
!The two arguments of the macro are identifiers associated with the x- and y-
!coordinates of the plot to be smoothed.  The value of the smoothing
!constant, f, is supplied interactively by the user.  On exit, the smoothed
```

```
!y-values are contained in the vector ysm_.  Any current model is abolished
!after use of this macro.
!
!macro arguments: %1 = vector of x-coordinates
!                 %2 = vector of y-coordinates
!
!input: value of smoothing constant, f, supplied interactively
!
!output: ysm_
!
!scalars used: %z1 - %z8
!
!vectors used: x_ y_ z1_ z2_ ysm_ ma_ w_ d_ f_
!
!macros called: sm_ dif_
!
$warn  $cal %z1=%coc  $out %poc  $var 1 f_  $data f_
$pr : 'Enter value of smoothing constant (e.g. 0.5):' :  $din %pic  $out $
$cal x_=%1 : y_=%2  $sort z1_ 1 x_  : z2_ 1 z1_  : x_ x_ z2_  : y_ y_ z2_
$cal %z2=%tr(%nu*f_)  : %z2=%z2+((%nu*f_)-%z2>0.5)  : %z3=%z4=1
$while %z3 sm_  $sort ysm_ ysm_ z1_  $out %z1  $pr  :  $plot ysm_ %2 %1 '.*'
$pr : 'Note: current model abolished' :  $del x_ y_ z1_ z2_ ma_ w_ d_ f_
$warn
$$endmac

$macro sm_
$cal d_=x_(%z4)-x_  : d_=%sqrt(d_*d_)  $sort d_  $cal %z5=1-(d_(1)=0)
: %z6=1 : %z7=2  $while %z6 dif_  $cal %z8=d_(%z8+1)
: ma_=(x_-x_(%z4))/%z8  : ma_=%sqrt(ma_*ma_)  : w_=(1-ma_*ma_*ma_)**3
: w_=w_*(ma_<1)  $yvar y_  $err n  $wei w_  $fit x_  $cal ysm_(%z4)=%fv(%z4)
: %z4=%z4+1  : %z3=(%z4<=%nu)
$$endmac

$macro dif_
$cal %z8=%z7-1  : %z5=%z5+(d_(%z7)>d_(%z8))  : %z7=%z7+1
: %z6=(%z5<=%z2)*(%z7<=%nu)
$$endmac
```

C.4 GLIM MACRO FOR COMPUTING THE $\Delta\hat{\beta}$-STATISTIC

```
$macro dbeta
!
!This macro computes the delta-beta statistic for assessing the impact of
!each observation on a parameter estimate, after fitting a linear logistic
!model.  The number of the parameter, obtained from $dis e, is supplied
!interactively by the user.  On exit, the values of the delta beta statistic
!are contained in the vector db_.  This macro can be used after prior weights
!of 0 or 1 have been defined to omit particular observations.
!
!macro arguments: none
!
!input: parameter number supplied interactively
!
!output: db_ (bd_ pw_)
!
!scalars used: %z1 - %z4
!
!vectors used: yv_ fv_ ww_ hj_ pw_ wt_ j_ z_ db_
!
!macros called: db1_ db2_ err_ pwt_
!
$warn  $cal %z1=%coc  $out %poc  $ext %vl %vc  $var 1 j_  $dat j_
$pr : 'Enter parameter number for beta coefficient:' :  $din %pic
```

```
$cal %z2=%if((j_<1)?(j_>%pl),1,2)  $swi %z2 err_ db1_
$del z_ fv_ yv_ ww_ hj_ j_ wt_  $warn
$$endmac

$macro db1_
$out  $cal pw_=1  $cal %z3=%sqrt(%vc(j_*(j_+1)/2))  $swi %pwf pwt_
$cal hj_=%vl*%wt/%sc  :  %z2=%z4=1  :  yv_=%yv  :  fv_=%fv  :  bd_=%bd
 :  wt_=%wt  :  ww_=wt_*pw_  $err n  $wei ww_  $whi %z2 db2_
$cal db_=pw_*db_*(yv_-fv_)/(1-hj_)/%z3  :  %yv=yv_
$err b bd_  $wei pw_  $f +  $out %z1
$$endmac

$macro db2_
$cal z_=0  :  z_(%z4)=1/wt_(%z4)  :  %yv=z_  $f +  $ext %pe
$cal db_(%z4)=%pe(j_)  :  %z4=%z4+1  :  %z2=%z4<=%nu
$$endmac

$macro err_  $pr : '**** error: index out of range ****' :  $$endmac

$macro pwt_  $cal pw_=%pw $$endmac
```

C.5 GLIM MACRO FOR COMPUTING THE ΔD-STATISTIC

```
$macro dstat
!
!This macro evaluates the delta D-statistic for assessing the impact of
!observation j on the fit at the remaining data points.  The value of j is
!supplied interactively by the user.  On exit, the values of the statistic
!are contained in the vector dd_.  This macro can be used after prior weights
!of 0 or 1 have been used to omit particular observations.
!
!macro arguments: none
!
!input: observation number supplied interactively
!
!output: dd_ (bd_ pw_)
!
!scalars used: %z1 %z2
!
!vectors used: yv_ fv_ ww_ hj_ pw_ wt_ chi_ var_ j_ z0_
!
!macros called: dst_ err_ pwt_
!
$warn  $cal %z1=%coc  $out %poc  $var 1 j_  $dat j_
$pr : 'enter index number of observation to be deleted (j):' :
$din %pic  $cal %z2=%if((j_<1)?(j_>%nu),1,2)  $swi %z2 err_ dst_
$del z0_ yv_ wt_ j_ chi_ hj_ var_ fv_ ww_  $warn
$$endmac

$macro dst_
$out  $cal yv_=%yv  :  fv_=%fv  :  wt_=%wt  :  bd_=%bd  :  pw_=1  $swi %pwf pwt_
$cal var_=%fv*(1-%fv/%bd)  :  z0_=0  :  z0_(j_)=1  :  %yv=z0_  :  ww_=wt_*pw_
$err n  $wei ww_  $f +  $cal hj_=pw_*%lp*%sqrt(wt_/wt_(j_))
 :  chi_=pw_*(yv_-fv_)/%sqrt(var_)  :  dd_=2*hj_*chi_*chi_(j_)/(1-hj_(j_))
 :  dd_=dd_+chi_(j_)*chi_(j_)*hj_*hj_/((1-hj_(j_))*(1-hj_(j_)))
$err b bd_  $cal %yv=yv_  $wei pw_  $f +  $out %z1
$$endmac

$macro err_  $pr : '**** error: index out of range ****' :  $$endmac

$macro pwt_  $cal pw_=%pw $$endmac
```

C.6 GLIM MACRO FOR COMPUTING THE C-STATISTIC

```
$macro cstat
!
!This macro computes the value of the C-statistic for assessing the influence
!of each observation on the classification of a future individual.  The
!values of the explanatory variables for the new individual are supplied
!interactively by the user in the order given in the output from $dis e.  On
!exit, the values of the C-statistic are contained in the vector c_.  This
!macro can be used after prior weights of 0 or 1 have been defined to omit
!particular observations.
!
!macro arguments: none
!
!input: value of x-variables for new individual supplied interactively
!
!output: c_ (bd_ pw_)
!
!scalars used: %z1 - %z3
!
!vectors used: yv_ bd_ ww_ pw_ wt_ x0_ c_ c1_ z_ hi_
!
!macros called: cst_ pwt_
!
$warn $ext %vl $cal %z1=%coc $out %poc $var %pl x0_ $dat x0_
$pr : 'Enter the values of x for the new individual:' : $din %pic $out
$cal pw_=1 $swi %pwf pwt_  $cal hi_=%vl*%wt/%sc : yv_=%yv : bd_=%bd
: wt_=%wt : ww_=wt_*pw_ : %z2=%z3=1 : c_=(yv_-%fv)/(1-hi_) : c1_=0
$err n  $wei ww_  $whi %z2 cst_  $cal c_=c1_*c_*pw_ : %yv=yv_  $err b bd_
$wei pw_  $f+  $del x0_ hi_ c1_ ww_ wt_ yv_ z_  $warn $out %z1
$$endmac

$macro cst_
$cal z_=0 : z_(%z3)=1/wt_(%z3) : %yv=z_  $f + $ext %pe
$cal c1_(%z3)=%cu(%pe*x0_) : %z3=%z3+1 : %z2=%z3<=%nu
$$endmac

$macro pwt_  $cal pw_=%pw $$endmac
```

C.7 GLIM MACRO FOR IMPLEMENTING THE WILLIAMS PROCEDURE FOR MODELLING OVERDISPERSION

```
$macro disp
!
!This macro estimates the heterogeneity factor, phi, and gives parameter
!estimates and s.e.'s adjusted for over-dispersion, in logistic regression.
!The macro should only be used after fitting the full model.  On exit, the
!directive $weight w_ is operative, and so sub-models can be fitted.  Weights
!can be cancelled using the weight directive.  This macro can be used after
!prior weights of 0 or 1 have been defined to omit particular observations.
!By changing a line in the subsidiary macro est_, called by disp, this macro
!can be used to fit a random effect using the approximate method.
!
!macro arguments: none
!
!output: (w_)
!
!scalars used: %z1 - %z6
!
!vectors used: w_ wv_ n1_ pw_
!
```

```
!macros called: est_ pwt_
!
$cal %z1=%coc  $out  $warn  $cal %z2=1  : %z3=%z4=0  : pw_=1
$swi %pwf pwt_  $cal w_=pw_  $while %z2 est_  $out %z1
$pr : 'Estimate of phi = ' %z3 ' after '*i %z4 ' iterations' :
: 'Deviance for full model, parameter estimates and standard errors'
: 'using a heterogeneity factor are as follows:' :  $wei w_  $f +  $dis e
$pr : 'Note: weight directive with weights in w_ is operative' :
$warn  $del wv_ n1_
$$endmac

$macro est_
$wei w_  $f +  $ext %vl  $cal wv_=%wt*%vl*%pw  : %z5=%z3
!
!To use the macro disp to fit a random effect using the approximate method,
!change the line below to read $cal n1_=(%bd-1)*%wt/%bd
!
$cal n1_=%bd-1
: %z3=(%x2-%cu(%pw*(1-wv_)))/%cu(n1_*%pw*(1-wv_))  : w_=(1+%z3*n1_)
: w_=pw_/w_  : %z2=%z5-%z3  : %z2=%sqrt(%z2*%z2)>=0.0001  : %z4=%z4+1
$$endmac

$macro pwt_  $cal pw_=%pw  $$endmac
```

References

Abdelbasit, K. M. and Plackett, R. L. (1981) Experimental design for categorized data. *International Statistical Review*, **49**, 111–126.

Abdelbasit, K. M. and Plackett, R. L. (1983) Experimental design for binary data. *Journal of the American Statistical Association*, **78**, 90–98.

Abramowitz, M. and Stegun, I. A. (1972) *Handbook of Mathematical Functions with Formulas, Graphs and Mathematical Tables*, U.S. Government Printing Office, Washington.

Adelusi, B. (1977) Carcinoma of the cervix uteri in Ibadan: coital characteristics. *International Journal of Gynaecology and Obstetrics*, **15**, 5–11.

Adena, M. S. and Wilson, S. R. (1982) *Generalized Linear Models in Epidemiological Research: Case-Control Studies*, INTSTAT Foundation, Sydney.

Agresti, A. (1990) *Categorical Data Analysis*, Wiley, New York.

Aitkin, M., Anderson, D. A., Francis, B. and Hinde, J. P. (1989) *Statistical Modelling in GLIM*, Clarendon Press, Oxford.

Anderson, D. A. (1988) Some models for overdispersed binomial data. *Australian Journal of Statistics*, **30**, 125–148.

Anderson, D. A. and Aitkin, M. (1985) Variance component models with binary response: interviewer variability. *Journal of the Royal Statistical Society*, **B47**, 203–210.

Anscombe, F. J. (1953) Contribution to the discussion of a paper by H. Hotelling. *Journal of the Royal Statistical Society*, **B15**, 229–230.

Aranda-Ordaz, F. J. (1981) On two families of transformations to additivity for binary response data. *Biometrika*, **68**, 357–363.

Armitage, P. and Berry, G. (1987) *Statistical Methods in Medical Research*, 2nd edn, Blackwell Scientific Publications, Oxford.

Ashton, W. D. (1972) *The Logit Transformation with Special Reference to its Uses in Bioassay*, Griffin, London.

Atkinson, A. C. (1981) Two graphical displays for outlying and influential observations in regression. *Biometrika*, **68**, 13–20.

Atkinson, A. C. (1982) Regression diagnostics, transformations and constructed variables (with discussion). *Journal of the Royal Statistical Society*, **B44**, 1–36.

Atkinson, A. C. (1985) *Plots, Transformations and Regression*, Clarendon Press, Oxford.

Azzalini, A., Bowman, A. W. and Härdle, W. (1989) On the use of nonparametric regression for model checking. *Biometrika*, **76**, 1–11.

Barnett, V. (1975) Probability plotting methods and order statistics. *Applied Statistics*, **24**, 95–108.

Barnett, V. (1982) *Comparative Statistical Inference*, 2nd edn, Wiley, London.

Barnett, V. and Lewis, T. (1985) *Outliers in Statistical Data*, 2nd edn, Wiley, London.

Berkson, J. (1944) Application of the logistic function to bio-assay. *Journal of the American Statistical Association*, **39**, 357–365.

Berkson, J. (1951) Why I prefer logits to probits. *Biometrics*, **7**, 327–339.

Berstock, D. A., Villers, C. and Latto, C. (1978) Diverticular disease and carcinoma of the colon. Unpublished manuscript.

Bishop, Y. M. M. (1969) Full contingency tables, logits and split contingency tables. *Biometrics*, **25**, 383–399.

Bishop, Y. M. M., Fienberg, S. E. and Holland, P. W. (1975) *Discrete Multivariate Analysis: Theory and Practice*, MIT Press, Cambridge, Massachusetts.

Bliss, C. I. (1935) The calculation of the dosage–mortality curve. *Annals of Applied Biology*, **22**, 134–167.

Bliss, C. I. (1940) The relation between exposure time, concentration and toxicity in experiments on insecticides. *Annals of the Entomological Society of America*, **33**, 721–766.

Blom, G. (1958) *Statistical Estimates and Transformed Beta-variables*, Wiley, New York.

Box, G. E. P. and Cox, D. R. (1964) An analysis of transformations (with discussion). *Journal of the Royal Statistical Society*, **B26**, 211–252.

Breslow, N. E., Day, N. E., Halvorsen, K. T., Prentice, R. L. and Sabai, C. (1978) Estimation of multiple relative risk functions in matched case–control studies. *American Journal of Epidemiology*, **108**, 299–307.

Breslow, N. E. and Day, N. E. (1980) *Statistical Methods in Cancer Research. 1: The Analysis of Case-Control Studies*, I.A.R.C., Lyon.

Breslow, N. E. and Day, N. E. (1987) *Statistical Methods in Cancer Research. 2: The Design and Analysis of Cohort Studies*, I.A.R.C., Lyon.

Breslow, N. E. and Storer, B. E. (1985) General relative risk functions for case–control studies. *American Journal of Epidemiology*, **122**, 149–162.

Brown, B. W. (1980) Prediction analyses for binary data. In *Biostatistics Casebook* (eds R. J. Miller, B. Efron, B. W. Brown and L. E. Moses), Wiley, New York.

Brown, C. C. (1982) On a goodness of fit test for the logistic model based on score statistics. *Communications in Statistics – Theory and Methods*, **11**, 1087–1105.

Burr, D. (1988) On errors-in-variables in binary regression – Berkson case. *Journal of the American Statistical Association*, **83**, 739–743.

Busvine, J. R. (1938) The toxicity of ethylene oxide to *Calandra Oryzae, C. Granaria, Tribolium Castaneum*, and *Cimex Lectularius*. *Annals of Applied Biology*, **25**, 605–632.

Carroll, R. J., Spiegelman, C. H., Lan, K. K. G., Bailey, K. T. and Abbott, R. D. (1984) On errors-in-variables for binary regression models. *Biometrika*, **71**, 19–25.

Chambers, E. A. and Cox, D. R. (1967) Discrimination between alternative binary response models. *Biometrika*, **54**, 573–578.

Chatfield, C. (1988) *Problem Solving: A Statistician's Guide*, Chapman and Hall, London.

Chatfield, C. and Collins, A. J. (1980) *Introduction to Multivariate Analysis*, Chapman and Hall, London.

Chatterjee, S. and Hadi, A. S. (1986) Influential observations, high leverage points, and outliers in linear regression. *Statistical Science*, **1**, 379–416.

Cleveland, W. S. (1979) Robust locally weighted regression and smoothing scatterplots. *Journal of the American Statistical Association*, **74**, 829–836.

Cochran, W. G. (1954) Some methods for strengthening the common χ^2 tests. *Biometrics*, **10**, 417–451.

Cochran, W. G. (1973) Experiments for nonlinear functions. *Journal of the American Statistical Association*, **68**, 771–781.

Collett, D. and Jemain, A. A. (1985) Residuals, outliers and influential observations in regression analysis. *Sains Malaysiana*, **14**, 493–511.

Collett, D. and Roger, J. H. (1988) Computation of generalised linear model diagnostics. *GLIM Newsletter*, **16**, 27–32. (Corrigendum in *GLIM Newsletter*, **17**, 3.)

Cook, R. D. and Weisberg, S. (1982) *Residuals and Influence in Regression*, Chapman and Hall, London.

Copas, J. B. (1983) Plotting *p* against *x*. *Applied Statistics*, **32**, 25–31.

Copas, J. B. (1988) Binary regression models for contaminated data (with discussion). *Journal of the Royal Statistical Society*, **B50**, 225–265.

Copenhaver, T. W. and Mielke, P. W. (1977) Quantit analysis: a quantal assay refinement. *Biometrics*, **33**, 175–186.

Cornfield, J. (1962) Joint dependence of the risk of coronary heart disease on serum cholesterol and systolic blood pressure: a discriminant function analysis. *Proceedings of the Federation of the American Society for Experimental Biology*, **21**, 58–61.

Cox, D. R. (1972) Regression models and life tables (with discussion). *Journal of the Royal Statistical Society*, **B74**, 187–220.

Cox, D. R. and Hinkley, D. V. (1974) *Theoretical Statistics*, Chapman and Hall, London.

Cox, D. R. and Oakes, D. (1984) *Analysis of Survival Data*, Chapman and Hall, London.

Cox, D. R. and Snell, E. J. (1968) A general definition of residuals (with discussion). *Journal of the Royal Statistical Society*, **B30**, 248–275.

Cox, D. R. and Snell, E. J. (1989) *Analysis of Binary Data*, 2nd edn, Chapman and Hall, London.

Crowder, M. J. (1978) Beta-binomial anova for proportions. *Applied Statistics*, **27**, 34–37.

Crowder, M. J. and Hand, D. J. (1990) *Analysis of Repeated Measures*, Chapman and Hall, London.

Crowder, M. J., Kimber, A. C., Smith, R. L. and Sweeting, T. J. (1991) *Statistical Analysis of Reliability Data*, Chapman and Hall, London.

Dales, L. G. and Ury, H. K. (1978) An improper use of statistical significance testing in studying covariables. *International Journal of Epidemiology*, **7**, 373–375.

Davison, A. C. (1988) Contribution to the discussion of a paper by J. B. Copas. *Journal of the Royal Statistical Society*, **B50**, 258–259.

Dempster, A.P., Laird, N. M. and Rubin, D. B. (1977) Maximum likelihood from incomplete data via the EM algorithm (with discussion). *Journal of the Royal Statistical Society*, **B39**, 1–38.

Dobson, A. J. (1990) *An Introduction to Generalized Linear Models*, Chapman and Hall, London.

Draper, C. C., Voller, A. and Carpenter, R. G. (1972) The epidemiologic interpretation of serologic data in malaria. *American Journal of Tropical Medicine and Hygiene*, **21**, 696–703.

Draper, N. R. and Smith, H. (1981) *Applied Regression Analysis*, 2nd edn, Wiley, New York.

Essenberg, J. M. (1952) Cigarette smoke and the incidence of primary neoplasm of the lung in the albino mouse. *Science*, **116**, 561–562.

Everitt, B. S. (1977) *The Analysis of Contingency Tables*, Chapman and Hall, London.

Everitt, B. S. (1987) *Introduction to Optimisation Methods*, Chapman and Hall, London.

Fieller, E. C. (1940) The biological standardisation of insulin. *Journal of the Royal Statistical Society Supplement*, **7**, 1–64.

Fieller, E. C. (1954) Some problems in interval estimation. *Journal of the Royal Statistical Society*, **B16**, 175–185.

Fienberg, S. E. (1980) *The Analysis of Cross-Classified Categorical Data*, 2nd edn, MIT Press, Cambridge, Massachusetts.

Finney, D. J. (1947) *Probit Analysis*, 1st edn, Cambridge University Press, Cambridge, UK.

Finney, D. J. (1971) *Probit Analysis*, 3rd edn, Cambridge University Press, Cambridge, UK.

Finney, D. J. (1978) *Statistical Method in Biological Assay*, 3rd edn, Griffin, London.

Fisher, R. A. (1922) On the mathematical foundations of theoretical statistics. *Philosophical Transactions of the Royal Society of London*, **A222**, 309–368. [Reprinted in Fisher (1950)].

Fisher, R. A. (1925) Theory of statistical estimation. *Proceedings of the Cambridge Philosophical Society*, **22**, 700–725. [Reprinted in Fisher (1950)].

Fisher, R. A. (1950) *Contributions to Mathematical Statistics*, Wiley, New York.

Fisher, R. A. and Yates, F. (1963) *Statistical Tables for Biological, Agricultural and Medical Research*, 6th edn, Oliver and Boyd, Edinburgh.

Fleiss, J. L. (1981) *Statistical Methods for Rates and Proportions*, Wiley, New York.

Follmann, D. A. and Lambert, D. (1989) Generalizing logistic regression by nonparametric mixing. *Journal of the American Statistical Association*, **84**, 295–300.

Fowlkes, E. B. (1987) Some diagnostics for binary logistic regression via smoothing. *Biometrika*, **74**, 503–515.

Fuller, W. A. (1987) *Measurement Error Models*, Wiley, New York.

Giltinan, D. M., Capizzi, T. P. and Malani, H. (1988) Diagnostic tests for similar action of two compounds. *Applied Statistics*, **37**, 39–50.

Golding, J., Limerick, S. and Macfarlane, A. (1985) *Sudden Infant Death: Patterns, Puzzles and Problems*, Open Books, Somerset.

Gordon, T. and Kannel, W. B. (1970) The Framingham, Massachusetts, study twenty years later. In *The Community as an Epidemiologic Laboratory: a Casebook of Community Studies* (eds. I. I. Kessler and M. L. Levin) John Hopkins Press, Baltimore.

Grizzle, J. E., Starmer, C. F. and Koch, G. G. (1969) Analysis of categorical data by linear models. *Biometrics*, **25**, 489–504.

Guerrero, M. and Johnson, R. A. (1982) Use of the Box–Cox transformation with binary response models. *Biometrika*, **65**, 309–314.

Haberman, S. J. (1978) *Analysis of Qualitative Data, 1: Introductory Topics*, Academic Press, New York.

Haseman, J. K. and Kupper, L. L. (1979) Analysis of dichotomous response data from certain toxicological experiments. *Biometrics*, **35**, 281–293.

Healy, M. J. R. (1988) *GLIM: An Introduction*, Clarendon Press, Oxford.

Hewlett, P. S. and Plackett, R. L. (1950) Statistical aspects of the independent joint action of poisons, particularly insecticides. II. Examination of data for agreement with the hypothesis. *Annals of Applied Biology*, **37**, 527–552.

Hewlett, P. S. and Plackett, R. L. (1964) A unified theory for quantal responses to mixtures of drugs: competitive action. *Biometrics*, **20**, 566–575.

Hinde, J. P. (1982) Compound Poisson regression models. In *GLIM 82* (ed. R. Gilchrist), Springer, New York, 109–121.

Hoblyn, T. N. and Palmer, R. C. (1934) A complex experiment in the propagation of plum rootstocks from root cuttings: season 1931–1932. *Journal of Pomology and Horticultural Science*, **12**, 36–56.

Holford, T. R., White, C. and Kelsey, J. L. (1978) Multivariate analysis for matched case–control studies. *American Journal of Epidemiology*, **107**, 245–256.

Holloway, J. W. (1989) A comparison of the toxicity of the pyrethroid trans-cypermethrin, with and without the synergist piperonyl butoxide, to adult moths from two strains of *Heliothis Virescens*, Final year dissertation, Department of Pure and Applied Zoology, University of Reading, UK.

Hosmer, D. W. and Lemeshow, S. (1980) Goodness of fit tests for the multiple logistic regression model. *Communications in Statistics–Theory and Methods*, **A9**, 1043–1069.

Hosmer, D. W. and Lemeshow, S. (1989) *Applied Logistic Regression*, Wiley, New York.

Jennings, D. E. (1986) Outliers and residual distributions in logistic regression. *Journal of the American Statistical Association*, **81**, 987–990.

Johnson, W. (1985) Influence measures for logistic regression: another point of view. *Biometrika*, **72**, 59–65.

Johnson, N. L. and Kotz, S. (1969) *Distributions in Statistics: Discrete Distributions*, Houghton Mifflin, Boston.

Jones, B. and Kenward, M. G. (1989) *Design and Analysis of Cross-Over Trials*, Chapman and Hall, London.

Jowett, G. H. (1963) The relationship between the binomial and *F* distributions. *The Statistician*, **13**, 55–57.

Kalbfleisch, J. D. and Prentice, R. L. (1980) *The Statistical Analysis of Failure Time Data*, Wiley, New York.

Kay, R. and Little, S. (1986) Assessing the fit of the logistic model: a case study of children with the haemolytic uraemic syndrome. *Applied Statistics*, **35**, 16–30.

Kay, R. and Little, S. (1987) Transformations of the explanatory variables in the logistic regression model for binary data. *Biometrika*, **74**, 495–501.

Kedem, B. (1980) *Binary Time Series*, Marcel Dekker, New York.

Kelsey, J. L. and Hardy, R. J. (1975) Driving of motor vehicles as a risk factor for acute herniated lumbar intervertebral disc. *American Journal of Epidemiology*, **102**, 63–73.

Kendall, M. G. and Stuart, A. (1979) *The Advanced Theory of Statistics*, Volume 2, 4th edn, Griffin, London.

Kenward, M. G. and Jones, B. (1987) A log-linear model for binary cross-over data. *Applied Statistics*, **36**, 192–204.

Krzanowski, W. J. (1988) *Principles of Multivariate Analysis: A Users Perspective*, Clarendon Press, Oxford.

Landwehr, J. M., Pregibon, D. and Shoemaker, A. C. (1984) Graphical methods for assessing logistic regression models (with discussion). *Journal of the American Statistical Association*, **79**, 61–83.

Lawless, J. F. (1982) *Statistical Models and Methods for Lifetime Data*, Wiley, New York.

Lee, A. H. (1988) Assessing partial influence in generalized linear models. *Biometrics*, **44**, 71–77.

Lilienfeld, A. M. and Lilienfeld, D. E. (1980) *Foundations of Epidemiology*, 2nd edn, Oxford University Press, New York.

Lindley, D. V. and Scott, W. F. (1984) *New Cambridge Elementary Statistical Tables*, Cambridge University Press, Cambridge, UK.

McCullagh, P. and Nelder, J. A. (1989) *Generalized Linear Models*, 2nd edn, Chapman and Hall, London.

MacMahon, B. and Pugh, T. F. (1970) *Epidemiology: Principles and Methods*, Little Brown, Boston.

Mantel, N. and Haenszel, W. (1959) Statistical aspects of the analysis of data from retrospective studies of disease. *Journal of the National Cancer Institute*, **22**, 719–748.

Mead, R. (1988) *The Design of Experiments*, Cambridge University Press, Cambridge, UK.

Minkin, S. (1987) Optimal designs for binary data. *Journal of the American Statistical Association*, **82**, 1098–1103.

Montgomery, D. C. and Peck, E. A. (1982) *Introduction to Linear Regression Analysis*, Wiley, New York.

Moolgavkar, S. H., Lustbader, E. D. and Venzon, D. J. (1985) Assessing the adequacy of the logistic regression model for matched case-control studies. *Statistics in Medicine*, **4**, 425–435.

Morgan, B. J. T. (1983) Observations on quantit analysis. *Biometrics*, **39**, 879–886.

Morgan, B. J. T. (1985) The cubic logistic model for quantal assay data. *Applied Statistics*, **34**, 105–113.

Nelder, J. A. (1985) GLIM 3.77 macros for univariate optimisation of an arbitrary function. *GLIM Newsletter*, **11**, 12–13.

Nelder, J. A. and Wedderburn, R. W. M. (1972) Generalized linear models. *Journal of the Royal Statistical Society*, **A135**, 370–384.

Payne, C. D. (ed.) (1986) *The GLIM Manual, Release 3.77*. Numerical Algorithms Group, Oxford.

Pierce, D. A. and Sands, B. R. (1975) *Extra-Bernoulli Variation in Binary Data*, Technical Report 46, Department of Statistics, Oregon State University, Oregon, USA.

Pierce, D. A. and Schafer, D. W. (1986) Residuals in generalized linear models. *Journal of the American Statistical Association*, **81**, 977–986.

Plackett, R. L. (1981) *The Analysis of Categorical Data*, 2nd edn, Griffin, London.

Pregibon, D. (1980) Goodness of link tests for generalized linear models. *Applied Statistics*, **29**, 15–24.

Pregibon, D. (1981) Logistic regression diagnostics. *Annals of Statistics*, **9**, 705–724.

Pregibon, D. (1982) Resistant fits for some commonly used logistic models with medical applications. *Biometrics*, **38**, 485–498.

Pregibon, D. (1984) Data analytic methods for matched case-control studies. *Biometrics*, **40**, 639–651.

Prentice, R. L. (1976) A generalisation of the probit and logit methods for dose response curves. *Biometrics*, **32**, 761–768.

Saunders-Davies, A. P. and Pontin, R. M. (1987) A centrifugation method for measuring the relative density (specific gravity) of planktonic rotifers (Rotifera), with values for the relative density of *Polyarthra major* Burckhardt and *Keratella cochlearis* (Gosse). *Hydrobiologia*, **147**, 379–381.

Schafer, D. W. (1987) Covariate measurement error in generalized linear models. *Biometrika*, **74**, 385–391.

Schlesselman, J. J. (1982) *Case-Control Studies: Design, Conduct, Analysis*. Oxford University Press, New York.

Searle, S. R. (1971) Topics in variance component estimation. *Biometrics*, **27**, 1–76.

Smith, P. G., Pike, M. C., Hill, A. P., Breslow, N. E. and Day, N. E. (1981) Algorithm AS 162: multivariate conditional logistic analysis of stratum matched case-control studies. *Applied Statistics*, **30**, 190–197.

Smith, W. (1932) The titration of antipneumococcus serum. *Journal of Pathology*, **35**, 509–526.

Snedecor, G. W. and Cochran, W. G. (1980) *Statistical Methods*, 7th edn, Iowa State University Press, Iowa, USA.

Stefanski, L. A. and Carroll, R. J. (1985) Covariate measurement error in logistic regression. *Annals of Statistics*, **13**, 1335–1351.

Stefanski, L. A. and Carroll, R. J. (1987) Conditional scores and optimal scores for generalized linear measurement-error models. *Biometrika*, **74**, 703–716.

Steel, R. G. D. and Torrie, J. H. (1980) *Principles and Procedures of Statistics: a Biometrical Approach*, McGraw-Hill, New York.

Stern, R. D. and Coe, R. D. (1984) A model fitting analysis of daily rainfall data (with discussion). *Journal of the Royal Statistical Society*, **A147**, 1–34.

Stiratelli, R., Laird, N. M. and Ware, J. H. (1984) Random-effects models for serial observations with binary response. *Biometrics*, **40**, 961–971.

Strand, A. L. (1930) Measuring the toxicity of insect fumigants. *Industrial and Engineering Chemistry: analytical edition* **2**, 4–8.

Stukel, T. A. (1988) Generalized logistic models. *Journal of the American Statistical Association*, **83**, 426–431.

Thisted, R. A. (1988) *Elements of Statistical Computing*, Chapman and Hall, London.

Truett, J., Cornfield, J. and Kannel, W. (1967) A multivariate analysis of the risk of coronary heart disease in Framingham. *Journal of Chronic Diseases*, **20**, 511–524.

Wang, P. C. (1985) Adding a variable in generalized linear models. *Technometrics*, **27**, 273–276.

Wang, P. C. (1987) Residual plots for detecting nonlinearity in generalized linear models. *Technometrics*, **29**, 435–438.

Weil, C. S. (1970) Selection of the valid number of sampling units and a consideration of their combination in toxicological studies involving reproduction, teratogenesis or carcinogenesis. *Food and Cosmetics Toxicology*, **8**, 177–182.

Wetherill, G. B. (ed.) (1986) *Regression Analysis with Applications*, Chapman and Hall, London.

Whitehead, J. R. (1983) *The Design and Analysis of Sequential Clinical Trials*, Ellis Horwood, Chichester.

Whitehead, J. R. (1989) The analysis of relapse clinical trials, with application to a comparison of two ulcer treatments. *Statistics in Medicine*, **8**, 1439–1454.

Whitehead, J. R. and Brunier, H. (1989) *PEST 2.0 Operating Manual*, University of Reading, UK.

Williams, D. A. (1975) The analysis of binary responses from toxicological experiments involving reproduction and teratogenicity. *Biometrics*, **31**, 949–952.

Williams, D. A. (1982a) Extra-binomial variation in logistic linear models. *Applied Statistics*, **31**, 144–148.

Williams, D. A. (1982b) Contribution to the discussion of a paper by A. C. Atkinson. *Journal of the Royal Statistical Society*, **B44**, 33.

Williams, D. A. (1983) The use of the deviance to test the goodness of fit of a logistic-linear model to binary data. *GLIM Newsletter*, **6**, 60–62.

Williams, D. A. (1984) Residuals in generalized linear models. *Proceedings of the 12th International Biometrics Conference*, Tokyo, 59–68.

Williams, D. A. (1987) Generalized linear model diagnostics using the deviance and single case deletions. *Applied Statistics*, **36**, 181–191.

Woodbury, M. A. and Manton, K. G. (1983) Framingham: an evolving longitudinal study, in *Encyclopaedia of Statistical Sciences*, (Eds S. Kotz and N. L. Johnson), Wiley, New York, **3**, 198–201.

Index of examples

For each illustrative example used in this book, the number of the example is given, followed by the page number in parentheses. When a complete section has been devoted to the analysis of a particular data set, the section number is given in boldface type.

INDEX